T0290967

Bioremediation

Bioremediation

Green Approaches for a Clean and Sustainable Environment

Edited by

Ram Naresh Bharagava

Department of Microbiology (DM), Babasaheb Bhimrao Ambedkar University (A Central University), Lucknow (U.P.), India

Sandhya Mishra

SERB-National Post Doctoral Fellow, Environmental Technologies Division CSIR-National Botanical Research Institute, Lucknow, Uttar Pradesh, India

Ganesh Dattatraya Saratale

Department of Food Science and Biotechnology, Dongguk University, South Korea

Rijuta Ganesh Saratale

Research Institute of Biotechnology & Medical Converged Science, Dongguk University, South Korea

Luiz Fernando Romanholo Ferreira

Waste and Effluent Treatment Laboratory, Institute of Technology and Research (ITP), Tiradentes University, Aracaju/Sergipe, Brazil

and

Graduate Program in Process Engineering, Tiradentes University (UNIT), Aracaju/Sergipe, Brazil

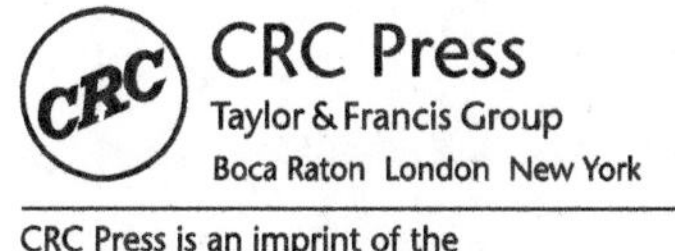

CRC Press is an imprint of the
Taylor & Francis Group, an informa business

First edition published 2022
by CRC Press
6000 Broken Sound Parkway NW, Suite 300, Boca Raton, FL 33487-2742

and by CRC Press
2 Park Square, Milton Park, Abingdon, Oxon, OX14 4RN

CRC Press is an imprint of Taylor & Francis Group, LLC

Library of Congress Cataloging-in-Publication Data
Names: Bharagava, Ram Naresh, editor.
Title: Bioremediation : green approaches for a clean and sustainable environment /
edited by Ram Naresh Bharagava,
Department of Environmental Microbiology (DEM), Babasaheb Bhimrao Ambedkar Central University, Lucknow (U.P.), India,
Sandhya Mishra, State Key Laboratory for Conservation and Utilization of Subtropical Agro-bioresources, Integrative Microbiology Research Centre (IMRC), South China Agricultural University (SCAU), Guangzhou, China,
Ganesh Dattatraya Saratale, Department of Food Science and Biotechnology, Dongguk University, South Korea,
Rijuta Ganesh Saratale, Research Institute of Biotechnology & Medical Converged Science, Dongguk University, South Korea and
Luiz Fernando Romanholo Ferreira, Institute of Technology and Research, Tiradentes University, Aracaju/Sergipe, Brazil.
Other titles: Bioremediation (CRC Press 2022)
Description: First edition. | Boca Raton, FL : CRC Press, 2022. | Includes
bibliographical references and index.
Identifiers: LCCN 2021053501 | ISBN 9781032019703 (hbk) | ISBN 9781032019710 (pbk) | ISBN 9781003181224 (ebk)
Subjects: LCSH: Bioremediation. | Green chemistry.
Classification: LCC TD192.5 .B5564 2022 | DDC 628.5—dc23/eng/20220112
LC record available at https://lccn.loc.gov/2021053501

ISBN: 9781032019703 (hbk)
ISBN: 9781032019710 (pbk)
ISBN: 9781003181224 (ebk)

DOI: 10.1201/9781003181224

Typeset in Times
by codeMantra

Contents

Preface

Environmental issues have been at the forefront of sustainable development and have become a serious matter of concern in the twenty-first century. Environmental sustainability with rapid industrialization is one of the major challenges of the current scenario worldwide. Industries are the key drivers of the world economy, but they are also the major polluters of environment due to the discharge of partially treated/untreated toxic and hazardous wastes containing organic and inorganic pollutants, which cause severe environmental (soil and water) pollution and toxic effects in living beings. Among the different sources of environmental pollution, industrial wastes are considered as the major source of environmental pollution because industries use cheap and poorly or non-biodegradable chemicals to obtain products of good quality within a short period of time and in an economic way; however, their toxicity is usually ignored. The untreated/partially treated wastewaters discharged from various industries contain potentially toxic and hazardous organic and inorganic pollutants, which cause the pollution of soil and aquatic resources including groundwater and severe toxic effects in humans, animals and plants.

The governments around the globe are strictly advocating the development of efficient treatment technologies to mitigate the environmental pollution and health threats. Hence, the adequate treatment of industrial wastes to degrade/detoxify pollutants is of utmost importance for environmental safety and for promoting the sustainable development of our society with low environmental impacts. The removal of a wide range of pollutants from contaminated sites requires our increasing understanding of different degradation pathways and regulatory networks to carbon flux for their degradation and detoxification. Therefore, this book will provide a comprehensive knowledge about the fundamental, practical and purposeful utilization of different treatment technologies for the treatment/management of industrial wastes. This book will describe the microbiological, biotechnological, biochemical and molecular aspects of various treatment technologies, including the use of 'omics' for the development of effective treatment technologies for industrial wastes/pollutants to combat the forthcoming challenges.

This book is mainly focused on the various treatment technologies such as integrated treatment approaches, in situ/ex situ bioremediation, nano-bioremediation, biofilm formation, composting, microbe-assisted bioremediation, role of advanced biotechnological tools and techniques in bioremediation process and application of nanomaterials in the removal/treatment of organic/inorganic pollutants from industrial wastewaters/contaminated sites along with their ecotoxicological effects on the environment as well as human/animal health.

For this book, many relevant topics have been contributed by the experts from different universities, research laboratories and institutes across the globe in relevant fields. Researchers working in the field of waste treatment/management and related fields for a clean and green environment will find this compilation most useful for further study to learn about the subject matter.

This book will be of great value to researchers, environmental chemists and scientists, microbiologists and biotechnologists, ecotoxicologists, waste treatment engineers and managers, environmental science managers, administrators and policy makers, industry persons and students at bachelor's, master's and doctoral levels in the relevant field. Thus, in this book, readers will find the updated information as well as the future directions in the field of waste treatment/management for a clean and green environment.

Editors

Dr. Ram Naresh Bharagava received his BSc (1998) in Zoology, Botany and Chemistry from University of Lucknow, Lucknow, Uttar Pradesh (U.P.), India, and MSc (2004) in *Molecular Biology and Biotechnology* from Govind Ballabh Pant University of Agriculture & Technology (GBPUAT), Pantnagar, Uttarakhand (U.K.), India. He obtained his PhDs in Microbiology in 2010 from CSIR-Indian Institute of Toxicology Research (CSIR-IITR), Lucknow, and Pt. Ravishankar Shukla University, Raipur, Chhattisgarh, India. He has published 13 edited books, 01 author book, 75 book chapters and more than 95 research papers in national and international journals of repute. He has an h-index of 40, i10-index of 85 and total citations of 5715 to his credit. He is also serving as a reviewer for various national and international journals in his field. Currently, Dr. Bharagava is working as a Sr. Assistant Professor and is actively engaged in teaching and research activities in government-sponsored projects on biodegradation and bioremediation of environmental pollutants at Department of Environmental Microbiology (DEM), Babasaheb Bhimrao Ambedkar Central University, Lucknow (U.P.), India. His major areas of research interest are biodegradation and bioremediation of environmental pollutants, Ecotoxicology, metagenomics and wastewater microbiology. He is a life member of the Indian Science Congress Association (ISCA), India, Association of Microbiologists of India (AMI), Biotech Research Society (BRSI) and Academy of Environmental Biology (AEB). He can be reached at bharagavarnbbau11@gmail.com and bharagavabiotech77@gmail.com.

Dr. Sandhya Mishra obtained her PhD (2019) in the area of Environmental Microbiology from the Department of Environmental Microbiology, Babasaheb Bhimrao Ambedkar (A Central) University (BBAU), Lucknow, Uttar Pradesh (U.P.), India. She did her PhD in microbial detoxification of chromium from tannery wastewater for environmental safety. She has published many research/review papers in national and international journals of high impact factor. Mainly, she has also written chapters for national and international edited books. She has also served as a potential reviewer for several national and international journals in her respective areas of the research. Presently, she is working as a Post-Doctoral Fellow at State Key Laboratory for Conservation and Utilization of Subtropical Agro-bioresources, Integrative Microbiology Research Centre (IMRC), South China Agricultural University (SCAU), Guangzhou, China. Her major areas of research interest are biodegradation and bioremediation of environmental pollutants such as heavy metals and pesticides, metagenomics, quorum sensing and wastewater microbiology. She is a member of the Association of Microbiologists of India (AMI) and Indian Science Congress Association (ISCA), India. She can be reached at sandhyamanshi@gmail.com and sandhyamishra@scau.edu.cn.

Dr. Ganesh Dattatraya Saratale currently works as an Associate Professor in the Department of Food Science and Biotechnology, Dongguk University, South Korea (https://food.dongguk.edu/?staff=staff11). He obtained his PhD degree in Biochemistry in 2006 and conducted postdoctoral research in Taiwan, South Korea and India. He has received numerous prestigious domestic and international academic awards, including meritorious Departmental research fellowship (during PhD, India), Young Scientist award (Govt. of India) and Brain Korea-21 (BK-21) (South Korea), and he also completed two major research projects related to lignocellulosic biorefineries. He has more than 185 publications in peer-reviewed scientific journals with total citations of about 8655, and he has an h-index of 51 (Google Scholar) to his credit. He edited two books and also contributed 16 book chapters. His name was featured in Stanford University's global list of top 2% scientists for 2019 and 2020 in the Environmental Engineering field. He served as an Editorial board member of many journals, including *Polymers, Applied Nano, Bioengineering International,* and *Frontiers in Bioscience-Landmark and Nanomanufacturing,* and also worked as Guest Editor for *International Journal of Hydrogen Energy and Bioresource Technology*

Reports (Elsevier). His research focuses majorly on the development of lignocellulosic biorefineries for biopolymers, hydrogen, ethanol and value-added biochemicals production. His research interest is well devoted to algal biofuels production; environmental biotechnology; bioelectrochemistry; and nanotechnology and its application towards energy, environment, agriculture and medical fields. He also delivered keynote speeches and invited talks in various conferences, seminars and workshops.

Dr. Rijuta Ganesh Saratale obtained a PhD in Biotechnology (India) in 2010. She did postdoctoral studies in various international universities/institutes such as University of Ostrava, Czech Republic, National Cheng Kung University, Taiwan, University of Pune, India, and Ewha Womans University, South Korea. Presently, she is working as an Assistant Professor in Research Institute of Biotechnology & Medical Converged Science, Dongguk University, South Korea. Her main fields of expertise are environmental biotechnology; wastewater treatment; algal biofuels; and nanotechnology for energy, environment and medical sector. She received the Young Scientist award by the Department of Science and technology, Govt. of India, and she also completed two major research projects related to bioremediation of environmental contaminants. Her name was featured in Stanford University's global list of top 2% scientists for 2019 and 2020 in the Environmental Engineering field. She has published more than 100 research publications in peer-reviewed journals (SCI/SCIE), five book chapters and one book. Her current h-index is 36 (Google Scholar) with more than 6350 citations. She presented her research work in various national and international conferences and has supervised about 21 master's theses.

Dr. Luiz Fernando Romanholo Ferreira is an Associate Professor at Tiradentes University and a researcher at the Institute of Technology and Research in Aracaju/Sergipe, Brazil. He earned his PhD in microbiology from the University of São Paulo. He has more than 120 scientific contributions in the form of research, reviews, book chapters and editorial-type scientific articles in various fields of science and engineering. He has experience as a visiting researcher at the Department of Molecular Biosciences and Bioengineering at the University of Hawaii at Manoa with Professor Samir Kumar Khanal. He is a level 2 researcher in technological development in environmental sciences at CNPq (DT-2) and serves on the editorial board of the *World Journal of Microbiology and Biotechnology* and is a Review Editor for *Frontiers in Microbiology*. He is a member of the Research Advisory Committee of the Foundation of Support to Research and Technological Innovation of the State of Sergipe [FAPITEC/ SE/BRAZIL] and a reviewing advisor of research projects of several research foundations in Brazil, such as FAPESP, the Brazilian National Council for Scientific and Technological Development (CNPq). He has an h-index of 15 with more than 700 citations and serves as a scientific reviewer in numerous peer-reviewed journals of high impact factor. He has a collaborative network with national and international institutes/universities across the globe. He is currently a member of the Society for Green Environment, India. Further, he has published more than 60 research articles with high impact and book chapters in leading international and national journals or books.

Contributors

Jayanthi Abraham
Microbial Biotechnology Laboratory, School of Biosciences and Technology
VIT University
Vellore, India

Lucia Maria Carareto Alves
Department of Agricultural, Livestock and Environmental Biotechnology, School of Agricultural and Veterinarian Sciences
São Paulo State University (UNESP)
Jaboticabal, Brazil

Iqbal Azad
Department of Chemistry
Integral University
Lucknow, India

Ram Naresh Bharagava
Laboratory of Bioremediation and Metagenomics Research (LBMR), Department of Microbiology
Babasaheb Bhimrao Ambedkar University (A Central University)
Lucknow, India

Shashi Lata Bharati
Department of Chemistry
North Eastern Regional Institute of Science and Technology
Nirjuli (Itanagar), India

Abhishek Kumar Bhardwaj
Department of Environmental Science
Veer Bahadur Singh Purvanchal University
Jaunpur, India

Sayed Muhammad Ata Ullah Shah Bukhari
Department of Microbiology
Quaid-i-Azam University
Islamabad, Pakistan

Bárbara Bonfá Buzzo
Department of Agricultural, Livestock and Environmental, School of Agricultural and Veterinarian Sciences
São Paulo State University (UNESP)
Jaboticabal, Brazil

Ankita Chatterjee
Microbial Biotechnology Lab, School of Bioscience and Technology
VIT University
Vellore, India

Pankaj Kumar Chaurasia
P.G. Department of Chemistry
L.S. College, B.R.A. Bihar University
Muzaffarpur, India

Chabungbam Victoria Devi
Department of Social Science, College of Horticulture and Forestry
Central Agricultural University
Pasighat, India

Geeta Dhania
Department of Environmental Science
M.D. University
Rohtak, India

Pranab Dutta
School of Crop Protection, College of Post-Graduate Studies in Agricultural Sciences
Central Agricultural University
Meghalaya, India

Luiz Fernando Romanholo Ferreira
Graduate Program in Process Engineering
Tiradentes University (UNIT)
Aracaju-Sergipe, Brazil
and
Waste and Effluent Treatment Laboratory
Institute of Technology and Research (ITP)
Tiradentes University (UNIT)
Aracaju-Sergipe, Brazil

Rajnish Gautam
Institute of Sustainable Industries and Liveable Cities, College of Engineering and Science
Victoria University
Melbourne, Australia

Ved Prakash Giri
Academy of Scientific and Innovative Research (AcSIR)
Ghaziabad, India
and
Department of Botany
Lucknow University
Lucknow, India

B. N. Hazarika
College of Horticulture and Forestry
Central Agricultural University
Pasighat, India

Punabati Heisnam
Department of Natural Resource Management, College of Horticulture and Forestry
Central Agricultural University
Pasighat, India

Muhsin Jamal
Department of Microbiology
Abdul Wali Khan University
Mardan, Pakistan

Rupali Kaur
Centre of Biotechnology
University of Allahabad
Prayagraj, India

Tahmeena Khan
Department of Chemistry
Integral University
Lucknow, India

Arun S Kharat
School of Life Sciences
Jawaharlal Nehru University
New Delhi, India

Roop Kishor
Laboratory of Bioremediation and Metagenomics Research (LBMR), Department of Environmental Microbiology
Babasaheb Bhimrao Ambedkar University (A Central University)
Lucknow, India

Madhuree Kumari
Division of Microbial Technology
CSIR-National Botanical Research Institute
Lucknow, India
Academy of Scientific and Innovative Research (AcSIR)
Ghaziabad, India
and
Department of Biochemistry
Indian Institute of Science
Bengaluru, India

Aradhana Mishra
Division of Microbial Technology
CSIR-National Botanical Research Institute
Lucknow, India
and
Academy of Scientific and Innovative Research (AcSIR)
Ghaziabad, India

Abhinash Moirangthem
Department of Horticulture, College of Agriculture
Central Agricultural University
Imphal, India

Nandkishor More
Department of Environmental Science
Babasaheb Bhimrao Ambedkar University (A Central University)
Lucknow, India

Ram Naraian
Department of Biotechnology
Veer Bahadur Singh Purvanchal University
Jaunpur, India

Dimuth Navaratna
Institute of Sustainable Industries and Liveable Cities, College of Engineering and Science
Victoria University
Melbourne, Australia

Muhammad Asif Nawaz
Department of Biotechnology
Shaheed Benazir Bhutto University
Sheringal, Pakistan

Shipra Pandey
Division of Microbial Technology
CSIR-National Botanical Research Institute
Lucknow, India
Academy of Scientific and Innovative Research (AcSIR)
Ghaziabad, India
and
Department of Chemical Engineering
Indian Institute of Technology
Bombay, India

Sidra Pervez
Department of Biochemistry
Shaheed Benazir Bhutto Women University
Peshawar, Pakistan

Bhanu Pratap
Department of Environmental Sciences
Babasaheb Bhimrao Ambedkar University (A Central University)
Lucknow, India

Daniel Delgado Queissada
UniAGES - University Center
Paripiranga, Brazil

Abdul Rahman Khan
Department of Chemistry
Integral University
Lucknow, India

Vivek Rana
Water Quality Management Division
Central Pollution Control Board
Delhi, India

Sana Raza
Department of Microbiology
Abdul Wali Khan University
Mardan, Pakistan

Redaina
Department of Microbiology
Abdul Wali Khan University
Mardan, Pakistan

Iraí Tadeu Ferreira de Resende
Instituto Federal de Sergipe
Lagarto, Brazil

Renan Lieto Alves Ribeiro
Centro Nacional de Pesquisa e Conservação de Mamíferos Carnívoros
Instituto Chico Mendes de Conservação da Biodiversidade (ICMBio/CENAP)
Atibaia, Brazil

Rakesh K. Sahoo
Institute of Physics
Bhubaneswar, India

Sushila Saini
Department of Botany
JVMGRR College
Charkhi Dadri, India

Cinara Ramos Sales
Department of Agricultural, Livestock and Environmental, School of Agricultural and Veterinarian Sciences
São Paulo State University (UNESP)
Jaboticabal, Brazil

Vanessa Cruz dos Santos
UniAGES - University Center
Paripiranga, Brazil

Luciana Maria Saran
Department of Agricultural, Livestock and Environmental, School of Agricultural and Veterinarian Sciences
São Paulo State University (UNESP)
Jaboticabal, Brazil

Liloma Shah
Department of Microbiology
Abdul Wali Khan University
Mardan, Pakistan

Jyoti Sharma
Water Quality Management Division
Central Pollution Control Board
Delhi, India

Ajeet Singh
Nanomaterials and Sensor Research Laboratory, Department of Physics
Babasaheb Bhimrao Ambedkar University (A Central University)
Lucknow, India

Arpita Singh
Department of Environmental Microbiology (DEM)
Babasaheb Bhimrao Ambedkar University (A Central University)
Lucknow, India

Saroj Kumar Singh
CSIR-Institute of Minerals and Materials Technology
Bhubaneswar, India

Yengkhom Disco Singh
Department of Post-Harvest Technology, College of Horticulture and Forestry
Central Agricultural University
Pasighat, India

Jesiel Alves da Silva
UniAGES - University Center
Paripiranga, Brazil

Débora da Silva Vilar
Graduate Program in Process Engineering
Tiradentes University (UNIT)
Aracaju-Sergipe, Brazil

Shanthy Sundaram
Centre of Biotechnology
University of Allahabad
Prayagraj, India

Anushree Suresh
School of Biosciences and Technology
VIT University
Vellore, India

Shikha Varma
Institute of Physics
Bhubaneswar, India

Anjali Verma
Department of Environmental Science
Babasaheb Bhimrao Ambedkar University (A Central University)
Lucknow, India

Bal Chandra Yadav
Nanomaterials and Sensor Research Laboratory, Department of Physics
Babasaheb Bhimrao Ambedkar University (A Central University)
Lucknow, India

1

Existing and Emerging Treatment Technologies for the Degradation and Detoxification of Textile Industry Wastewater for the Environmental Safety

Ajeet Singh, Roop Kishor, Ram Naresh Bharagava, and Bal Chandra Yadav
Babasaheb Bhimrao Ambedkar University (A Central University)

CONTENTS

1.1 Introduction

Textile industries (TIs) are spread globally, having a market size of ≈1 trillion dollars, and India contributes to ≈7% of the total world exports. Globally, TIs offer employment to ≈35 million workers and are the fifth largest source of foreign currency (Kaur et al., 2018; Tara et al., 2019; Kishor et al., 2020). India is the second largest exporter of dyes after China. But, unfortunately, TIs are a major source of environmental pollution because they release huge volumes of coloured wastewater into valuable water resources (Bener et al., 2019; Kishor et al., 2021a). Textile production is a complex process, which consists of sizing, desizing, bleaching, scouring, mercerizing, dyeing, printing, washing and finishing stages (Kadam et al., 2018; Sen et al., 2019; Kishor et al., 2021b). These stages use large volumes of freshwater and a large number of different chemicals (Kishor et al., 2020).

For example, TIs consume ≈1.6 million L of groundwater for the production of 8,000 kg of textile fabrics per day (Khan and Malik, 2017; Kishor et al., 2018) and ≈20% of wastewater is discharged into environment (Kishor et al., 2021a). Several chemicals such as acids, bases, surfactants, salts, dispersants, dyes and finishing agents are used at different stages of textile production (Bener et al., 2019; Kishor et al., 2021c). Among these chemicals, the dyes are the major source of environmental pollution (Chandanshive et al., 2020; Kishor et al., 2021c). Dyes are employed in textile, cosmetic, leather, printing, paper and medicine industries as a colouring agent (Haq et al., 2018; Kishor et al., 2021c). Besides,

DOI: 10.1201/9781003181224-1

azo dyes are the largest used synthetic dye, about 70% of all dye production per year (Kaur et al., 2018; Kishor et al., 2020).

They contain at least one azo group (–N=N–) as a chromophore as well as sulphonic (SO3–) and hydroxyl (OH–) groups (Kishor et al., 2021a). TIWW is characterized by its intensive colour, high pH, temperature, BOD, COD, TSS, total nitrogen, total solids and toxic metals (Cao et al., 2019; Kishor et al., 2021b). TIWW causes serious threats in water and soil ecologies. In a water ecosystem, it reduces photosynthetic activity and dissolved oxygen (DO) content, leading to anoxic conditions, which ultimately affects fauna and flora. In soil ecologies, it reduces soil fertility due to the accumulation of recalcitrant pollutants and metals (Cao et al., 2019; Bener et al., 2019; Kishor et al., 2021b). TIWW is highly toxic to plants (Kishor et al., 2021b). It also causes severe threats to human beings (Sen et al., 2019; Kishor et al., 2021d). Hence, the treatment of TIWW is urgently needed for the protection of environment and public health.

Different physico-chemical, advanced treatment and biological methods have been reported for the treatment of TIWW (Kaur et al., 2018; Kishor et al., 2018; Ceretta et al., 2020). The physico-chemical methods are not feasible due to the production of sludge and high costs (Kishor et al., 2018; Cao et al., 2019). AOPs and biological process effectively degrade pollutants into non-toxic and inorganic compounds (Kishor et al., 2021a). The biological treatments use archaea, bacteria, fungi, yeasts, algae and plants to transform and degrade pollutants (Khandare and Govindwar, 2015; Kadam et al., 2018; Kishor et al., 2021c). These biological agents are used in bioreactors, with adequate agitation and aeration. These agents are also used in wetland treatment. Biological treatments are able to degrade, transform and detoxify pollutants into non-toxic and mineralized compounds (Garg et al., 2020; Kishor et al., 2021a). Bioprocesses have many limitations. They can degrade only biodegradable compounds and take long time for the complete degradation of pollutants. In addition, the biological agents may be inhibited/prevented by toxic compounds during the treatment process (Cao et al., 2019; Bener et al., 2019; Kishor et al., 2021a).

Besides, AOPs such as ozonation and photocatalytic, photo-Fenton, electrocatalytic and electrochemical oxidation can degrade wastewater pollutants. In AOPs, various oxidizing agents such as H_2O_2 and O_3; many catalysts such as CdS, Fe_2O_3, TiO_2, GaP, ZnO and ZnS; and also high-energy radiations such as UV light are utilized (Kaur et al., 2018; Bener et al., 2019; Kishor et al., 2021a). AOPs are well reported as effective and efficient, but may not be suitable due to high cost, incomplete mineralization and the toxic products generated (Kaur et al., 2018). Nowadays, a combination of AOPs and biological process is an alternative solution for the treatment of recalcitrant compounds. In a combination system, AOPs can break down complex recalcitrant and persistent pollutants into more easily biodegradable compounds by free radicals, which are further completely mineralized by biological agents into water and carbon dioxide.

Therefore, this chapter provides detailed knowledge about wastewater generation, its nature and pollutants. It deals with various toxic effects on the environment and living organisms. It also presents different existing and emerging treatment techniques for the treatment of TIWW, key issues and challenges.

1.2 Characteristics and Pollutants of Textile Industry Wastewater

Textile manufacturing is a complex process, which involves many steps such as sizing, desizing, bleaching, mercerizing, dyeing, printing, washing and finishing for the production of textiles (Garg et al., 2020; Kishor et al., 2021a). These steps consume large volumes of freshwater and a wide range of chemicals (Kishor et al., 2018, 2020). For example, about 20% of wastewater is released from the dyeing and finishing stages (Khan and Malik, 2017). Further, around 200 L of potable water is consumed to produce 1 kg textile products. TIWW is characterized by its intensive colour, high pH, temperature, BOD, COD, TDS, TOC, TSS, SS, total nitrogen, phosphate, sulphate, chloride and various toxic metals, as shown in Table 1.1 (Waghmode et al., 2019; Garg et al., 2020; Kishor et al., 2021b). Various researchers have reported the textile wastewater characteristics, as shown in Table 1.1.

TIWW has alkalis, binders, phthalates, perfluorinated compounds, bleaching agents, organic compounds, volatile organic compounds, dioxins, perfluorooctanoic acid, surfactants, dispersing agents, brominated flame retardants, fixing agents and finishing agents (Bener et al., 2019; Kishor et al., 2021a, 2021c). Dyes such as methylene blue, methyl orange, Congo red, methyl red, reactive red, crystal violet, Remazol Brilliant Blue R and azure B dye are reported in TIWW (Cao et al., 2019; Garg et al., 2020;

TABLE 1.1

Physico-Chemical Characteristics of Textile Industry Wastewater

References	pH	EC	COD	BOD_5	TDS	TSS	TS	TOC	Cl^-	Alk.	NO_3	SO_4^{-2}	PO_4^{-3}	TN	Cr	Cd	Pb	Phenol	Zn	Cu	Sulphide	As
Hubadillah et al. (2020)		1614	416												1.92	0.45			0.90	1.05		
Ağtaş et al. (2020)		9,758	580					161														
Chandanshive et al. (2020)	10.9	0.89	1,794	1,350	5,143	1,900									2.93	0.11	0.80					2.89
Hussain et al. (2019)	7.5	5.2	689	248	3,367	235	4,732	307	2,586			275	13.7	22.5	0.19	1.05		0.38				0.48
Bener et al. (2019)	8.5	8.96	295			34		220	0.1			<1			0.028			0.355	0.09			
El-Mekkawi et al. (2019)	7.5		1,150	507		206							7.53	39.2								
Núñez et al. (2019)	11.6	51,200	800																			
Oktem et al. (2019)	8.1		505	128				527	5,500			311	16.4	28.7				0.85				
Tara et al. (2019)	8.2	8.8	513	283	5,251	324	5,420	201	1,383						9.67	0.88						
Kaur et al. (2018)	10		1,156	196	9,640	8,430																
Kadam et al. (2018)	9.5		1,438	1,230	8,230	5,175									4.20	0.09	.070					2.05
GilPavas and Correa-Sanchez (2018)	9.3	4,010	720	115				164														
Arcanjo et al. (2018)	9	1,608	78						42.3	264	6.2	136	0.4									
Watharkar et al. (2018)	8.6		940	768	3,361	3,172									3.20							1.80
Chandanshive et al. (2017)	10.7		1,328	1,140	9,562	7,280									2.91	0.07	0.42					2.12
Amare et al. (2017)	10.43	5.69	90,666	15,190	2993.44						226	1393.33	3,310	389.65	0.025	0.028	0.047		0.130	0.042		
Guadie et al. (2017)	8.5	1.3			757	248	1,005		74.2	528			12	6.96								
Khan and Malik (2017)	8.04				7,970				ND						1.533	0.088	0.199		2.694			
Okareh et al. (2017)	8.60	10.36	898.15	222.08	6,210	9,530	15,740		718.25													
Tomei et al. (2016)	9		1,117			0.54		158	38.6		3.8	4.5	3.2									

Except pH, all the parameters are expressed in mg/L, but conductivity is expressed in µmho/cm.

Alk., alkalinity; BOD, biochemical oxygen demand; COD, chemical oxygen demand; EC, electrical conductivity; TDS, total dissolved solids; TOC, total organic carbon; TS, total solids; TSS, total suspended solids.

Kishor et al., 2021a). Different toxic metals such as Cr, Sb, Cu, Hg, Zn, Pb, Cd and Ni are also reported in TIWW (Table 1.1) (Waghmode et al., 2019; Kishor et al., 2021a).

1.3 Toxicity Profile of Textile Industry Wastewater

TIs discharge large volumes of highly toxic wastewater, which causes serious threats to water and soil ecologies. In water resources, it causes colouration of water, leading to reduction in penetration of sunlight and decrease in photosynthetic activity and dissolved oxygen content, which results in anoxic conditions and thus causes severe toxic threats to aquatic fauna and flora (Kishor et al., 2018; Cao et al., 2019). In soil ecologies, it accumulates over long time duration and disturbs microbial communities and thus ultimately causes soil pollution and salinity.

Different metals can accumulate in living tissues through the food chain, which causes severe threats to public health, such as diarrhoea, neuromuscular disorders, liver problems, haemorrhage, dermatitis, central nervous system disorder and kidney dysfunction (Khan and Malik, 2017; Kishor et al., 2021a). TIWW may reduce seed germination and biomass production of plants (Bener et al., 2019; Kishor et al., 2021a). The nonylphenol ethoxylates (NPEs) are reported as endocrine-disrupting chemicals and can lead to hormonal imbalance in all forms of life (Khandare and Govindwar, 2015; Kishor et al., 2021a). TIWW causes carcinogenic, allergenic and mutagenic effects in rats, molluscs, plants, fish and mammalian cells.

1.4 Existing and Emerging Treatment Technologies

Different existing and emerging treatment methods have been reported by various workers for the degradation of industrial wastewater. These methods may be used individually or in combination for the enhanced and efficient treatment of TIWW. The classification of various treatment methods reported by various workers is shown in Figure 1.1.

1.4.1 Biological Treatment

Biological treatment is a cost-effective, green, eco-friendly and effective method for the degradation of textile dyes and wastewater (Cao et al., 2019; Kishor et al., 2020). It uses bacteria, fungi, yeasts, algae and plants to remediate wastewater pollutants. It can degrade, decolourize and mineralize the pollutants into less toxic and inorganic compounds (Bharagava et al., 2018; Kishor et al., 2021a). Different biological agents were used by various researchers for the degradation of textile industry wastewater.

Bacterial treatment (BT) is an efficient approach to degrading and decolourizing the textile industrial wastewater (Haq et al., 2018; Garg et al., 2020; Kishor et al., 2021b). Bacteria produce enzymes, which are involved in the degradation of industrial wastewater pollutants into less toxic/non-toxic and mineralized form (Khandare and Govindwar, 2015; Kishor et al., 2018; Cao et al., 2019). Bacteria may utilize the pollutants as an energy source (C/N). Different species such as *Aeromonas, Arthrobacter, Alcaligenes, Bacillus* and *Pseudomonas* are reported in the mineralization of various dyes and wastewater pollutants (Kishor et al., 2018, 2021a). *Bacillus cohnii* is able to degrade and decolourize the Congo red and TIWW within 12 and 48 hours (Kishor et al., 2021b).

Fungal/yeast (F/Y) treatment is able to degrade and detoxify recalcitrant industrial wastewater pollutants (Sen et al., 2016; Kishor et al., 2020). Many F/Y species are able to degrade/convert a wide number of textile dyes and wastewater pollutants into less toxic/non-toxic products (Sen et al., 2016; Kishor et al., 2018). Many F/Y species produce different ligninolytic enzymes such as laccase, lignin peroxidase (LiP) and manganese peroxidase (MnP) during the treatment of dyes and wastewater pollutants (Haq et al., 2018; Kishor et al., 2021a).

For example, *Phanerochaete chrysosporium* decolourized various dyes by producing LiP, MnP and laccase enzyme (Sathishkumar et al., 2019). F/Y treatment has many disadvantages such as requirement of long growth phase, need for nitrogen restrictive enzymes, unreliable enzymes production and large reactor size due to the long holding time for complete degradation.

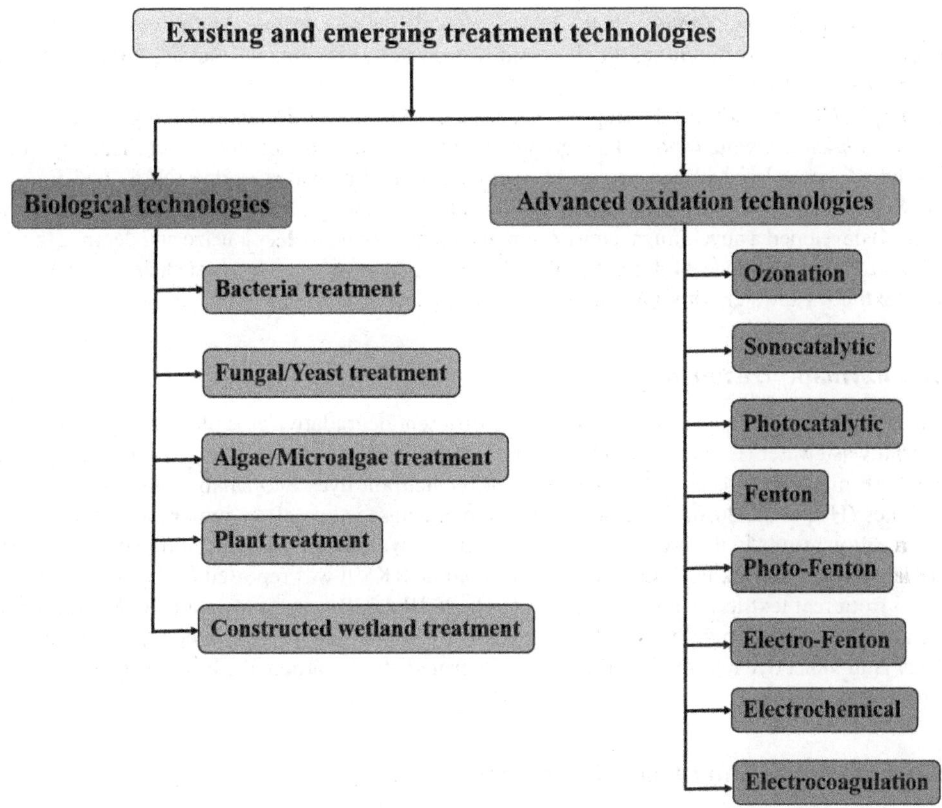

FIGURE 1.1 Classification of different treatment methods used for the degradation of textile industry wastewater.

Algae are the most diverse groups of photosynthesis organisms, which are commonly found in freshwater, marine water and wastewater (Shahid et al., 2020; Kishor et al., 2021a). Microalgae rapidly grow in wastewater because they provide the bulk of biomass and biofuel production, provide cheaper growth media with ample nutrients (nitrogen or phosphorus) and provide the possibility of integrating algal cultivation with the existing infrastructure of wastewater remediation (Shahid et al., 2020). Algae are made of lipids (70%), carbohydrates (60%) and proteins and essential amino acids, respectively (Shahid et al., 2020). Microalgae have many useful properties such as highest carbon fixation, higher biomass productivity, and good harvest index and have a little growth cycle as compared to terrestrial plants (Sinha et al., 2016; Kishor et al., 2018; Shahid et al., 2020).

Microalgae play a major role in carbon mitigation and bioremediation of toxic pollutants (Kishor et al., 2020). They are used as a bioindicator to detect climate change in an aquatic environment. Microalgae are able to degrade recalcitrant chemicals and also uptake nitrogen and phosphorus (80%–100%) from the wastewater for higher productivities of biomass and valuable products such as poultry, biofertilizers, medicine, cosmetics and green fuels, for example biodiesel and bioalcohol (Sinha et al., 2016; Shahid et al., 2020).

As discussed above, different microbial agents were applied as a pure/single culture for the degradation of industrial wastewater. But, these agents are ineffective against complex and toxic compounds and also require a long time. However, for effective treatment, several researchers have reported that microbial consortium is more effective compared to the use of pure/single culture (Cao et al., 2019; Kishor et al., 2021a). In a consortium system, the pure/single strain might attack dye molecules at different sites. Moreover, the degradation products, which appeared due to the metabolic activity of one strain, may be utilized as a substrate by another (Kishor et al., 2021a). Many workers have developed bacteria–bacteria, bacteria–algal, bacteria–fungal, bacteria–yeast, bacteria–plant, fungal–fungal,

fungal–yeast, fungal–algae, fungal–plant, algal–algal and algal–yeast consortium for the decolourization and degradation of different textile dyes and wastewater pollutants (Cao et al., 2019; Garg et al., 2020; Kishor et al., 2021a).

For example, Cao et al. (2019) developed a bacterial consortium of *Bacillus*, *Betaproteobacteria* and *Gammaproteobacteria*, which showed a high potential for the degradation and mineralization of direct blue 2B dye. A bacteria–algal consortium of nitrifier-enriched activated sludge (NAS) and *Chlorella vulgaris* was developed for the remediation of wastewater (Sepehri et al., 2020). In addition, Mishra and Malik (2014) developed a novel fungal consortium, which was found to decolourize and degrade the Acid Blue 161 and Pigment Orange 34 dyes. It is also able to remediate heavy metals such as Cr and Cu from synthetic textile wastewater within 48 hours.

1.4.1.1 Enzymatic Treatment

Bacteria, fungi, yeasts and other microbes produce different degradative enzymes during the treatment of industrial wastewater (Haq et al., 2018; Cao et al., 2019; Kishor et al., 2021a). Azoreductase, laccase, LiP and MnP may participate in the conversion of recalcitrant dyes into small, simple and non-toxic intermediates (Haq et al., 2018; Cao et al., 2019). Azoreductase is one of the important enzymes whose catalytic reaction results in the breakdown of azo bond and synthesizes aromatic amine into a colourless water (Kishor et al., 2021b). For example, *Bacillus cohnii* (RKS9) was reported to degrade and remove Congo red from real textile wastewater (Kishor et al., 2021b). LiP enzyme was able to decolourize and degrade azure B dye within 48 hours (Haq et al., 2018). *Aeromonas hydrophila* degraded and mineralized the crystal violet dye into small and less toxic intermediates by producing laccase and LiP enzymes (Bharagava et al., 2018).

1.4.1.2 Phytoremediation (Plant Treatment)

Plant treatment is a green, solar energy-driven, eco-friendly and cost-effective method for the clean-up of environmental pollutants. The plant can remove or convert many toxic recalcitrant pollutants such as textile dyes, pesticides, surfactants, polyaromatic hydrocarbons, chlorinated solvents, heavy metals, polyaromatic biphenyls and even various gases (Chandanshive et al., 2020; Kishor et al., 2021a). Phytoextraction, phytostabilization, phytovolatilization, phytofiltration and phytotransformation mechanisms are involved in the remediation of recalcitrant pollutants. *Salvinia molesta* was reported in the degradation of Rubine GEL (Chandanshive et al., 2017). *Vetiveria zizanioides* is able to decolourize and remove ADMI (colour) (68%), COD (75%), BOD (73%), TDS (77%) and TSS (34%) from textile wastewater. It also removed different heavy metals such as As, Cr, Cd and Pb from wastewater (Chandanshive et al., 2020).

1.4.2 Advanced Oxidation Processes (AOPs)

In 1980s, for the first time AOPs were proposed for drinking water treatment, but are now largely used in the treatment of different types of wastewaters (Gupta et al., 2021a). In AOPs, different species such as hydroxyl radicals (OH*), sulphate radicals (SO_4^-), superoxide radical anions (O_2^{-*}), hydroperoxyl radical (HO_2*), hydrogen peroxide (H_2O_2), electrons (e^-) and hydrogen radical (H*) are generated in appropriate quantity to remove pollutants and purify wastewater at near-ambient temperature (Kishor et al., 2021a; Gupta et al., 2021b). Researchers focus on AOPs for the reduction of recalcitrant compound and toxic elements in wastewater, and further research is going on to increase the efficiency of existing the AOPs and developing newer AOPs (Ceretta et al., 2020; Gupta et al., 2021a). Various AOPs such as ozonation and sonocatalytic, photocatalytic, Fenton, photo-Fenton, electro-Fenton, electrochemical and electrocoagulation treatments were reported in the treatment of TIWW.

Ozonation has widely been used in wastewater treatment (Bener et al., 2019). The redox potential of ozone is 2.07 V, which being a powerful oxidant can degrade organic pollutants using two mechanisms: (i) direct electrophilic attack by molecular ozone and (ii) indirect attack by OH* radicals produced through ozone decomposition. Both mechanisms of ozonation split organic bonds and dissociated aromatic rings in recalcitrant organic compounds (Kishor et al., 2021a).

In recent decades, the ultrasound-based sonochemical process has widely been used for the treatment of wastewater (Kishor et al., 2021a). In this process, a high temperature and high pressure are created in the solution by the ultrasound waves, which generate bubbles. The degradation of the dye pollutant through the sonocatalytic process enhanced in the presence of a suitable catalyst, and this is likely due to a synergistic effect between ultrasonic irradiation and the solid semiconductor catalyst.

The photocatalytic process is one of the most significant techniques in the list of AOPs (Gupta et al., 2021a). It has a high efficiency and has received great attention for the removal of recalcitrant organic pollutants and dissolved solids (Gupta et al., 2021a; Kishor et al., 2021a). Many semiconductors such as ZnO, CdS, TiO_2, SnO_2, CeO_2 and NiO are used to generate highly reactive species such as hydroxyl radicals (OH*), hydroperoxyl radicals (HO_2) and superoxide radical anions (O_2^{*-}). These species are highly effective in the mineralization of wastewater pollutants (Hien et al., 2020; El-Mekkawi et al., 2020; Kishor et al., 2021a; Gupta et al., 2021b).

Recently, Ceretta et al. (2020) have reported a ZnO/polypyrrole composite in the treatment of real textile wastewater.

$$ZnO + hv \rightarrow ZnO\left(e_{CB}^{-} + h_{VB}^{+}\right) \tag{1.1}$$

$$ZnO\left(h_{VB}^{+}\right) + H_2O \rightarrow ZnO + H^{+} + OH^{-} \tag{1.2}$$

$$ZnO\left(h_{VB}^{+}\right) + OH^{-} \rightarrow ZnO + OH^{*} \tag{1.3}$$

$$ZnO\left(e_{CB}^{-}\right) + O_2 \rightarrow ZnO + O_2^{*-} \tag{1.4}$$

$$O_2^{*-} + H^{+} \rightarrow ZnO + HO_2^{*} \tag{1.5}$$

$$HO_2^{*} + HO_2^{*} \rightarrow H_2O_2 + O_2 \tag{1.6}$$

The process has many advantages such as being fast, high stability and little chemical consumption.

The Fenton treatment (Fe^{2+}/H_2O_2) is one of the most widely studied AOPs for the treatment of wastewater. In this process, H_2O_2 oxidants dissociate by Fe^{2+} catalyst into hydroxyl radicals HO*, which is a strong and non-selective oxidant accomplished of oxidizing highly recalcitrant organic pollutants. The process occurred at a suitable amount of catalyst and oxidant, pH value (2–4), temperature and treatment time (Saratale et al., 2018; Kishor et al., 2020). The possible reaction in the Fenton process is as follows:

$$Fe^{2+} \cdot H_2O_2 \rightarrow OH^{-} \cdot OH * \tag{1.7}$$

$$Fe^{3+} + H_2O_2 \rightarrow Fe^{2+} \cdot HO_2^{*} \tag{1.8}$$

The photo-Fenton (UV/Fe^{2+}/H_2O_2) process is an emerging and highly effective process for the treatment of TIWW. It is a combination of Fenton reagents (H_2O_2 and Fe^{2+}) and UV–Vis radiation ($\lambda < 600$ nm). In this process, the mixture of ferrous sulphate and H_2O_2 produces hydroxyl radicals (OH*), which is able to oxidize the organic pollutants in wastewater (Rosa et al., 2020; Gupta et al., 2021a; Kishor et al., 2021a). Additionally, the formation of hydroxyl radicals (OH*) could be increased by the UV light. In this way, the concentration of Fe^{2+} and overall reaction is accelerated. This extra radical is generated by two additional reactions: (i) photo-reduction of Fe^{3+} to Fe^{2+} ions and (ii) peroxide photolysis via shorter wavelength ($\lambda < 600$ nm), as shown in Equations 1.9–1.10.

$$Fe(OH)^{2+} \cdot hv \rightarrow OH * \lambda < 580 nm \tag{1.9}$$

$$H_2O_2 + hv \rightarrow OH * \cdot \lambda < 310 nm \tag{1.10}$$

Recently, Tian et al. (2020) have developed a $ZnAl_2O_4/BiPO_4$ composite for the treatment of textile wastewater pollutants, which is found to be able to remove colour and COD from real textile wastewater pollutants.

The electro-Fenton (EF) process is known as Fenton-based electrochemical advanced oxidation process. In this process, the hydroxyl radicals are produced by the reaction of electrogenerated hydrogen peroxide and ferrous ions that are either added to the solution or released by an iron source (GilPavas and Correa-Sanchez, 2019). The formation of hydroxyl radicals can be shown as follows:

$$O_2 + 2H^+ + 2e^- \rightarrow H_2O_2 \text{ (in cathode)} \tag{1.11}$$

$$Fe^{2+} + H_2O_2 + H^+ \rightarrow Fe^{3+} + HO^* + H_2O \text{ (in solution)} \tag{1.12}$$

The main advantages of the EF process are better control of the process, avoidance of the storage and transport of H_2O_2, continuous regeneration of Fe^{2+} at the cathode leading to better degradation rate of organic pollutants, and minimized sludge production. The disadvantages of the EF process are that it is optimal only under strong acidic conditions, the catalyst (Fe^{2+}) dissolved in wastewater is non-recyclable, and it consumes high energy (GilPavas et al., 2018; Kishor et al., 2021a).

Electrochemical oxidation (EO) is a relatively new technique for the treatment of toxic and non-biodegradable wastewater. In this process, electrodes and electrolyte are used for the generation of hydroxyl radicals (OH*), which broadly decolourize and mineralize different toxic pollutants dissolved in wastewater. In this process, many types of electrodes such as Ti/SnO_2, $TiBO_2$, BDD, $TiPbO_2$ and graphite and Ti/RuO_2 and SnO_2 are used as anode in the treatment of wastewater (Kaur et al., 2018, Viana et al., 2018; Kishor et al., 2021a).

Many studies of the EO process showed the decolourization, COD removal, TOC and mineralization of toxic pollutants in wastewater. For example, Viana et al. (2018) showed 100% decolourization and 73.77% COD removal within 180 minutes by using $Ti(RuO_2)_{0.8}$-$(Sb_2O_3)_{0.2}$ electrodes under optimal conditions (current density 0.62–12.38 mA/cm^2 and electrolyte (NaCl) concentration 0.00062–0.1238 M).

Electrocoagulation (EC) is an electrochemical process widely used for the treatment of dyes, heavy metals, phenols, surfactants and pesticides. In an EC set-up, the cathode and anode (iron or aluminium) are connected to an external monopolar or bipolar power supply. Characteristically, these electrodes (iron (Fe) and aluminium (Al)) are easily available, are of low cost, and have high removal efficiency. Some factors such as electrode type, distance between the electrodes, applied current density, pH, conductivity of the electrolyte and treatment time affect the EC process. At cathode, hydroxyl ions are generated, which remove toxic pollutants from the wastewater, and metal ions generated from the anode act as subverting agents and neutralize the electric charge of contaminants, resulting in the removal of organic pollutants from wastewater.

For example, Bener et al. (2019) reported the treatment of real textile wastewater and its reuse in agriculture irrigation.

1.4.3 Combination of AOPs and Biological Treatment Methods

TIWW has various recalcitrant/xenobiotic chemicals such as dyes, pentachlorophenol (PCP), volatile organic compounds (VOCs), phthalates, surfactants and different heavy metals (Cao et al., 2019; Kishor et al., 2021a). Therefore, several treatment methods have been reported in the degradation of TIWW. But these methods are not efficient and effective for complete mineralization of TIWW pollutants. For example, AOPs are fast, effective and efficient, but these are not suitable due to their cost and generation of various toxic, carcinogenic and mutagenic substances (Waghmode et al., 2019; Ceretta et al., 2020). Further, biological methods require long time and are ineffective against recalcitrant and toxic pollutants. In the present time, the combination of AOPs and biological treatments is an alternative approach to the effective degradation and mineralization of chemicals (Waghmode et al., 2019; Sun et al., 2020). In AOPs, the highly potential and non-reactive species such as sulphate and hydroxyl radicals are produced during the treatment process (Kishor et al., 2021a).

These radicals can convert or mineralize complex molecular structures of pollutants into simpler intermediates and more degradable products (Waghmode et al., 2019; Kishor et al., 2021b). These products are further mineralized by microorganisms by involving degradative enzymes. For example, Waghmode et al. (2019) used a sequential photocatalysis and biological process for the degradation and mineralization of methyl red. In this process, photocatalysis can break down the complex structure of methyl red into more easily degradable products. Further, these products are mineralized into simpler and inorganic compounds by the biological method. A combination of plasma advanced oxidation process (PAOP) and microbial fuel cells (MFCs) was developed for the mineralization of methylene blue (MB) dye. In this system, PAOP was able to degrade 97.7% of MB. But, it was only able to mineralize 23.2% of MB dye. MFCs mineralized 63% of MB dye. The combined system enhanced the mineralization of MB dye with an improved power density of 519 mW/m^2 (Sun et al., 2020). Various combined treatment approaches have been reported by many workers for the treatment of TIWW/dyes and are shown in Table 1.2.

TABLE 1.2

Combined Treatment Approaches for TIWW/Dyes

Combined Treatment Approaches	Textile Wastewater and Dyes	Optimum Parameters	Treatment Efficiency	References
Biological + photocatalysis treatment	Real textile wastewater	Biological = pH = 7, T = 25°C, 125 rpm and 24 hours; photocatalysis = ZnO/PPy = 25:1, simulated dyes wastewater = 50 mg/L	Colour (95.7%) and TOC (99.8%)	Ceretta et al. (2020)
Plasma oxidation (PO) + microbial fuel cell (MFC)	Methylene blue (MB) dye	PO = diameter = 25 and 45 mm, height = 120 and 360 mm, dye = 300 mg/L and voltage = 25–40. MFC = chamber long/diameter = 4/3, dye = 300 mg/L, pH = 7, T = 450°C and 30 minutes	97.7%	Sun et al. (2020)
Biological + photocatalysis treatment	Textile wastewater	Photocatalysis = catalyst = 0.6 g/L, UV radiation, T = 20–25°C and 2 hours; biological = pH = 7, T = 30°C, 120 rpm and 2 hours	98% colour and 63% TOC	da Silva et al. (2019)
Biological + electrochemical oxidation treatment	Congo red and textile wastewater	Electrochemical = current = 20 mA/cm^2, NaCl = 2 g/L, pH = 7; biological = T = 27°C and 120 rpm	98% and 93%	Sathishkumar et al. (2019)
Photocatalysis + biological treatment	Methyl red (MR) dye	Photocatalysis = pH = 7, T = 30°C and 2 hours; biological = pH = 7, T = 30°C, 120 rpm and 2 hours	100%	Waghmode et al. (2019)
Biological + electrochemical oxidation treatment	Acid Blue 161 and Procion Red MX-5B dye	Electrochemical = current = 50 mA/cm^2, dye = 100 mg/L, electrolyte = (Na_2SO_4 0.02 mol/L), and 1 hour; biological = T = 30°C and 12 hours	95%	Almeida et al. (2019)
Adsorption + photocatalysis	Textile wastewater	Photocatalysis = dye = 5.0 mg/L, rpm = 140, voltage = 500 W, pH = 6–7, T = 25°C and T = 60 minutes. Adsorption = dye = 5.0 mg/L, pH = 6.1, T = 27°C, rpm = 140 and 30 minutes	99.20%	Fazal et al. (2019)
Ozonation + biological treatment	Industrial textile wastewater	–	98%	Paździor et al. (2016)
Biological + chemical treatment	Textile wastewater	pH 3 for 100% sample	Colour (> 92%) and COD (87%)	Hayat et al. (2015)

1.5 Challenges and Recommendations

TIs are one of the major sources of global economy in many countries, but unfortunately, they are also the major source of environmental pollution. The treatment of textile wastewater is a major challenge for TIs. Besides it, TIs are also facing many challenges from government sectors, such as lack of specific and dedicated industrial areas for the positioning of textile industries, increasing demands for various types of textile fabrics, increasing cost of raw materials, lack of financial support from the government and lack of advanced processing techniques and waste treatment technologies in developing countries.

The mitigation of these challenges requires large-scale financial supports from the government for proper functioning of TIs. TIs should also use environmentally friendly natural colouring agents of biological origin instead of synthetic agents as they may be helpful in reducing the treatment cost. TIs should adopt recycling/reuse of treated wastewater to minimize the use of fresh ground water for economic and environmental benefits.

1.6 Conclusions

TIWW is highly toxic to the environment and livings beings. TIWW has various dyestuffs, pesticides, surfactants, phthalates, chlorinated solvents, polyaromatic biphenyls, finishing agents and metals such as Cr, Cd, As, Pb, Sb, Ni and Zn. Therefore, different treatment methods such as physico-chemical processes, AOPs and biological processes have been reported for the treatment of textile wastewater. Physico-chemical methods can convert pollutants from one phase to another phase or concentrate them within one phase. AOPs and biological processes effectively degrade/mineralize pollutants into non-toxic and inorganic compounds. In the present time, the combination of AOPs and biological treatments is an alternative approach to the effective degradation of chemicals. In AOPs, potential and non-reactive species such as sulphate and hydroxyl radicals are produced during the treatment process. These radicals can convert or mineralize complex molecular structures of pollutants into simpler intermediates and more degradable products. These products are further mineralized by microorganisms by involving degradative enzymes.

Acknowledgement

Mr. Ajeet Singh is grateful to Council of Scientific & Industrial Research (CSIR), Government of India, for financial support in the form of Senior Research Fellowship (F.No.16–9(June2017)/2018 NET/CSIR).

REFERENCES

Ağtaş, M., Yılmaz, Ö., Dilaver, M., Alp, K., and Koyuncu, I. 2020. Hot water recovery and reuse in textile sector with pilot scale ceramic ultrafiltration/nanofiltration membrane system. *J. Clean. Product.* 256, 120359.

Amare, E., Kebede, F., and Mulat, W. 2017. Analysis of heavy metals, physicochemical parameters and effect of blending on treatability of wastewaters in Northern Ethiopia. *Int. J. Environ. Sci. Technol.* 14, 1679–1688.

Arcanjo, G.S., Mounteer, A.H., Bellato, C.R., da Silva, L.M.M., Dias, S.H.B., and da Silva, P.R. 2018. Heterogeneous photocatalysis using TiO_2 modified with hydrotalcite and iron oxide under UV–visible irradiation for color and toxicity reduction in secondary textile mill effluent. *J. Environ. Manage.* 211, 154–163.

Bener, S., Bulca, Ö., Palas, B., Tekin, G., Atalay, S., and Ersöz, G. 2019. Electrocoagulation process for the treatment of real textile wastewater: Effect of operative conditions on the organic carbon removal and kinetic study. *Process Saf. Environ. Prot.* 129, 47–54.

Bharagava, R.N., Mani, S., Mulla, S.I., and Saratale, G.D. 2018. Degradation and decolorization potential of an ligninolytic enzyme producing *Aeromonas hydrophila* for crystal violet dye and its phytotoxicity evaluation. *Ecotoxicol. Environ. Saf.* 156, 166–175.

Cao, J., Sanganyado, E., Liu, W., Zhang, W., and Liu, Y. 2019. Decolorization and detoxification of direct blue 2B by indigenous bacterial consortium. *J. Environ. Manag.* 242, 229–237.

Ceretta, M.B., Vieira, Y., Wolski, E.A., Foletto, E.L., and Silvestri, S. 2020. Biological degradation coupled to photocatalysis by ZnO/polypyrrole composite for the treatment of real textile wastewater. *J. Water Pro. Eng.* 35, 101230.

Chandanshive, V., Kadam, S., Rane, N., Jeon, B.H., Jadhav, J., and Govindwar, S. 2020. In situ textile wastewater treatment in high-rate transpiration system furrows planted with aquatic macrophytes and floating phytobeds. *Chemosphere* 252, 126513.

Chandanshive, V.V., Rane, N.R., Tamboli, A.S., Gholave, A.R., Khandare, R.V., and Govindwar, S.P. 2017. Co-plantation of aquatic macrophytes *Typha angustifolia* and *Paspalum scrobiculatum* for effective treatment of textile industry effluent. *J. Hazard. Mater.* 338, 47–56.

da Silva, L.S., Gonçalves, M.M.M., and Raddi de Araujo, L.R. 2019. Combined photocatalytic and biological process for textile wastewater treatments. *Environ. Res.* 91(11), 1490–1497.

El-Mekkawi, D.M., Abdelwahab, N.A., Mohamed, W.A., Taha, N.A., and Abdel-Mottaleb, M.S.A. 2020. Solar photocatalytic treatment of industrial wastewater utilizing recycled polymeric disposals as TiO2 supports. *J. Clean. Prod.* 249, 119430.

Fazal, T., Razzaq, A., Javed, F., Hafeez, A., Rashid, N., Amjad, U.S., Rehman, M.S., Faisal, A., and Rehman, F. 2019. Integrating adsorption and photocatalysis: A cost effective strategy for textile wastewater treatment using hybrid biochar-TiO_2 composite. *J. Hazard. Mater.* 10, 121623.

Garg, N., Garg, A., and Mukherji, S. 2020. Eco-friendly decolorization and degradation of reactive yellow 145 textile dye by *P. aeruginosa* and *T. pantotropha. J. Environ. Manag.* 263, 110383.

GilPavas, E. and Correa-Sanchez, S. 2019. Optimization of the heterogeneous electro-Fenton process assisted by scrap zero-valent iron for treating textile wastewater: Assessment of toxicity and biodegradability. *J. Water Process. Eng.* 32, 100924.

GilPavas, E., Dobrosz-Gómez, I., and Gómez-García, M.Á. 2018. Optimization of sequential chemical coagulation-electro-oxidation process for the treatment of an industrial textile wastewater. J. Water Process Eng 22, 73–79.

Guadie, A., Tizazu, S., Melese, M., Guo, W., Ngo, H.H., and Xia, S. 2017. Biodecolorization of textile azo dye using *Bacillus* sp. strain CH12 isolated from alkaline lake. *Biotechnol. Rep.* 15, 92–100.

Gupta, S.K., Gupta, A.K., Yadav, R.K., Singh, A., and Yadav, B.C. 2021b. Highly efficient S-g-CN/Mo-368 catalyst for synergistically NADH regeneration under solar light. *Photochem. Photobiol.* https://doi.org/10.1111/php.13484.

Gupta, S., Yadav, R.K., Gupta, A.K., Yadav, B., Singh, A., and Pandey, B.K. 2021a. One-pot highly efficient synthesis of n-enrich graphene quantum dots as a photocatalytic platform for NAD+/NADP+ reduction. *Photochem. Photobiol.* https://doi.org/10.1111/php.13460.

Haq, I., Raj, A., and Markandeya. 2018. Biodegradation of azure-B dye by *Serratia liquefaciens* and its validation by phytotoxicity, genotoxicity and cytotoxicity studies. *Chemosphere* 196, 58–68.

Hayat, H., Mahmood, Q., Pervez, A., Bhatti, Z.A., and Baig, S.A. 2015. Comparative decolorization of dyes in textile wastewater using biological and chemical treatment. *Sep. Purif. Technol.* 154, 149–153.

Hien, N. T., Nguyen, L. H., Van, H. T., Nguyen, T. D., Nguyen, T. H., Chu, T. H., Nguyen, T. V., Vu, X. H., and Aziz, K. H. 2020. Heterogeneous catalyst ozonation of direct black 22 from aqueous solution in the presence of metal slags originating from industrial solid wastes. *Sep. Purif. Technol.* 233, 115961.

Hubadillah, S.K., Othman, M.H.D., Tai, Z.S., Jamalludin, M.R., Yusuf, N.K., Ahmad, A., Rahman, M.A., Jaafar, J., Kadir, S.H.S.A., and Harun, Z. 2020. Novel hydroxyapatite-based bio-ceramic hollow fiber membrane derived from waste cow bone for textile wastewater treatment. *Chem. Eng. J.* 379, 122396.

Kadam, S.K., Chandanshive, V.V., Rane, N.R., Patil, S.M., Gholave, A.R., Khandare, R.V., Bhosale, A.R., Jeon, B.H., and Govindwar, S.P. 2018. Phytobeds with *Fimbristylis dichotoma* and *Ammannia baccifera* for treatment of real textile effluent: An in-situ treatment, anatomical studies and toxicity evaluation. *Environ. Res.* 160, 1–11.

Kaur, P., Kushwaha, J.P., and Sangal, V.K. 2018. Electrocatalytic oxidative treatment of real textile wastewater in continuous reactor: Degradation pathway and disposability study. *J. Hazard. Mater.* 346, 242–252.

Khan, S. and Malik, A. 2017. Toxicity evaluation of textile effluents and role of native soil bacterium in biodegradation of a textile dye. *Environ. Sci. Pollut. Res.* 25(5), 4446–4458.

Khandare, R. V., and Govindwar, S. P. 2015. Phytoremediation of textile dyes and effluents. Current scenario and future prospects. Biotechnol. Adv. 33, 1697–1714.

Kishor, R., Bharagava, R.N., Ferreira, L.F.R., Bilal, M., and Purchase, D. 2021d. Molecular techniques used to identify perfluorooctanoic acid degrading microbes and their application in a wastewater treatment reactor/plant. In Shah M. and Rodriguez-Couto S. (ed) *Wastewater Treatment Reactors*, Elsevier. 253–271.

Kishor, R., Purchase, D., Ferreira, L.F., Mulla, S.I., Bilal, M., and Bharagava, R.N. 2020. Environmental and health hazards of textile industry wastewater pollutants and its treatment approaches. In: Hussain, C.M. (ed.) *Environmental Resource Management*, Springer Nature Switzerland. https://doi.org/10.1007/978-3-319-58538-3_230-1.

Kishor, R., Purchase, D., Saratale, G.D., Ferreira, L.F.R., Bilal, M., Iqbal, H.M., and Bharagava, R.N. 2021b. Environment friendly degradation and detoxification of Congo red dye and textile industry wastewater by a newly isolated *Bacillus cohnni* (RKS9). *Environ. Technol. Innov.* 22, 101425.

Kishor, R., Purchase, D., Saratale, G.D., Saratale, R.G., Ferreira, L.F.R., Bilal, M., Chandra, R., and Bharagava, R.N. 2021a. Ecotoxicological and health concerns of persistent coloring pollutants of textile industry wastewater and treatment approaches for environmental safety. *J. Environ. Chem. Eng.* 105012.

Kishor, R., Saratale, G.D., Saratale, R.G., Ferreira, L.F.R., Bilal, M., Iqbal, H.M., and Bharagava, R.N. 2021c. Efficient degradation and detoxification of methylene blue dye by a newly isolated ligninolytic enzyme producing bacterium *Bacillus albus* MW407057. *Colloid. Surface. Biointerface.* 111947.

Kishor, R., Bharagava, R.N., and Saxena, G. 2018. Industrial wastewaters: The major sources of dye contamination in the environment, ecotoxicological effects, and bioremediation approaches. In: Bharagava, R.N. (ed.) *Recent Advances in Environmental Management.* CRC Press/Taylor & Francis Group, Boca Raton, FL, pp. 1–25.

Mishra, A. and Malik, A. 2014. Novel fungal consortium for bioremediation of metals and dyes from mixed waste stream. *Bioresour. Technol.* 171, 217–226.

Núñez, J., Yeber, M., Cisternas, N., Thibaut, R., Medina, P., and Carrasco, C. 2019. Application of electrocoagulation for the efficient pollutants removal to reuse the treated wastewater in the dyeing process of the textile industry. *J. Hazard. Mater.* 371, 705–711.

Okareh, O. T., Ademodi, T. F., and Igbinosa, E. O. 2017. Biotreatment of effluent from Adire textile factories in Ibadan, Nigeria. *Environ. Monit. Assess.* 189(12), 629.

Oktem, Y.A., Yuzer, B., Aydin, M.I., Okten, H.E., Meric, S., and Selcuk, H. 2019. Chloride or sulfate? Consequences for ozonation of textile wastewater. *J. Environ. Manag.* 247, 749–755.

Paździor, K., Bilińska, L., and Ledakowicz, S. 2019. A review of the existing and emerging technologies in the combination of AOPs and biological processes in industrial textile wastewater treatment. *Chem. Eng. J.* 376, 120597.

Rosa, J.M., Tambourgi, E.B., Vanalle, R.M., Gamarra, F.M.C., Santana, J.C.C., and Araújo, M.C., 2020. Application of continuous H_2O_2/UV advanced oxidative process as an option to reduce the consumption of inputs, costs and environmental impacts of textile effluents. *J. Clean. Prod.* 246, 119012.

Saratale, R.G., Ghodake, G.S., Shinde, S.K., Cho, S.-K., Saratale, G.D., Pugazhendhi, A., and Bharagava, R.N. 2018. Photocatalytic activity of $CuO/Cu(OH)_2$ nanostructures in the degradation of Reactive Green 19A and textile effluent, phytotoxicity studies and their biogenic properties (antibacterial and anticancer). *J. Environ. Manag.* 223, 1086–1097.

Sathishkumar, K., AlSalhi, M.S., Sanganyado, E., Devanesan, S., Arulprakash, A., and Rajasekar, A. 2019. Sequential electrochemical oxidation and bio-treatment of the azo dye Congo red and textile effluent. *J. Photochem. Photobiol. B Biol.* 200, 111655.

Sen, S.K., Patra, P., Das, C.R., Raut, S., and Raut, S. 2019. Pilot-scale evaluation of bio-decolorization and biodegradation of reactive textile wastewater: An impact on its use in irrigation of wheat crop. *Water Resour. Ind.* 21, 100106.

Sen, S.K., Raut, S., Bandyopadhyay, P., and Raut, S., 2016. Fungal decolouration and degradation of azo dyes: a review. Fungal Biol. Rev. 30 (3): 112–133.

Sepehri, A. Sarrafzadeh, M.H., and Avateffazeli M. 2020. Interaction between Chlorella vulgaris and nitrifying-enriched activated sludge in the treatment of wastewater with low C/N ratio, J. Clean. Prod. 247, 20119164.

Shahid, A., Malik, S., Zhu, H., Xu, J., Nawaz, M.Z., Nawaz, S., Alam, M.A., and Mehmood, M.A. 2020. Cultivating microalgae in wastewater for biomass production, pollutant removal, and atmospheric carbon mitigation; a review. *Sci. Total Environ.* 704, 135303.

Sinha, S., Singh, R., Chaurasia, A.K., and Nigam, S. 2016. Self-sustainable *Chlorella pyrenoidosa* strain NCIM 2738 based photobioreactor for removal of direct red-31 dye along with other industrial pollutants to improve the water-quality. *J. Hazard. Mater.* 306, 386–394.

Sun, Y., Cheng, S., Lin, Z., Yang, J., Li, C., and Gu, R. 2020. Combination of plasma oxidation process with microbial fuel cell for mineralizing methylene blue with high energy efficiency. *J. Hazard. Mater.* 384, 121307.

Tara, N., Arslan, M., Hussain, Z., Iqbal, M., Khan, Q.M., and Afzal, M. 2019. On-site performance of floating treatment wetland macrocosms augmented with dye-degrading bacteria for the remediation of textile industry wastewater. *J. Clean. Prod.* 217, 541–548.

Tian, Q., Ran, M., Fang, G., Ding, L., Pan, A., Shen, K., and Deng, Y. 2020. $ZnAl_2O_4/BiPO_4$ composites as a heterogeneous catalyst for photo-Fenton treatment of textile and pulping wastewater. *Sep. Purif. Technol.* 239, 116574.

Tomei, M. C., Soria Pascual, J., and Mosca Angelucci, D. 2016. Analysing performance of real textile wastewater bio-decolourization under different reaction environments. *J. Clean. Prod.* 129, 468–477.

Viana, D.F., Salazar-Banda, G.R., and Leite, M.S. 2018. Electrochemical degradation of reactive black 5 with surface response and artificial neural networks optimization models. *Sep. Sci. Technol.* 53(16), 2647–2661.

Waghmode, T.R., Kurade, M.B., Sapkal, R.T., Bhosale, C.H., Jeon, B.H., and Govindwar, S.P. 2019. Sequential photocatalysis and biological treatment for the enhanced degradation of the persistent azo dye methyl red. *J. Hazard. Mater.* 371, 115–122.

Watharkar, A.D., Khandare, R.V., Waghmare, P.R., Jagadale, A.D., Govindwar, S.P., and Jadhav, J.P. 2015. Treatment of textile effluent in a developed phytoreactor with immobilized bacterial augmentation and subsequent toxicity studies on *Etheostoma olmstedi* fish. *J. Hazard. Mater.* 283, 698–704.

2

Nanotechnology: A Valuable Asset for Contribution to Positive Impact on Environment

Ved Prakash Giri
Academy of Scientific and Innovative Research (AcSIR)
Lucknow University

Shipra Pandey
CSIR-National Botanical Research Institute
Academy of Scientific and Innovative Research (AcSIR)
Indian Institute of Technology

Madhuree Kumari
CSIR-National Botanical Research Institute
Academy of Scientific and Innovative Research (AcSIR)
Indian Institute of Science

Aradhana Mishra
CSIR-National Botanical Research Institute
Academy of Scientific and Innovative Research (AcSIR)

CONTENTS

DOI: 10.1201/9781003181224-2

2.1 Introduction

Nanoparticles being very small (10^{-9}m) differ from their bulk material counterpart in many of their physical and chemical properties. The core of nanotechnology lies in the fact that they will reduce the use of their chemical counterpart (Jeevanandam et al., 2018). It can reduce the cost involved by many folds because of their unique properties such as improving productivity and product quality, reducing postharvest losses, as well as increasing the competitiveness of agricultural producers and market access (Barbosa-Cánovas et al. 2017). Advances in nanotechnology may present new chances of reducing the environmental risk. Because of small particle size, large surface-to-volume ratio, higher water solubility, high penetration power and easy anchoring capacity on solid surfaces, nanoparticles play an important role in improving the current crop management techniques and combating pathogen attacks (Gehrke et al. 2015). Similarly, they can be revealed as next-generation low-cost water filters, food additives, smart delivery vehicles or quick pollutant-detecting sensors (Lammersa 2013).

Although the path of nanotechnology unleashes enormous possibilities for the poor, it is full of obstacles. While most investment in case of nanotechnology research and development activities is carried by the USA and the European Union, the developing countries are still lagging far behind in this kind of research and development activities where this technology is needed the most by the environment for pure water and cheaper grains. 'Nano-divide' (Beumer 2016), which refers to the gap between the rich and poor countries because of the high cost of the nanomaterials and technology, needs to be resolved. A vast number of collaborations between government ministries, multinational companies, agencies and research institutions worldwide are needed to boost the research and development activities and invest the capital required for growth and benefits of this new technology in developing countries.

This chapter concludes the potential benefits and broadening of nanotechnology for building certain nanotechnological factors and methods. It can provide an impactful solution to several environmental problems and move towards resource saving by preventing the environment from pollution, risk factors and agricultural damages. It may reveal a valuable asset to a positive impact on environment and reduce the constraints faced during employing the technology by developing nations.

2.2 Nanotechnology in Potential Benefits for Environment

About three-fourth of the total population of India resides in rural area (Census India SRS Bulletins' Registrar General of India, Govt of India, 2011) and relies largely on agriculture to run their livelihood. Although the demand for food has constantly increased, the land available for agriculture has reduced, limiting the farmers to search for alternatives or to pay heavily for new technologies. In such a situation, nanotechnology can help in a cost-effective manner by producing more grains using nano-pesticides and nano-fertilizers increasing shelf life of food (Zahedi et al. 2020) and connecting them to market. Similarly, the problem of water crisis is also one of the major problems generated annually in the environment *(India's Water Economy: Bracing for a Turbulent Future, World Bank).* Salamanca-Buentello et al. (2005) correlated the top ten applications of nanotechnology for the developing countries based on UN Millennium Development Goals such as enhancement of agricultural production, treatment and remediation of water, and food storage and processing. Though in initial stage, there are many nanotechnology-based products that can prove very fruitful for agriculture practices, nutrition, medicine, food, water safety, etc. The applications of nanomaterials and their potential benefits to the environment are shown in Figure 2.1.

Pheromones are chemical signals released by insects to attract opposite sex. In agricultural practices, they are widely used in integrated pest management. Currently, pheromones are sprayed by mechanical means, which results in uneven distribution and requires heavy dosages. Nanofibres composed of biodegradable polymers such as polyamide and cellulose acetate were used to release pheromones, which not only reduced the cost of application, but also resulted in even distribution (Bisotto-de-oliveira et al. 2015).

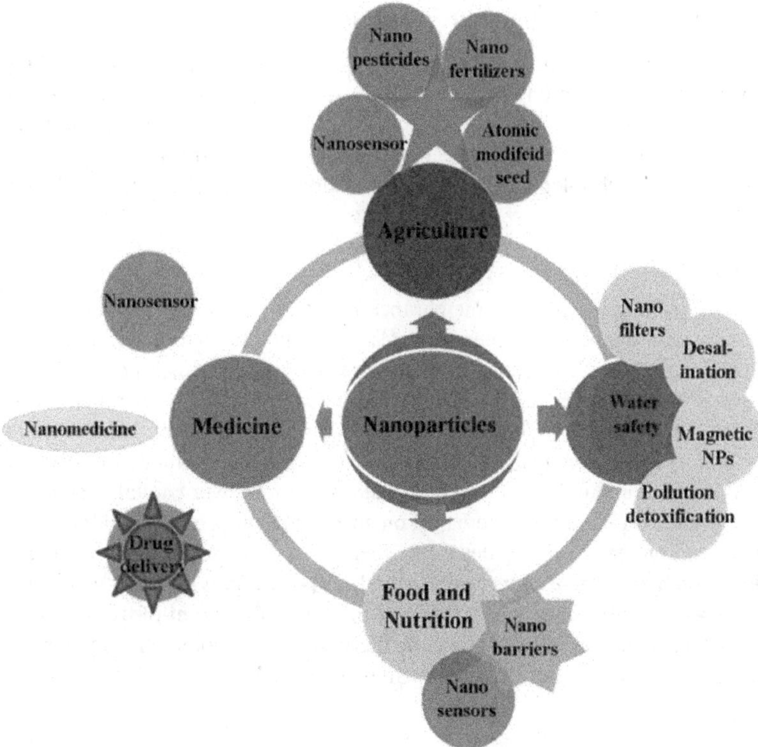

FIGURE 2.1 Application of nanomaterials and their potential benefits.

Nanotechnology applications have now reached the core of seeds, such as atomically modified seeds. The greatest advancement of this technique is the ability to modify the trait of interest without going into details of genetic manipulations. In March 2004, ETC group carried out a nanotechnology research on atomically modified characteristics of local rice varieties of Thailand. They inserted nitrogen atoms into the membrane of rice cells by drilling a hole and reported the rearrangement of their characteristics (Jha et al. 2011). Similarly, nanoparticles-based techniques are also helping the researchers in delivering the gene of interest to target cell through transformation. Francois Tourney Brian Tsceoyn and colleagues at Iowa State University described the delivery foreign genetic material into plant cells with the help of silica nanoparticles (Jha et al. 2011).

2.2.1 Nanotechnology in Water Safety, Treatment and Remediation

"Water shortage would be the cause of third world war" (Sunder Lal Bahuguna).

Access to clean water and sanitation is a basic resource to live, but is not available to everyone. Approximately 3.4 million people die each year from causes related to water, sanitation and hygiene. Nearly all deaths, that is 99%, occur in the developing world. World Health Organization (Prüss-Üstün and WHO, 2008) reported a work titled *Safer Water, Better Health: Costs, Benefits And Sustainability Of Interventions To Protect And Promote Health*. The water disinfection crisis causing hardships of lives through sickness than any war claims mortality through weapons.

All these data clearly demonstrate the need to develop new technologies to provide everyone with access to clean water without making it a costly affair. In such a context, several nanotechnology applications may help to improve the quality of water; some of them are as follows.

2.2.1.1 Nano-Enabled Water Filtration Techniques

Nanotechnology bears a great potential to clean water, desalinate it, detect water pollutants at a very early stage and detoxify it. The National Research Council (NRC) report (2009) on emerging technologies focused on several aspects of *Nanoscience* that will benefit farmers in Africa in an economical and effective way for water purification. Pradeep and Anshup (2009) from India are also continuously working on affordable nanotechnologies for water purification and cheap filtration units of nanoparticles (Sarkar et al. 2016).

2.2.1.1.1 Nanofilters

Nanofilters apply the techniques of nano-membranes made up of carbon nanotubes, nano-polymers, nanoporous ceramics and magnetic nanoparticles. Carbon nanotube channels can be utilized to expel pollutants from drinking water. Researchers at Rensselaer Polytechnic Institute and the Banaras Hindu University (Varanasi, India) teamed up to study the structure of a carbon nanotube channel to expel microbes from water and published a report in the GEO year book 2007 entitled *Emerging Challenges, Nanotechnology and the Environment.* Nanofilters simply higher operational, easier integration in research and resulting into a simpler unit operation along with a lower capital expenditure which can filter viruses as well. Carbon nanotubes are based on size exclusion principle, and the strong capability to adsorb chemical and biological molecules ranks them on a much higher position than conventional filtration techniques such as reverse osmosis (Das et al. 2014). Biopolymer-reinforced synthetic granular nanocomposites were synthesized and optimized for affordable point-of-use water purification by Sankar et al. (2013). Due to the better adsorption and thermodynamic potential, polymer nanocomposites have emerged as a cost-effective water filtration technology (Pandey et al. 2017).

2.2.1.1.2 NanoCeram

It is another type of water channel media demonstrating the advantages of standard pleated filters with a forward leap in nano-innovation. The channels have evacuation abilities along with the extent of regular filtration advancements and also have the ability to expel a wide scope of waterborne contaminants including microscopic organisms and colloidal particles. Karim et al. (2009) structured an inexpensive electropositive cartridge NanoCeram filter, which proficiently expelled enteroviruses and noroviruses from huge volumes of water.

2.2.1.1.3 Magnetic Nanoparticles for Filtration

Monodisperse magnetite (Fe_3O_4) nanocrystals have a strong and irreversible connection with arsenic while holding their magnetic properties. Earlier reports exhibited that the magnetic field improved the removal of cobalt and iron from simulated groundwater (Iranmanesh and Hullige, 2017). The magnetic field-improved filtration/sorption process varies altogether from the magnetic partition process utilized in the processing of minerals and water treatment. Iron oxides have been utilized for the treatment of fluid wastes containing radioactive and unsafe metals. These procedures include adsorption, precipitation and chemical and physical strategies. Similarly, a radioactive wastewater precipitation process incorporates ferric hydroxide floc to remove radioactive contaminants such as americium, plutonium and uranium (Yang et al. 2017).

2.2.1.1.4 Desalination

Ninety-seven per cent of the total water available on earth resides in seas and oceans. The saline water of oceans can be desalinated to obtain fresh water for common uses. The enormous scope of desalination normally requires a lot of vitality, and specific and costly infrastructure, making it very expensive as compared to the use of fresh water of rivers. Nanomaterials have a number of unique physical and chemical properties that make them specifically attractive as filtration media for water desalination. Four classes of nanoscale materials are being assessed as practical materials for water purification: (i) metal-containing nanoparticles, (ii) carbonaceous nanomaterials, (iii) zeolites and (iv) dendrimers. Zhao et al. (2013) synthesized thermo-responsive magnetic nanoparticles (MNPs) that attracted solute to separate water from saline or seawater by forward assimilation and change in temperature. Similarly, Ge et al.

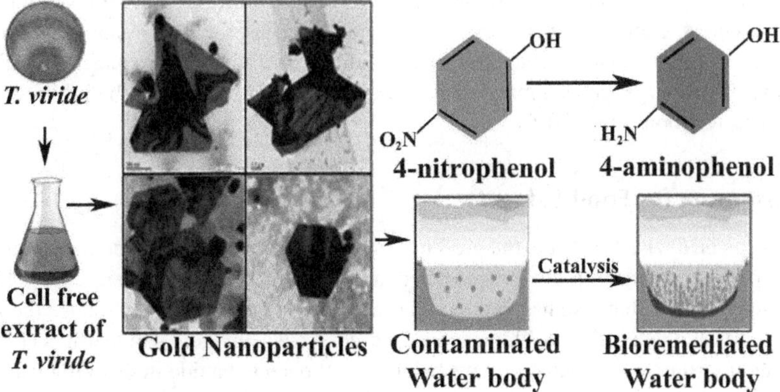

FIGURE 2.2 Degradation of water pollutants by diverse shapes and size of biosynthesized gold nanoparticles and its ability to degrade 4-nitropdhenol into 4-aminophenol. (Image adopted from Kumari et al. 2016.)

(2016) used hydroacid MNPs in forward osmosis for seawater desalination. Nanotechnology would give novel chances to develop more proficient and cost-effective nanostructured reactive membranes for water filtration, purification and desalination.

2.2.1.2 Prevention and Detoxification of Water Pollutants

Nanosensors can be used to detect pollutants in water at very early stage and remove them from contaminated sites (Steffens et al. 2017). Nanoscale zeolites, metal oxides, titanium dioxide, carbon nanotubes and fibres, enzymes and various noble metals, mainly bimetallic nanoparticles (Pandey et al. 2015), have been used for bioremediation. For the removal of arsenic, zinc and copper oxide nanoparticles can be used as a point-of-source purification materials (McDonald et al. 2015). Kumari et al. (2016a,b) reported the catalytic activity of biosynthesized gold nanoparticles (GNPs) for the conversion of 4-nitrophenol into 4-aminophenol (Figure 2.2). The process was strongly influenced by the GNPs' structure and dimension. Commonly, biodegradation practices are traditional, very expensive and time-consuming and require large amount of raw material. Diverse shapes and sizes of GNPs could revolutionize the biodegradation processes that can reveal a positive impact on the current scenario of the environment by removing the pollutants.

2.3 Nanotechnology – an Eco-Friendly Manner for Food Safety

A large proportion of people of developing countries live in rural areas and don't have access to satisfactory nutritional sources, safe food and screening procedures for checking their health status. The lack of healthy nutrition adds to the greater part of deaths of children under five in less industrialized nations. The utilization of a few economical nanotechnologies can possibly diminish this death rate (Buentello et al. 2005), including nanocomposites in food packaging; decontamination of food equipment, or food; nanotechnology-based antigen-detecting biosensors for identifying pathogen contamination; nanosensors for pest identification; and nanoparticles as pesticides, insecticides, etc. (Gruère et al. 2011).

2.3.1 Nanobarriers for Food Safety

Nanobarriers are nanoparticles-based coating that reduces the permeability of carbon dioxide, oxygen and other gases so that microbial access will be prevented. Nano-polymers, nanofilms, nanoclays and nanocomposites have successfully been used for this purpose (Saurabh et al. 2019). Polymer nanocomposites (PNCs) have enhanced polymer barrier properties. They are also stronger, are more flame resistant and possess better thermal properties (e.g. melting point, degradation and glass transition temperatures) than control polymers. Nanoclays have also been used to reduce the bioavailability of aflatoxin and deoxynivalenol and zearalenone mycotoxins. Because of the antimicrobial properties of silver nanopar ticles, they have also emerged as nanobarriers to increase the shelf life of food by inhibiting the growth of food pathogens (Carbone et al. 2016).

2.3.2 Nanosensors for Food Safety

Nanomaterials can be devised to detect the presence of gases, aromas, chemical contaminants and pathogens, or to respond to changes in environmental conditions (Wang and Duncan, 2017). This not only will assure quality control, but also has the potential to improve food safety and reduce the frequency of foodborne illnesses at a reduced cost. Fluorescence-based assay using gold nanoclusters (Liu et al. 2010) and a nanoscale liposome-based detector have been used to detect cyanide in drinking water at concentrations as low as 2 nM and contamination of drinking water with pesticides, respectively (Vamvakaki and Chaniotakis, 2007).

Similarly, nanosensors have widely been used to detect pathogenic microorganisms at a very early stage. Several protein-based bacteria at picomolar (pM) levels have been detected using antibody-labelled luminescent quantum dots. Easy-to-read colorimetric metal nanoparticle detectors have also been developed for detecting numerous other small molecules (Zhang et al. 2009a, b), proteins (Hong et al. 2009) and metal ions. There are examples that express the importance of nanomaterials, specifically magnetic nanomaterials, as vehicles for the simultaneous isolation and optical or magnetic detection of microorganisms. Gilardoni et al. (2016) assessed that MNPs can be used to remove *Mycobacterium avium* spp. paratuberculosis from contaminated whole milk.

2.4 Nanotechnology for Resource Saving in Pharmaceutical Industries

Medical research always demands cutting-edge technology to keep it with the pace of time. This makes the field of medicine a very expensive affair that cannot be afforded by the poor. Nanomedicine majorly contributes to the healthcare area by providing new and improved therapies as well as new diagnostic methods, which are faster and more affordable than the currently available methods.

2.4.1 Smart Drug Delivery Vehicle

Recent advances in nanotechnology have reignited the interest in drug delivery (Lammersa 2013). Liposomes, gold colloidal nanoparticles and iron oxide nanoparticles have proved their candidature to be used in cancer treatment. Polymeric nanoparticles owing to their unique features and cost-effectiveness have marked their role as an efficient smart drug delivery vehicle; some examples of smart drug delivery vehicles approved by FDA are mentioned in Table 2.1.

Douglas et al. (2012) demonstrated the role of nanorobots loaded with combinations of antibody fragments used in different types of cell signalling stimulation for cell targeting tasks. Similarly, the union of quantum dots is giving new bits of knowledge into the mechanism of non-viral nucleic acid delivery by the convergent characterization of delivery barriers. It has the potential to accelerate the design of improved carriers to understand the capability of nucleic acid therapeutics and gene medication (Grigsby et al. 2012).

2.4.2 Nanosensors for Pharmaceutical Industries

To detect pathogens at an early stage and at low cost has always been a mammoth task in medical research. Nanosensors because of their high catalytic property provide a low-cost and more accurate alternative to the current techniques available for sensing the pathogen and other chemicals. Glucose nanosensors have been used for the detection of increased glucose level, and triglyceride nanosensors are able to detect increased level of fat (Ansary and Faddah 2010). A nanoparticle hybridization assay

TABLE 2.1

Therapeutic Agents Used in Drug Formulation

S. No.	Therapeutic Name	Drug Formulation	Indication	References
1.	Myocet	A liposome-containing chemotherapeutic agent doxorubicin	Metastatic breast cancer	Netterwald (2013)
2.	Doxil	A liposome-containing chemotherapeutic agent doxorubicin	AIDS-related Kaposi sarcoma; ovarian cancer (if first-line treatment fails); and also multiple myeloma	Netterwald (2013)
3.	BIND-014	Cancer drug docetaxel, covered in ligands that interact with a common tumour antigen known as prostate-specific membrane antigen (PSMA)	Advanced or metastatic solid tumours	Netterwald (2013)
4.	Lipusu	Liposomal PTX	Ovarian cancer and non-small cell lung cancer	Zhang et al. (2009a,b)
5.	Marqibo Kit	Liposomal vincristine	Acute lymphoblastic leukaemia	Silverman and Deitcher (2013)
6.	CosmoFer, Dexferrum	Iron dextran colloid	Iron-deficient anaemia	Anselmo and Mitragotri (2016)
7.	Diafer	Five per cent iron isomaltoside	Iron-deficient anaemia	Anselmo and Mitragotri (2016)
8.	Estrasorb™	Micellar estradiol	Menopausal therapy	Bobo et al. (2016)
9.	Abraxane®/ABI-007	Albumin-bound paclitaxel nanoparticles	Breast cancer and pancreatic cancer	Bobo et al. (2016)
10.	NanoTherm®	Iron oxide	Glioblastoma	Ulbrich et al. (2016)
11.	CALAA-01	Polymer-based nanoparticle containing small interfering RNA (siRNA)	Cancer cell imaging	Zuckerman et al. (2014)
12.	AuroShell	Photothermal gold nanoshell	Tumour cell imaging	Schwartz (2009)

has been developed, which involves ubiquitous and specific magnetic DNA probes targeting bacterial 16S rRNAs, to detect amplified target DNAs using a miniaturized NMR device (Chung et al. 2013). Kim et al. (2017) evaluated the potential of metal nanoparticles as the catalysts to establish high-performance sensor arrays for the pattern recognition of biomarkers to monitor their ability as chemosensitive biosensors. Banerjee et al. (2017) emphasized the importance of sensing and detecting pathogenic bacteria before the food reaches our dining table.

2.5 Nanotechnology in Medicine

Nanomedicines are the nanoparticles themselves acting as potent medicines for diseases or a combination of medicine, natural products or drugs to enhance the efficacy of the traditional medicine or drugs. Kumari et al. (2017a–c) demonstrated the antimicrobial and antioxidative role of metal nanoparticles. Kang et al. (2016) developed a blood-cleansing device (biospleen) for the removal of pathogens from infected blood with the help of nanotechnology. They formulated streptavidin-coated superparamagnetic nanobeads conjugated with mannose-binding lectin, a broad-spectrum opsonin for capturing and eliminating pathogens. A number of publications suggest nanoparticles loaded with a natural product and drug to treat some difficult diseases. Tiwari et al. (2014) demonstrated the potency of curcumin-loaded nanoparticles to cure reverse cognitive deficits in Alzheimer's disease model in an economical manner. Similarly, Landis et al. (2017) showed the role of cross-linked polymer-stabilized nanocomposites for the

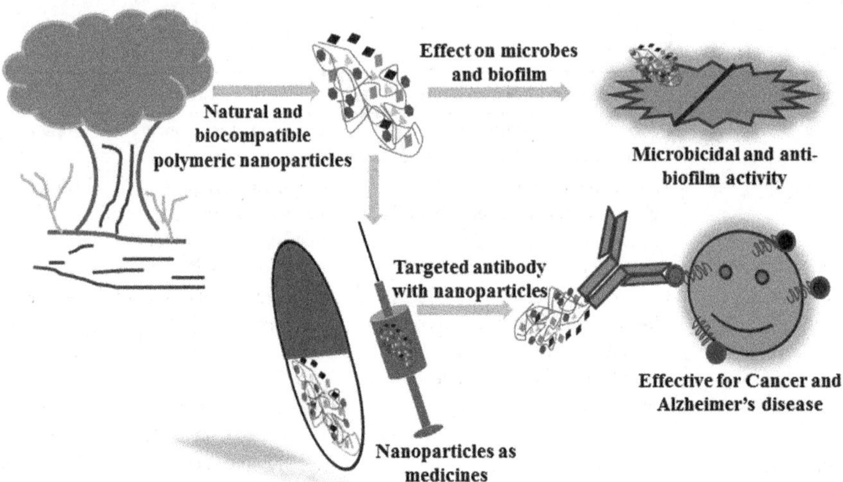

FIGURE 2.3 Nanotechnology applied in pharmaceuticals industries as nano-medicine.

treatment of bacterial biofilms. Nanoparticles have also been used synergistically with antibiotics to curb multiple drug resistance of pathogens (Kumari et al. 2016a,b) (Figure 2.3).

2.6 Impact of Nanotechnology in Agriculture

In an agriculture-based developing country like India, there is an urgent need to develop technologies that can combat the adverseeffects without hampering the budget of farmers and improved crop productivity. The investigation has been proceeded to achieve advancements in technology and a major role in shaping sustainable agricultural development goals (SDGs). Seventeen goals of SDGs have been subjected in agriculture, among them goal two is 'zero hunger' with the aim of 'end hunger' (Bebbington and Unerman 2018). Nanotechnology is now used as an emerging tool in agriculture with a myriad of applications, *viz.* nano-pesticides, nano-fertilizers and nanosensors, to enhance food security, improve nutrition and promote sustainable agriculture (Mir et al. 2017). In the previous decades, green nanotechnology had the potential to be the reason for next '*Green Revolution in India*'. Nanoparticles are broadly defined as particles having at least one dimension that belongs to size 1–100 nm. There are two types of approaches that can be used in the synthesis of nanomaterials: 'top-down' and 'bottom-up approaches'. They provide nanoscale materials by the fabrication of nanostructure (Biswas et al. 2012). The nanotechnology has the ability to influence the bioavailability of biologically active agents by their solubility and permeability. A broad range of nanomaterials are being used in agriculture for a myriad of applications, including antimicrobial activity, nutrition acquisition, pesticidal activity, pesticide detection could achieve in a sustainable manner (Kumari et al. 2019; Ashrafi et al. 2020).

Several researches have previously been done for the synthesis and application of nanomaterials in different perspectives; some of them are mentioned in Table 2.2. To monitor and regulate the food supply, nanosensors play an important role in hazardous materials and early disease detection (Yusof and Isha 2020). Single-walled carbon nanotubes (SWCNTs) are able to detect H_2O_2 in *Arabidopsis* plant that is key factor for ROS generation (Wu et al. 2020). Nanomaterials-based products have the potential to enhance plant growth by promoting seed germination, vigour index, productivity, etc. (Yusefi-Tanha et al. 2020). The most salient feature of nanomaterials is their precise and targeted activity, which takes a leap over chemical fertilizers and pesticides. The chitosan-loaded Zn/B nano-fertilizer has been applied to enhance the uptake of macro- and micronutrients (Yatim et al. 2019).

TABLE 2.2

Applications of Nanomaterials in Agriculture

Nanomaterials	Size Distribution	Targeted Plant	Functions	References
Fe_3O_4, TiO_2 and carbon	30–50 nm	Cucumber	Seed germination	Mushtaq (2011)
Cu-Zn NPs and carbon nanofibres	50–60 nm	*Cicer arietinum*	Controlled release of micronutrients	Kumar et al. (2018)
Single-walled CNTs	-	*Arabidopsis thaliana*	H_2O_2 detection	Wu et al. (2020)
Carbon nanodots	3–5 nm	Cherry tomato	Pesticide detection	Ashrafi Tafreshi et al. (2020)
Chitosan-loaded Zn/B nano-fertilizer	200–2,000 nm	Coffee seedling	Enhanced uptake of Zn, nitrogen and phosphorus	Wang and Nguyen (2018)
ZnO	38–500 nm	Soybean	Enhanced seed yield and antioxidant activity	Yusefi-Tanha et al. (2020)
Chitosan	539 nm	Maize	Delivery of Cu and salicylic acid	Sharma et al. (2020)
Cu nanowires	-	Alfalfa	Enhanced growth at physiological and molecular levels	Cota-Ruiz et al. (2020)
CNTs	-	Catharanthus seeds	Enhanced seed germination and stress tolerance	Pandey et al. (2019)
Nanoclay polymer composite	-	Plant/soil	Controlling nitrification in soil	Saurabh et al. (2019)
Mentha longifolia oil nanoemulsion	14–36 nm	*Ephestia kuehniella*	Pesticidal effect	Louni et al. (2018)
Chitosan NPs	89.8 nm	*Fusarium oxysporum* f.sp. *ciceris*, *Pyricularia grisea* and *Alternaria solani*	Antifungal activity	Sathiyabama and Parthasarathy (2016)
Cellulose acetate + clay + oregano oil	-	*Alternaria alternata*, *Geotrichum candidum* and *Rhizopus stolonifer*	Antifungal activity	Pola et al. (2016)
Graphene oxide	-	*Xanthomonas oryzae* pv. *oryzae*	Antibacterial activity	Chen et al. (2013)
Selenium–arabinogalactan	67 nm	*Clavibacter michiganensis*	Antibacterial activity	Papkina et al. (2015)
Graphene oxide–silver nanocomposite	10–35 nm	*Fusarium graminearum*	Antifungal activity	Chen et al. (2016)
Chitosan–Cu-Zn nanocomposite	1.7–23.7 nm	*A. alternata*, *R. solani* and *B. cinerea*	Antifungal activity	Al-Dhabaan et al. (2017)
Fe_3O_4/ZnO/AgBr nanocomposites		*Fusarium* spp.	Antifungal activity	Hoseianzadeh et al. (2016)
Silver/chitosan nanocomposite		*Rhizoctonia solani*, *Aspergillus flavus* and *Alternaria alternata*	Seed-borne fungal pathogens	Kaur et al. (2012)

(Continued)

TABLE 2.2 (*Continued*)

Applications of Nanomaterials in Agriculture

Nanomaterials	Size Distribution	Targeted Plant	Functions	References
Chitosan–copper nanocomposites	4–15 nm	*Sclerotium rolfsii* and *Rhizoctonia solani*	Antifungal activity	Rubina et al. (2017)
Silver+chitin nanofibres	10 nm	*Alternaria alternata*, *A. brassicae*, *A. brassicicola*, *Bipolaris oryzae*, *Botrytis cinerea* and *Penicillium digitatum*	Antifungal activity	Ifuku et al. (2015)

2.7 Status of Nanotechnology in Research and Development

The use of nanotechnology has given birth to serious concerns over ethical, economic, environmental, legal and social issues (NE3LS) (Endo et al. 2011). In particular, the application of nanotechnology in second and third world countries requires broadening of our imagination, belief and support on socio-economic front. Maximum research in the field of nanotechnology is carried out in developed countries. Advanced developed countries such as the USA, Japan and the member states of the European Union are investing billions of dollars annually in nanoscale research to build the scientific foundations for nanotechnology commercialization, such as 'The Nanoscale Science, Engineering, and Technology (NSET) Washington, DC'. Among developing countries, China, Brazil and India have heavily invested in nanotechnology, while smaller developing countries lag behind. To have the agricultural, food and water nanotechnology accessed at the grass-roots level, basic capital, infrastructure and sufficient economic funding are needed, which are very hard to find in third world countries. Although research initiatives have been taken in countries such as South Africa, Thailand, Vietnam, Jordan, Kazakhstan, Pakistan, Uzbekistan and Venezuela, there is a need for proper policy making, support of public sector and financial findings to make nanotechnology do wonders for poor people and the environment.

2.8 Costs and Access to Nanotech Applications

Patents and copyrights have always been used as a weapon to make money out of research. Recently, a number of patents have been filed and awarded in the USA claiming innovation and application in nanotechnology (ETC Group 2005). These patents not only establish the monopoly of the developed world on the market, but also enhance the cost of technology by many folds, making it a dream for poor people. Moreover, the absence of any form of IPR protection also discourages private companies to invest in developing countries and reduces investment in agricultural technology, which would ultimately keep useful technologies away from those who would benefit from them the most. Licensing agreement, market segmentation and searching alternatives for patent are some measures that could help these technologies reach the bottom line.

The term 'nano-divide' is generally used to denote the increasing gap between the rich and the poor countries due to nanotechnology, and this needs to be solved. Different kinds of nano-divide including access, profit, benefit and control of nano-divides need to be addressed with the involvement of organizations such as UNESCO for the ethics and policies of nanotechnologies (Beumer 2016).

2.9 Conclusions and Future Aspects

Nanotechnology has given hope to over millions of people living in developing countries for better access to food and water. It has the potential to increase crop yield and improve food and water safety. Nano-enabled technologies help in the detection of contaminants and pollutants at a very early stage without affecting the economy and environment. However, this technology has given a new hope to developing countries along with positive impact on environment. Most of the research and development carried out in the field of this science aims at the developed world and protects the natural environment. There is a wide information gap associated with the risks involved in this technology and the fear of unknown, which needs to be solved. This is the time to shift the direction of the research towards achieving useful goals for agricultural, pharmaceutical and sustainable development. There is necessity for taking into account the present and future constraints associated with global warming and bioenergy. Governance options, intellectual property rights and the role of national, regional and international regulatory frameworks need to be modified, by which the outcomes of research could not remain confined to the premises of laboratory only, but can reach the hands of last man in the queue.

REFERENCES

Al-Dhabaan, F.A., Shoala, T., Ali, A.A. et al. 2017. Chemically-produced copper, zinc nanoparticles and chitosan–bimetallic nanocomposites and their antifungal activity against three phytopathogenic fungi. *International Journal of Agricultural Technology* 13(5):753–69.

Ansary, A., and Faddah, L.M. 2010. Nanoparticles as biochemical sensors. *Nanotechnology, Science and Applications* 3:65–76.

Anselmo, A.C., and Mitragotri, S. 2016. Nanoparticles in the clinic. *Bioengineering & Translational Medicine* 1:10–29.

Ashrafi Tafreshi, F., Fatahi, Z., Ghasemi, S.F. et al. 2020. Ultrasensitive fluorescent detection of pesticides in real sample by using green carbon dots. *PLos One* 15:e0230646.

Banerjee, T., Shelby, T., and Santra, S. 2017. How can nanosensors detect bacterial contamination before it ever reaches the dinner table? *Future Microbiology* 12:97–100.

Barbosa-Cánovas, G.V., Pastore, G.M., Candoğan. et al. 2017. *Global Food Security and Wellness.* Springer, New York.

Bebbington, J., and Unerman, J. 2018. Achieving the United Nations sustainable development goals. *Accounting, Auditing & Accountability Journal* 31:2–24.

Beumer, K. 2016. Broadening nanotechnology's impact on development. *Nature Nanotechnology* 11:398–400.

Bisotto-de-oliveira, R, Morais, R.M., Roggia, I. et al. 2015. Polymers nanofibers as vehicles for the release of the synthetic sex pheromone of *Grapholita molesta* (lepidoptera, tortricidae). *Revista Colombiana de Entomologia* 41(2):262–269.

Biswas, A., Bayer, I.S., Biris, A.S. et al. 2012. Advances in top–down and bottom–up surface nanofabrication: techniques, applications & future prospects. *Advanced Colloid Interface Science* 170:2–27.

Bobo, D., Robinson, K.J., Islam, J. et al. 2016. Nanoparticle-based medicines: a review of FDA-approved materials and clinical trials to date. *Pharmacology Research* 33:2373–87.

Buentello, F.S, Persad, D.L., Court, E.B. et al. 2005. Nanotechnology and the developing world. *PLoS Med* 2:97.

Carbone, M., Donia, D.T., Sabbatella, G. et al. 2016. Silver nanoparticles in polymeric matrices for research food packaging. *Journal of King Saud University Science* 28:273–279.

Census India SRS Bulletins' Registrar General of India, Govt of India, 2011.

Chen, J., Sun, L., Cheng, Y., et al. 2016. Graphene oxide-silver nanocomposite: novel agricultural antifungal agent against *Fusarium graminearum* for crop disease prevention. *ACS Applied Materials & Interfaces* 8(36):24057–24070.

Chen, J., Wang, X., and Han, H. 2013. A new function of graphene oxide emerges: inactivating phytopathogenic bacterium *Xanthomonas oryzae* pv. *Oryzae*. *Journal of Nanoparticle Research* 15(5):1658.

Chung, H.J., Castro, C.M., Im, H. et al. 2013. A magneto-DNA nanoparticle system for rapid detection and phenotyping of bacteria. *Nature Nanotechnology* 8:369–75.

Cota-Ruiz, K., Ye, Y., and Valdes, C. 2020. Copper nanowire as nanofertilizers for alfalfa plants: understanding nano-bio systems interactions from microbial genomics, plant molecular responses and spectroscopic studies. *Science of the Total Environment* 742:140572.

Das, R., Ali, Md. E., Hamid, S.B.A. et al. 2014. Carbon nanotube membranes for water purification: a bright future in water desalination. *Desalination* 336:97–109.

Douglas, M., Bachelet, I., and Church, G.M. 2012. A logic-gated nanorobot for targeted transport of molecular payloads. *Science* 335:831–4.

Duncan, T.V. 2011. Applications of nanotechnology in food packaging and food safety: barrier materials, antimicrobials and sensors. *Journal of Colloid Interface Science* 363:1–24.

Endo, C.A., Emond, C., Battista, R., et al. 2011, July. The Ne3LS Network, Québec's initiative to evaluate the impact and promote a responsible and sustainable development of nanotechnology. *Journal of Physics: Conference Series* 304(1):012090.

ETC Group. 2004. Down on the farm: the impact of nano-scale technologies on food and agriculture. Action Group on Erosion, Technology, and Conservation. November, Ottawa, Canada.

ETC Group. 2005. The potential impacts of nano-scale technologies on commodity markets: the implications for commodity dependent developing countries. Trade-Related Agenda Development and Equity research Papers. Geneva, Switzerland.

Ge, Q., Yang, L., and Cai, J. 2016. Hydroacid magnetic nanoparticles in forward osmosis for seawater desalination and efficient regeneration via integrated magnetic and membrane separations. *Journal of Membrane Science* 520:550–559.

Gehrke, I., Geiser, A. and Somborn-Schulz, A. 2015. Innovations in nanotechnology for water treatment. *Nanotechnology, Science and Applications* 8: 1.

Gilardoni, L.R., Fernández, B., and Morsella, C. et al. 2016. *Mycobacterium paratuberculosis* detection in cow's milk in Argentina by immunomagnetic separation-PCR 47. *Brazilian Journal of Microbiology* 47:506–512.

Grigsby, C.L., Ho, Y.P., and Leong, K.W. 2012. Understanding nonviral nucleic acid delivery with quantum dot-FRET nanosensors. *Nanomedicine* 4:565–577.

Gruère, G., Narrod, C., and Abbott, L. 2011. *Agricultural, Food, and Water Nanotechnologies for the Poor.* International Food Policy Research Institute, Washington, DC.

Hong, S., Choi, I., Lee, S. et al. 2009. Sensitive and colorimetric detection of the structural evolution of superoxide dismutase with gold nanoparticles. *Analytical Chemistry* 81:1378.

Hosseinzadeh, H., and Khoshnood, N. 2016. Removal of cationic dyes by poly (AA-co-AMPS)/montmorillonite nanocomposite hydrogel. *Desalination and Water Treatment* 57(14):6372–6383.

Ifuku, S., Tsukiyama, Y., Yukawa, T., et al. 2015. Facile preparation of silver nanoparticles immobilized on chitin nanofiber surfaces to endow antifungal activities. *Carbohydrate Polymers* 117:813–817.

Iranmanesh, M., and Hullige, J. 2017. Magnetic separation: its application in mining, waste purification, medicine, biochemistry and chemistry. *Chemical Society Reviews* 46:5925–5934.

Jeevanandam, J., Barhoum, A., Chan, Y.S., et al. 2018. Review on nanoparticles and nanostructured materials: history, sources, toxicity and regulations. *Beilstein Journal of Nanotechnology* 9(1): 1050–1074.

Jha, Z., Behar, N., Sharma, S.N. et al. 2011. Nanotechnology: prospects of agricultural advancement. *Nano Vision* 1: 88–100.

Kang, J.H., Super, M., and Ingber, D.E. 2016. *A BioSpleen Blood Cleansing Device for Sepsis Therapy.* Seoul National University.

Karim, M.R., Rhodes, E.R., Brinkman, N. et al. 2009. New electropositive filter for concentrating enteroviruses and noroviruses from large volumes of water. *Applied Environmental Microbiology* 75:2393–2399.

Kaur, P., Thakur, R., and Choudhary, A. 2012. An in vitro study of the antifungal activity of silver/chitosan nanoformulations against important seed borne pathogens. *International Journal of Scientific Technology and Research* 1(7):83–86.

Kim, S.J., Choi, S.J., Jang, J.S., et al. 2017. Innovative nanosensor for disease diagnosis. *Accounts of Chemical Research* 50:1587–1596.

Kumar, R., Ashfaq, M., and Verma, N. 2018. Synthesis of novel PVA–starch formulation-supported Cu–Zn nanoparticle carrying carbon nanofibers as a nanofertilizer: controlled release of micronutrients. *Journal of Material Science* 53:7150–7164.

Kumari, M., Giri, V.P., Pandey, S., et al. 2019. An insight into the mechanism of antifungal activity of biogenic nanoparticles than their chemical counterparts. *Pesticide Biochemistry and Physiology* 157:45–52.

Kumari, M., Mishra, A., Pandey, S., et al. 2016a. Physico-chemical condition optimization during biosynthesis lead to development of improved and catalytically efficient gold nano particles. *Scientific Reports* 6(1):1–14.

Kumari, M., Pandey, S., Bhattacharya, A., et al. 2017a. Protective role of biosynthesized silver nanoparticles against early blight disease in *Solanum lycopersicum*. *Plant Physiology and Biochemistry* 121:216–225.

Kumari, M., Pandey, S., Giri, V.P., et al. 2016b. Tailoring shape and size of biogenic silver nanoparticles to enhance antimicrobial efficacy against MDR bacteria. *Microbial Pathogenesis* 105:346–355.

Kumari, M., Pandey, S., Mishra, S.K., et al. 2017b. Effect of biosynthesized silver nanoparticles on native soil microflora via plant transport during plant-pathogen-nanoparticles interaction. *3 Biotech* 7:345.

Kumari, M., Shukla, S., Kumari, M., et al. 2017c. Enhanced cellular internalization: a bactericidal mechanism more relative to biogenic nanoparticles than chemical counterparts. *ACS Applied Material Interfaces* 9:4519–4533.

Lammersa, T. 2013. SMART drug delivery systems: back to the future vs. clinical reality. *International Journal of Pharmacology* 454:527–529.

Landis, R.F., Gupta, A., Lee, Y.W., et al. 2017. Cross-linked polymer-stabilized nanocomposites for the treatment of bacterial biofilms. *ACS Nanotechnology* 24:946–952.

Liu, Y., Ai, K., and Cheng, X. 2010. Gold-nanocluster-based fluorecent sensors for highly sensitive and selective detection of cyanide in water. *Advanced Functional Materials* 20:951.

Louni, M., Shakarami, J. and Negahban, M. 2018. Insecticidal efficacy of nanoemulsion containing Mentha longifolia essential oil against Ephestia kuehniella (Lepidoptera: Pyralidae). *Journal of Crop Protection* 7(2):171–182.

McDonald, K.J., Reynolds, B., and Reddy, K.J. 2015. Intrinsic properties of cupric oxide nanoparticles enable effective filtration of arsenic from water. *Science Reports* 5:11110.

Mir, S.A., Shah, M.A., Mir, M.M., et al. 2017. New horizons of nanotechnology in agriculture and food processing industry. In *Integrating Biologically-Inspired Nanotechnology into Medical Practice*. IGI Global, 230–258.

Mushtaq, Y.K. 2011. Effect of nanoscale Fe_3O_4, TiO_2 and carbon particles on cucumber seed germination. *Journal of Environmental Science and Health Part A* 46(14):1732–1735.

National Research Council, 2009. *Emerging Technologies to Benefit Farmers in sub-Saharan Africa and South Asia*. Washington, DC: The National Academies Press. https://doi.org/10.17226/12455.

Netterwald, J. 2013. Nano-vehicles for cancer drugs. *The Scientist*.

Pandey, K., Anas, M., and Hicks, V.K. 2019. Improvement of commercially Valuable traits of industrial crops by application of carbon-based nanomaterials. *Science Reports* 9(1):1–14.

Pandey, N., Shukla, S.K., and Singh, N.B. 2017. Water purification by polymer nanocomposites: an overview. *Nanocomposites* 3:47–66.

Pandey, S., Kumari, M., Singh, S.P. et al. 2015. Bioremediation via nanoparticles: an innovative microbial approach. In *Handbook of Research on Uncovering New Methods for Ecosystem Management through Bioremediation*, IGI Global. pp. 491–515.

Papkina, A.V., Perfileva, A.I., Zhivet'yev, M.A. et al. 2015. Complex effects of selenium-arabinogalactan nanocomposite on both phytopathogen *Clavibacter michiganensis* subsp. sepedonicus and potato plants. *Nanotechnologies in Russia* 10(5–6):484–491.

Pola, C.C., Medeiros, E.A., Pereira, O.L. et al. 2016. Cellulose acetate active films incorporated with oregano (*Origanum vulgare*) essential oil and organophilic montmorillonite clay control the growth of phytopathogenic fungi. *Food Packaging and Shelf Life* 9:69–78.

Pradeep, T., and Anshup. 2009. Noble metal nanoparticles for water purification: a critical review. *Thin Solid Films* 517:6441–6478.

Prüss-Üstün, A., and World Health Organization. 2008. Safer water, better health: costs, benefits and sustainability of interventions to protect and promote health. *World Health Organization*.

Rubina, M.S., Vasil'kov, A.Y., Naumkin, A.V. et al. 2017. Synthesis and characterization of chitosan–copper nanocomposites and their fungicidal activity against two sclerotia-forming plant pathogenic fungi. *Journal of Nanostructure in Chemistry* 7(3):249–258.

Salamanca-Buentello, F., Persad, D.L., Court, E.B. et al. 2005. Nanotechnology and the developing world. *PLoS Medicine* 2:383–386.

Sankar, M.U., Aigal, S., Maliyekkal, S.M. et al. 2013. Biopolymer-reinforced synthetic granular nanocomposites for affordable point-of-use water purification. *PNAS* 110:8459–8464.

Sarkar, D., Mahitha, M.K., and Som, A. 2016. Metallic nanobrushes made using ambient droplet sprays. *Advanced Materials* 28:2223–2228.

Sathiyabama, M., and Parthasarathy, R. 2016. Biological preparation of chitosan nanoparticles and its in vitro antifungal efficacy against some phytopathogenic fungi. *Carbohydrate Polymers* 151:321–325.

Saurabh, K., Kanchikeri, M., and Datta, S.C. 2019. Nanoclay polymer composites loaded with urea and nitrification inhibitors for controlling nitrification in soil. *Archive of Agronomy and Soil Science* 65(4):478–491.

Schwartz, J.A. 2009. Feasibility study of particle-assisted laser ablation of brain tumors in orthotopic canine model. *Cancer Research* 69(4):1659.

Sharma G, Kumar A, Devi KA 2020. Chitosan nanofertilizer to foster source activity in maize. *International Journal of Biology and Macromolecules* 145:226–234.

Silverman, J.A., and Deitcher, S.R. 2013. Marqibo® (vincristine sulfate liposome injection) improves the pharmacokinetics and pharmacodynamics of vincristine. *Cancer Chemotherapy Pharmacology* 71:555–64.

Steffens, C., Steffens, J., Graboski, A.M., et al. 2017. Nanosensors for detection of pesticides in water. *New Pesticides and Soil Sensors* 595–635.

Tiwari, S.K., Agarwal, S., Seth, B., et al. 2014. Curcumin-loaded nanoparticles potently induce adult neurogenesis and reverse cognitive deficits in Alzheimer's disease model via canonical wnt/β-catenin pathway. *ACS Nano* 28:76–103.

Ulbrich, K., Holá, K., Šubr, V., et al. 2016. Targeted drug delivery with polymers and magnetic nanoparticles: covalent and noncovalent approaches, release control, and clinical studies. *Chemical Reviews* 116:5338–5431.

Vamvakaki, V., and Chaniotakis, N.A. 2007. Pesticide detection with a liposome-based nano-biosensor. *Biosensors and Bioelectronics* 15: 2848–53.

Wang, Y., and Duncan, T.V. 2017. Nanoscale sensors for assuring the safety of food products. *Current Opinion Biotechnology* 44: 74–86.

Wang, S.L., and Nguyen, A.D. 2018. Effects of Zn/B nanofertilizer on biophysical characteristics and growth of coffee seedlings in a greenhouse. *Research on Chemical Intermediates* 44(8):4889–4901.

Watkins, K., 2006. Human Development Report 2006-Beyond scarcity: Power, poverty and the global water crisis. *UNDP Human Development Reports (2006).*

Wu, H., Nißler, R., and Morris, V. 2020. Monitoring plant health with near-infrared flouorescent H_2O_2 nanosensors. *Nano Letters* 20:2432–2442.

Yang, H.M., Sun, K., Chan, H., et al. 2017. Sodium-copper hexacyanoferrate-functionalized magnetic nanoclusters for the highly efficient magnetic removal of radioactive caesium from seawater. *Water Research* 15: 81–90.

Yatim, N.M., Shaaban, A., Dimin, M.F. et al 2019. Urea functionalized multiwalled carbon nanotubes as efficient nitrogen delivery system for rice. *Advances in Natural Science, Nano Science and Nanotechnology* 10(1):015011.

Yusefi-Tanha, E., Fallah, S., Rostamnejadi, A. 2020. Zinc oxide nanoparticles (ZnONPs) as novel nanofertilizer: influence on seed yield and antioxidant defense system in soil grown soybean (*Glycine max* cv. Kowsar). *Science of the Total Environment* 140240.

Yusof, N.A., and Isha, A. 2020. Nanosensors for early detection of plant diseases. In *Nanomaterials for Agriculture and Forestry Applications*. Elsevier, 407–419.

Zahedi, S.M., Karimi, M., and Teixeira da Silva, J.A. 2020. The use of nanotechnology to increase quality and yield of fruit crops. *Journal of Science Food and Agriculture* 100:25–31.

Zhang, J., Cao, X., and Wang, L. 2009a. Adenosine detection by using gold nanoparticles and designed aptamer sequences. *Analyst* 134:1355.

Zhang, Q., Huang, X.E., and Gao, L.L. 2009b. A clinical study on the premedication of paclitaxel liposome in the treatment of solid tumors. *Biomedicine and Pharmacotherapy* 63: 603–7.

Zhao, Q., Chen, N., and Zhao, D. 2013. Thermo responsive magnetic nanoparticles for seawater desalination. *ACS Applied Material Interfaces* 5–21.

Zuckerman, J.E., Gritli, I., and Tolcher, A. 2014. Correlating animal and human phase Ia/Ib clinical data with CALAA-01, a targeted, polymer-based nanoparticle containing siRNA. *Proceedings of the National Academy of Sciences USA* 111: 11449–11454.

3

Wastewater Treatment Using Biochar-Amended Constructed Wetland Systems

Vivek Rana and Jyoti Sharma
Central Pollution Control Board

CONTENTS

3.1 Introduction

The aquatic water bodies degrade inexorably due to burgeoning population, industrialization, urbanization and over-exploitation of natural water resources. The continuous discharge of various pollutants (organic as well as inorganic in nature) emanating from different municipal and industrial sources is a matter of concern globally. Till now, a lot of conventional methods have been employed to treat different types of wastewaters, but these techniques have some inherent limitations such as high capital cost and technical complexity, due to which applicability is limited. Hence, it is imperative to explore an alternative eco-friendly and cost-effective technology to treat the wastewater (Batool and

DOI: 10.1201/9781003181224-3

Zeshan 2017). Constructed wetlands (CWs) have been considered as a viable solution to the remediation of wastewaters loaded with various pollutants. The treatment process depends upon factors such as the potential of plants, microbial degradation pathways, substrate properties and treatment dynamics.

Plants used in the CWs also play a significant role in the uptake of various contaminants and metabolizing them. The trace metals such as Fe, Cu, Pb and Zn enhance the plant growth if present in acceptable concentration; however, a high concentration of metals can induce plant toxicity, which in turn leads to the disruption of metabolic activity. Thus, hyperaccumulator plant species have a high potential to tolerate the high concentration of metals in their tissues. The most commonly used hyperaccumulators are *Typha latifolia* L. and *Phragmites australis* (Cav.) Trin. ex Steud. due to their dense root system, rapid growth, high biomass and rapid detoxification (Kumari and Tripathi 2015). Zhu and Haynes (2010) used various species of plants for the removal of nutrients, such as *Canna indica* L., *Cassia tora* L., *Arundo donax* L. and *Imperata cylindrica* (L.) P. Beauv.

Various substrates used in the CWs are rice husk, palm tree mulch, steel slag, gravel and sand and limestone–coco peat mixture (Batool and Zeshan 2017). The removal efficiency of the substrate is assumed to depend mainly on its sorption capacity and the chemical and physical conditions of the CWs. Biochar has also been proved to be excellent an amendment to improve the fertility of the soil. The application of biochar as a substrate medium is mainly owing to its large surface area, high sorption ability and high nutrient removal potential. An extensive literature is available on the amendment of biochar in the soil to enhance the treatment potential of a soil for contaminants. However, limited attention has been paid to the implication of biochar as a filter medium in CWs for the depuration of pollutants from different sources (Kasak et al. 2018). This chapter highlights the efficiency of CWs to treat different types of wastewaters and the effect of biochar amendment on the efficiency of CWs.

3.2 Phytoremediation

An upsurge in the generation of municipal and industrial wastewater and their direct discharge into aquatic resources of water is an environmental concern. The traditional technologies to combat the problem of wastewater remediation are proved to be costly and have operational complexities. So, it is imperative to find a cost-effective and sustainable alternate for environmental remediation. Phytoremediation is a green, sustainable and cost-effective technology for the degradation of environmental pollutants (Rana and Maiti 2018a). The phytoremediation process involves the direct use of green plants to remediate the aquatic water resources loaded with environmental pollutants. One of the major advantages of the removal of pollutants using this technology is the interaction of plant rhizosphere with pollutants. The phytoremediation can be performed either in natural habitats or in CWs. Recently, a study carried out by Gong et al. (2018) has evaluated the efficacy of field-scale remediation techniques, their technical progress and developments and revealed that physical and chemical techniques used for the remediation of wastewater are not environment-friendly as they release secondary pollutants into the environment. However, biological methods, especially phytoremediation techniques, are cost-effective and energy-efficient as compared to conventional methods.

The concept of using plants for the remediation of pollutants has been adopted since the 1970s, but it got recognition from the government and commercial sector in the 1980s. Gradually, it has become a widely accepted and exploited technology over the years. Various plant species belonging to the families Ranunculaceae, Lemnaceae, Cyperaceae, Haloragaceae, Hydrocharitaceae, Potamogetonaceae, Typhaceae, Najadaceae, Pontederiaceae and Juncaceae have been explored for their phytoremediation potential (Prasad 2006). Researchers have explored many plants (such as *Wolffia globosa* (Roxb.), *Ceratophyllum demersum* L., *Lemna gibba* L., *T. latifolia* and *Eichhornia crassipes* (Mart.) for their potential to remove pollutants from the environment (Boonyapookana et al. 2002). The efficiency of water hyacinth for the removal of nitrogen and potassium was reported to be 60%–80% and 69%, respectively (Zhou et al. 2007).

3.2.1 Types of Phytoremediation

The efficacy of phytoremediation of plants mainly depends on the survival capacity of plant species to stress conditions. Other factors responsible for the removal of pollutants are pH, temperature, solar radiation, water salinity, size and biomass of aquatic plant, etc. Based on the application, different types of phytoremediation techniques are as follows:

i. Phytoextraction, which involves the uptake of the toxic metal contaminants from soil by plant roots and subsequent translocation of these pollutants to aerial parts such as shoots and leaves, followed by the removal of that plant from the contaminated site after complete growth for safe disposal of biomass accompanied with contaminants. The success of phytoextraction using plants depends upon the bioavailability of metals and plant tolerance to that level of metal.
ii. Phytovolatilization, which utilizes the transpiration process of plants by converting more toxic compounds from the soil into less toxic vapour and thus releasing them into the atmosphere.
iii. Phytofiltration, which involves the benefits of plant species to absorb pollutants from wastewater or aqueous system.
iv. Phytostabilization, in which plant species are employed to reduce the mobility of contaminants and stabilize them, thus preventing their dispersal into the nearby environment.
v. Phytodegradation, wherein the plants uptake organic xenobiotics followed by their degradation using enzymes.
vi. Rhizodegradation, which involves degradation of pollutants by rhizosphere microbe-assisted interaction.

3.2.2 Potential Plants for Phytoremediation

The use of aquatic plants is considered as an advanced strategy for the removal of both organic and inorganic pollutants from wastewater through various phytoremediation processes such as phytovolatilization, phytofiltration, phytodegradation and phytoextraction. However, removal of pollutants depends on the type of plant species, the concentration of pollutants, pH, temperature, etc. It is commendable to highlight that various aquatic plant species have catered for significant advances in phytoremediation of wastewater with noteworthy success (Akinbile et al. 2016). A lot of free-floating plants [such as *Pistia stratiotes* L., *Salvinia molesta* D. Mitch., *Azolla pinnata* R. Br., *Spirodela polyrhiza* (L.) Schleid., *E. crassipes and Riccia fluitans* L.], submerged plants [such as *Hygrophila corymbosa* Lindau, *Najas marina* L., *Ruppia maritima* L., *Hydrilla verticillata* (L.f.) Royle, *Egeria densa* Planch., 1849, *Vallisneria americana* Michx. and *Myriophyllum aquaticum* (Vell.) Verdc.] and emergent plants [such as *Distichlis spicata* (L.) Greene, *I. cylindrica, Iris virginica* L., *Justicia americana* L., *Diodia virginiana* L., *P. australis* and *T. latifolia*] have shown their potential for the removal of pollutants from wastewater (Ekperusi et al. 2019).

The phytoremediation potential of *P. stratiotes* was explored by Mukherjee et al. (2015) for the treatment of rice mill effluent. The removal efficiency for soluble COD, total nitrogen, ammoniacal nitrogen and soluble phosphorus was observed as 65%, 70%, 98% and 65%, respectively. Duckweed has emerged as a sustainable, reliable and cost-effective alternative for the treatment of industrial and municipal wastewaters. The highly explored species of Araceae family are *Lemna minor* L. and *L. gibba* (Mkandawire and Tauert 2004).

The *L. gibba* has proved its potential for the removal of TSS, BOD, total nitrogen, ammonium nitrogen and phosphate (Ekperusi et al. 2019). Zhang et al. (2009) reported that *W. globosa* can tolerate arsenic up to 400 mg/kg of dry weight. Moreover, *W. globosa* has been reported for the efficient accumulation of Cd and Cr (Boonyapookana et al. 2002), whereas *Phragmites communis* Trin. and *N. marina* have evidenced high accumulation of trace metals in their roots and leaves (Baldantoni et al. 2004). *E. crassipes* has shown its high potential for the removal of pollutants from domestic wastewater. A significant removal of nitrogen and potassium was observed using *E. crassipes* (Zhou et al. 2007). *E. crassipes* is a potential plant to treat wastewater laden with metals [Cd, Zn and Cr(VI)], nutrients and organic matter.

3.3 Constructed Wetlands

Constructed wetlands are engineered systems that are designed to utilize natural components such as soil, plants and microorganisms to degrade the inorganic/organic pollutants present in wastewater (Rana and Maiti 2018b). CWs are designed in such a way that they can use the same process as that of natural wetlands, but in a controlled environment. It was in the early 1950s when Kaithe Seidel did the experiment of wastewater treatment by using wetland plants in Germany at Max Planck Institute (Seidel 1955). Seidel performed a lot of experiments that aimed at the treatment of various industrial wastewaters using wetland plants (Seidel 1976). The wetlands used by Seidel in his experiments were of either vertical or horizontal subsurface flow, but for the first time, fully developed CWs with free surface water were used in 1967 in the Netherlands (De Jong 1976). The efficacy of a wastewater treatment system can be enhanced by combining various types of CWs. Generally, a hybrid system comprises of vertical and horizontal flow systems arranged in a staged manner, but all types of CWs can be combined to increase the removal efficiency of the treatment system. Various studies have been reported on the use of these artificial wetlands for the treatment of grey water, sewage water and industrial wastewaters (paper & pulp industry, dairy industry, pharmaceutical industry, glass industry and petrochemical industry).

The CWs have various advantages over a conventional treatment system as it is environment-friendly, economically sustainable and simple, has high removal efficiency and raises the aesthetic value of a wastewater treatment site (Thullner et al. 2018). Furthermore, they do not require any skilled labour, have low operational and maintenance cost and can be used under varied environmental conditions. Production of less quantity of sludge gives extra benefits to CWs over conventional technologies associated with the problem of sludge bulking. Hence, it could be considered as a viable option in both developed and developing countries for the treatment of a wide array of wastewaters. Flores et al. (2019) conducted a study on life cycle assessment of CWs and activated sludge process and concluded that these are the engineered systems which treat the contaminated sites in a natural manner without causing any harm to the environment.

The treatment potential of CWs can be improved by varying their operational parameters such as hydraulic retention time, hydraulic load, depth of water, substrate and the type of plants used. The hydraulic load measures the magnitude of force with which the volume of wastewater is supplied to the CWs surface per unit time. The hydraulic conditions of the CWs are influenced by influent loading rate and position of effluent or influent. Sharif et al. (2014) reported that a hydraulic loading rate of more than 5 cm/day influences the carbon loading rate, which in turn affects the pollutant removal in CWs. The hydraulic retention time of the CWs is also a crucial factor in pollutants removal efficiency. Kotti et al. (2010) reported that an adequate removal of suspended solids can be achieved at a HRT of 14 days, while a HRT of 20 days favoured the removal of BOD and phosphate. The feeding mode such as batch mode or continuous mode also affects the removal mechanism of pollutants in CWs. An efficient removal of total suspended solids and ammonium nitrogen was observed in CWs operated in batch mode as compared to continuous mode (Abdelhakeem et al. 2016). The water level is another significant parameter that affects the performance of CWs. The CWs operated at shallow water depth are usually preferred. Mburu et al. (2019) found that the deep system (0.8 m depth) is beneficial for the removal of TSS and ammoniacal nitrogen from slaughterhouse wastewater as compared to the shallow system (0.6 m depth). However, the average water depth of both kinds of CWs does not vary significantly, as both are shallow water system.

3.3.1 Types of Constructed Wetlands

Based on the pattern of flow of water, CWs have been categorized into two basic types: (i) surface flow (SF) and (ii) subsurface flow (SSF). However, several modifications are possible by merging two or more types of wetlands.

3.3.1.1 Free Surface Water Constructed Wetlands

These are natural modified lagoons characterized by shallow flowing water (30–40 cm deep) over the saturated substrate. The flow of water is maintained by shallow depth, low velocity and litter of wetland plants. The attached surface to the emergent vegetation is provided by an impermeable layer bearing porous material, i.e. sand or gravel, to control the leaching of contaminants. The plants used in the CWs play a vital role by acting as hyperaccumulators of heavy metals and nutrients, provide a surface to microbes, maintain hydraulic conditions, and act as a surface insulator (Zheng et al. 2016). An efficient removal of total nitrogen was observed in FSW-CWs in comparison with other types of CWs. The FSC-CWs are more prone to varied climatic conditions and harshly vulnerable to cold climates. These are highly used CWs in rural areas due to the low construction cost and less energy requirements associated with them. Shrikhande et al. (2014) reported that a minimum of 5 days of hydraulic retention time is required for the treatment of sewage by using *Canna* or *T. latifolia*.

3.3.1.2 Subsurface Flow Constructed Wetlands

These wetlands have 30–90 cm depth; roots and substrates are in contact with water flow, which provides propagation of biofilm and pollutants removal. They are also known as the vegetative submerged bed. The SSF-CWs are artificially constructed channels underlined by impervious structures (such as gravel, rock and stone), emerging plants and saturated substrate, through which the water flows either vertically or horizontally. The SSF-CWs are also target-specific; for example, VFSSF-CWs are preferred when a nitrification-enhanced process in a system is required. However, when a system requires denitrification more than nitrification, then HFSSF-CWs are preferred. The emerging plants used in SSF-CWs play a fundamental role in the transfer of oxygen to the root system in the subsurface area as it is usually an anaerobic zone. Mustapha et al. (2018) evaluated the heavy metal removal potential of three plant species, namely *T. latifolia, Cynodon dactylon* (L.) Pers. and *Cyperus alternifolius* Rottb., 1772, in pilot-scale vertical flow SSF-CWs. Unlike the FWS-CWs, the SSF-CWs are not limited to cold climates. However, nitrogen removal in this system is limited by the accessibility of carbon and oxygen because nitrogen removal in a submerged system mainly depends on the availability of oxygen to the root zone. The sand media provide a large surface area to the microbes and make oxygen available at the surface to enhance the removal process.

3.3.1.3 Horizontal Flow Constructed Wetlands

A horizontal flow CW system is characterized by horizontal flow of water through the granular bed with 30–90 cm depth and a flow rate of 0.05–0.1 m^3 (García 2011). It has a well-arranged system of pipes wherein inlet and outlet are provided with coarse gravel for the filtration purpose. The removal of chromium was studied by Papaevangelou et al. (2017) in both horizontal and vertical flow CWs planted with macrophytes. The results of the study revealed that the horizontal flow system showed a higher removal efficiency as compared to the vertical flow system.

3.3.1.4 Vertical Flow Constructed Wetlands

In this system, water flows in a vertical direction down through the substrate bed. The vertical flow system faces clogging of pipes as these are buried under the soil. The removal potential for ammonia, COD and metals was studied by Yalcuk and Ugurlu (2009) in a combined CW system of two vertical flow units with one horizontal flow unit. Ammonia was removed to a significant level in the vertical flow system.

3.3.1.5 Hybrid Constructed Wetland Systems

The hybrid CWs involve a combination of different types of wetland systems arranged either in series or in parallel at the filter stage. The first original hybrid system was built in the UK and France in the 1980s. Nowadays, it is a widely adapted strategy for the enhanced removal of pollutants from wastewater. The performance of a hybrid CW system can be optimized by employing various CWs in a single remedial design to improve the wastewater treatment process. The VF–HF system is the most frequently used hybrid system for the treatment of municipal and industrial wastewaters loaded with high ammoniacal nitrogen. However, the HF–VF system has some constraints over the VF–HF system; thus, it can only be employed for the treatment of municipal sewage. Vymazal and Březinová (2016) reported a removal efficiency of 83% for COD and 79% for nitrogen in a multistage hybrid CW system. Similarly, Ali et al. (2018) evaluated the performance of a full-scale hybrid CW system for the treatment of municipal sewage over the year under varied continuous flow and reported >90% removal of organic pollutants and pathogens. A significant reduction in the number of emerging contaminants, especially antibiotics, was reported by using a hybrid CW system. Recently, the removal efficiencies for antibiotics and antibiotic resistance genes in a hybrid CW system were reported as 87.4%–95.3% and 87.8%–99.1%, respectively (Chen et al. 2019).

3.3.2 Municipal Wastewater Treatment Using Constructed Wetlands

Constructed wetlands have been considered as a viable, low-cost, sustainable and alternate technology for the remediation of municipal wastewater. CWs have shown potential for the removal of contaminants from municipal wastewater; however, the removal efficiency varies among different types of CWs. CWs have catered for significant advances in the treatment of domestic wastewater mainly because of the low maintenance cost, easy operation and ecological benefits. Casierra-Martínez et al. (2017) assessed the efficacy of two plant species of the Colombian region, viz. *Cyperus ligularis* L. and *Echinochloa colona* (L.) Link, for the removal of nutrients and organic matter when cultivated for 4 months under a continuous loading of 42 L/day and 2.3 days HRT. The efficiency of *C. ligularis* and *E. colona* to remove COD was found to be 69% and 63%, respectively.

Burgos et al. (2017) achieved more than 50% removal of COD by using *C. papyrus* at an organic loading rate of 0.47–1.94 g COD/m^{2}day. Moreover, the removal efficiency increases with an increase in the organic loading rate. Further, a high C/N ratio also enhances the removal efficiency for COD.

Bohórquez et al. (2017) used *Heliconia psittacorum* L.f. under different configurations of vertical flow CWs with fine sand and medium gravel as substrate media and a feeding frequency of 20 pulses/day and 10 pulses/day, respectively. The removal efficiency for pollutants was found to be significant in the sand medium as compared to the gravel beds. The removal efficiency for COD, BOD, TSS and NH_4^+-N was found to be >91% and >39%, >96% and >47%, >85% and 0%, and >77% and >36% for sand and gravel, respectively. Abou-Elela (2017) carried out a pilot study to observe the effect of hydraulic loading rate and hydraulic retention time in subsurface horizontal flow CW cultivated with *Cyprus papyrus* L. for the treatment of municipal wastewater. The results revealed that the maximum performance of the CW was observed at a HRT of 4.7 days and HLR of 0.07 m^3/m^2. D. The reduction in COD, TSS and BOD was observed to be 86%, 80% and 87%, respectively.

Caselles-Osorio et al. (2017) conducted a pilot experiment in a horizontal subsurface flow CW planted with *C. articulates* to observe the removal potential for COD and nitrogenous compounds present in domestic wastewater. COD and NH_4^+-N removal efficiencies of 91% and 80%, and 85% and 40% were observed for a CW planted with *C. articulates* and an unplanted wetland, respectively.

3.3.3 Industrial Wastewater Treatment Using Constructed Wetlands

Constructed wetlands have been considered as a viable option for the remediation of industrial wastewater. A horizontal subsurface flow CW planted with *A. donax* and *Sarcocornia fruticosa* (L.) A. J. Scott was used for the treatment of tannery wastewater and showed removal efficiencies of 65%, 73%, 65%, 73% and 75% for COD, BOD, TSS, NH_4^+-N and TKN, respectively (Calheiros et al. 2012). Davies et al. (2005) investigated the treatment potential of vertical flow CW planted with *P. australis* for the removal of azo dye Acid Orange 7 (AO7) and reported the removal of TOC, COD and dye to be 71%, 64% and 74%, respectively. *P. australis* degraded aromatic amines along with the dye. Zhang et al. (2010) carried out a study on a combination of vertical CW (down-flow and up-flow) with a total area of 320 m^2, in which *C. indica* was planted in the down-flow unit, while *T. latifolia* in the up-flow unit. The system was maintained under a HLR and HRT of 22.5–33.8 cm/day and

TABLE 3.1

Municipal and Industrial Wastewater Treatment Potential of Constructed Wetlands

Location	Type of Wastewater	Plant Species	Treatment Efficiency	References
South Africa	Municipal wastewater	*Vetiveria zizanioides* (L.) Nash	COD (96.6%–97.2%), Sulphate (46.4%–88.7%) and TSS (98.3–99.7)	Badejo et al. (2017)
Colombia	Domestic wastewater	*Cyperus ligularis* L. and *Echinochloa colona* (L.) Link	COD (26%–63%)	Casierra-Martínez et al. (2017)
India	Textile wastewater	*Typha angustifolia* L. and *Paspalum scrobiculatum* L.	BOD (68%), colour (62%), COD (65%), TDS (45%), TSS (35%), As (60%), Cd (28%), Cr (59%) and Pb (45%)	Chandanshive et al. (2017)
Poland	Secondary domestic wastewater	*Phragmites australis* (Cav.) Trin. ex Steud., *Glyceria maxima* (Hartm.) Holmb. and *Sida hermaphrodita* (L.) Rusby	BOD (95%), COD (95%), TSS (95%), TN (94%) and TP (95%)	Marzec et al. (2018)
Greece	Secondary treated wastewater	*Juncus acutus* L.	Bisphenol A (76.2%) and ciprofloxacin (93.9%)	Christofilopoulos et al. (2019)
Brazil	Urban wastewater	*Canna x generalis, Equisetum sp., Chrysopogon zizanioides* (L.) Roberty, *Hymenachne grumosa* (Nees) Zuloaga and *Cyperus papyrus* L.	TP (94%), ammonium nitrogen (N–NH_3) (93.8%), TN (93.8%), DOC (80%), BOD (84%), COD (77%) and turbidity (99.7%)	Dell'Osbel et al. (2020)
Egypt	Domestic wastewater	*Phragmites australis* (Cav.) Trin. ex Steud.	COD (88%), BOD (88%) and TSS (88.5%)	Khalifa et al. (2020)

BOD, biochemical oxygen demand; COD, chemical oxygen demand; DOC, dissolved organic carbon; TN, total nitrogen; TP, total phosphorus; TSS, total suspended solids.

0.9–1.3 days, respectively. The removal efficiency was reported as 26% for COD, 58% for TSS, 48% for TN, 56% for BOD, and 34% for NH_4^+-N.

Shi et al. (2011) used an integrated system of vertical and horizontal flow CW systems planted with salt-tolerant species, namely *P. australis, Sporobolus alterniflorus* (Loisel.) P. M. Peterson & Saarela *and Scirpus mariqueter* T. Tang & F. T. Wang, to treat brackish wastewater. The CW system showed a high removal efficiency for TSS (66%) and TN (67%), whereas a low removal efficiency for COD (27%), nitrate (59%) and TP (24%) was reported. The treatment of wastewater emanating from a mountain cheese factory by a hybrid CW cultivated with *P. australis* was studied by Comino et al. (2011), and the treatment efficiency was reported as 72% for COD, 50% for TN, 80% for non-ionic surfactants, 60% for TSS, 55% for BOD and 37% for TP. A recent study carried out by Hamad (2020) has reported that *T. latifolia* removed 68.5% COD, 71% BOD, 70% TSS and 82.3% ammonia, whereas the removal efficiency of *P. australis* was reported as 85.5% for COD, 86.2% for BOD, 83.9% for TSS and 92.3% for ammonia. The application of CWs for the treatment of municipal and industrial wastewaters is summarized in Table 3.1.

3.4 Biochar: A Sustainable Eco-Tool

3.4.1 Characteristics of Biochar

Biochar is a carbon-rich by-product produced by the combustion of biomass under a limited supply of oxygen in a controlled environment through slow and fast pyrolysis, gasification and hydrothermal carbonization (Wu et al. 2017). Biochar is composed of volatile and condensed aromatic organic substances and inorganic elements. Biochar production is driven by factors such as (i) biomass properties (such as type of biomass, moisture content and particle size), (ii) reaction conditions (such as reaction temperature, reaction time and heating rate), (iii) surrounding environment (such as carrier gas type and flow rate of carrier gas) and (iv) other factors (such as catalyst and reactor type). Moreover, the process conditions (such as temperature, residence time, heating rate and particle size) also affect the properties of the biochar (Tripathi et al. 2016). The beneficial physical (high surface area, high porosity, high surface charge and high water-holding capacity) and chemical properties (high pH and high nutrient exchange) of biochar provide various environmental remediation services such as high complexation and immobilization of metals, sorption and partitioning of organic pollutants, and climate change mitigation (low greenhouse gas emission and high carbon sequestration).

3.4.2 Biochar Production Methods

The energy from the biomass is extracted using bioenergy conversion technologies that encompass (i) biochemical conversion, which comprises the use of biological catalysts and biological organisms to produce the energy from biomass, and (ii) thermochemical conversion, which involves heat and chemical catalysts to produce energy from biomass. The thermochemical conversion of biomass may take place through combustion, gasification or pyrolysis.

Combustion is a thermochemical process wherein the chemical energy stored in the biomass is extracted in the form of heat by directly burning it in the presence of oxygen (Nussbaumer 2003). In gasification, the carbonaceous contents of the biomass are converted into a gaseous fuel in the presence of a gaseous medium (such as oxygen, air, nitrogen, carbon dioxide, steam or some mixture of these gases) at high temperatures (700°C–900°C) (Chew et al. 2020). Pyrolysis involves the thermal degradation of biomass under an inert or very low stoichiometric oxygen atmosphere. The description of different thermochemical biomass conversion techniques is shown in Table 3.2.

Pyrolysis is a vastly adopted technique, which results in the formation of stable products and solid residue after heating the biomass to a temperature above its limit of thermal stability. Pyrolysis is considered advantageous in many ways: (i) it can reduce and convert biowaste into useful products such as bio-oil, bio-syngas and biochar; (ii) it can treat the biomass (dry, wet and hard/soft) and waste (sewage sludge or other industrial wastes) efficiently; (iii) this technique is flexible both with feedstock type and with operating conditions; (iv) the texture and properties of the product can be changed with a change in the pyrolysis conditions; and (v) this technique is eco-friendly as it mitigates global warming by emitting low sulphur and NO_x gases (Tripathi et al. 2016).

3.4.3 Effect of Biochar Application on Biophysical and Biochemical Properties of the Soil

The physicochemical properties of biochar, such as pH, composition, water-holding capacity, surface area, electrical conductivity and pore size distribution, are driven by pyrolysis conditions and vary

TABLE 3.2

Process Description Of Thermochemical Conversion Techniques

Parameters	Combustion	Gasification	Pyrolysis
Temperature (°C)	800–1,000	700–900	400–1,200
Air supply	Excess	Marginal	Nil
Pressure (MPa)	0.1	0.1	0.1–0.5
Resources	Solid biomass	Solid biomass	Solid biomass
Status	Commercial	Commercial	Developing
Pretreatment	Not required	Required	Required
Cost	Low	High	High
Harmful emission	High	Low	Low
Products	Heat	Bio-syngas, bio-oil and biochar	Biochar, bio-oil and gaseous products

Source: Adopted from Tripathi, M., Sahu, J.N., and Ganesan, P., Renew. Sustain. Energy Rev., 55, 467–481, 2016.

widely with the properties of feedstock. Biochar is prepared at high temperatures and typically has a large inner surface area with high porosity, organic C and adsorption capacity, along with high pH and cation exchange capacity. When added to the soil, it increases the soil's physico-chemical properties (such as cation exchange capacity, pore size distribution, soil structure, bulk density, hydraulic conductivity and soil water retention) and increases soil nutrient bioavailability. Biochar also increases soil C sequestration and thus reduces atmospheric greenhouse gas concentration. It also assists in the immobilization of toxic metals in contaminated soils. Table 3.3 summarizes the effect of biochar amendment on soil properties.

In biochar-amended soils, the N and P cycling rates are affected by the soil physicochemical properties. Biochars having a high surface area adsorb ionic forms of N and P at higher capacities. Biochars composed of low nutrient content tend to benefit the reduction of NO_3 and ortho-P leaching in nutrient-rich soil; however, the amendment of biochar in nutrient-rich soil could decrease the bioavailable content of N and P, which have consequences for microbially mediated reactions in the soil N and P cycles.

Biochar increases the cation exchange capacity of soil due to its large surface area, high pore volume and negative surface charge. The improved cation exchange capacity leads to higher nutrient retention capability and a lower nutrient loss, which favours microbial activity in the soil. Biochar provides nutrients (such as K, Na, Mg, N and P) to microorganisms in the soil through the sorption of nutrient cations and inorganic anions with its surface functional groups (oxygen-containing groups such as the carboxylate group) (Zhu et al. 2017).

TABLE 3.3

Effect of Biochar Application on Biophysical and Biochemical Properties of the Soil

Biochar Source	Soil Type	Effect on Soil Properties/Soil Quality Changes
Different types of feedstock	Different soil types	Increase in soil pH, CEC, available K, Ca and Mg, total N and available P; decrease in Al saturation of acid soils
Wood charcoal	Anthrosol and ferralsol	Increase in soil C content, pH value and available P; reduction in leaching of applied fertilizer N, Ca and Mg and lower Al contents
Eucalyptus logs; maize stover	Clay loam oxisol; silt loam	Increase in total N derived from the atmosphere by up to 78%; higher total soil N recovery with biochar addition
Charcoal site soil	Haplic Acrisols	Increase in total porosity from 46% to 51% and saturated soil hydraulic conductivity by 88% and reduction in bulk density by 9%
Peanut hulls, pecan shells and poultry litter	Loamy sand	Biochars produced at higher pyrolysis temperature increased soil pH, while biochar made from poultry litter increased available P and Na
Wood and peanut shell–chicken manure–wheat chaff	Sandy soils	Increase in P availability from 163% to 208%, but decreased AMF abundances in soils from 43% to 77%
Wood- and manure-derived biochars	Different soil types	Increase the soil's saturated hydraulic conductivity and plant's water accessibility, as well as boost the soil's total N concentration and CEC, improving soil field capacity, and reduce NH_4-N leaching
Manure, corn stover, woods and food waste	Alfisol	Tissue N concentration and uptake decreased with increasing pyrolysis temperature and application rate, but increased K and Na content
Different biochars	Different soil types	Sources increased crop yield and improved microbial habitat and soil microbial biomass, rhizobia nodulation, plant K tissue concentration, soil pH, soil P, soil K, total soil N and total soil C compared with control conditions
Peanut hull	Ultisols	Increased K, Ca and Mg in the surface soil (0–15 cm). Increased K was reflected in the plant tissue analysis
Simoca and activated wundowie	Loamy sand clay	Increased soil microbial activity more in clay than in loamy soil
Acacia whole tree green waste	Planosol	Increase in porosity via direct pore contribution, creation of accommodation pores or improved aggregate stability
Wheat straw	Fimi-Orthic anthrosols	Increase in soil pH, organic carbon and total nitrogen and reduction in yield-scaled N_2O emissions

Source: Adopted and modified from Agegnehu, G., Srivastava, A.K., and Bird, M.I., Appl. Soil Ecol. 119, 156–170, 2017.

3.5 Wastewater Treatment Potential of Biochar-Amended Constructed Wetlands

Biochar is being widely used for the removal of a wide array of pollutants from different types of wastewaters as listed in Table 3.4.

3.5.1 Nutrient Uptake in Biochar-Amended Constructed Wetlands

Biochar plays a vital role in the retention, transformation and export of N in constructed wetlands. The high porosity and large surface area of biochars increase microbial attachment sites, which leads to stimulation of N transformation. The release of labile organic carbon from biochar enhances the denitrification process, which results in the improvement in the removal efficiency of NO_3-N, NH_4^+-N and N and reduction in the accumulation of N_2O in constructed wetlands (El Hanandeh et al. 2017). Moreover, biochar also favours the growth of wetland macrophytes, which also increases NH_4^+-N assimilation by plants. The carbon source present in the CWs also drives the relative abundances of nitrifying, denitrifying and sulphate-reducing bacteria, which shows a positive correlation with the removal efficiencies for ammonia, nitrate and sulphate (Chen et al. 2015). Zhou et al. (2019) reported a higher reduction of total N in biochar-amended subsurface flow CWs. The influent C/N ratio also affects the nitrogen removal efficiencies of biochar-amended CWs. In a study conducted by Zhao et al. (2019), biochar amendment also showed an improvement in total N removal under low influent C/N ratios (<7). However, Jia et al. (2018) reported the NH_4^+-N removal efficiency to be independent of external carbon sources under different

TABLE 3.4

Enhancement of Pollutant Removal Efficiency in Biochar-Amended Constructed Wetlands

Location	Type of Wastewater	Type of Biochar	Pollutant Removal Efficiency	References
California, the USA	Dairy effluent	Biochar derived from a mixture of mixed hardwood shavings, including maple, aspen, choke cherry and alder	Biochar adsorbed (in vitro) up to 0.24 mg/g of PO_4^{3-} (50%) and 5.3 mg/g of NH^{4+} (18%) from dairy manure effluent	Sarkhot et al. (2013)
Rawalpindi, Pakistan	Dye wastewater	Rice husk-derived biochar	80% adsorption of dye in the presence of rice husk biochar	Saba et al. (2015)
Guangdong Province, China	Pesticide-contaminated water	Iron (Fe)-impregnated biochar derived from *Cyperus alternifolius* Rottb., 1772	Enhancement in pesticide removal, particularly diuron (primarily due to adsorption and microbial degradation)	Tang et al. (2016)
Liaoning Province, China	High nitrate-laden wastewater	Hydrophytes (reed and cattail)-based biochar modified by treatment with HCl	Increased adsorption of nitrate	Wang et al. (2017)
Tartu County, Estonia	Municipal wastewater	95% Charcoal from alder (*Alnus* sp.) with small proportions of charcoals from birch (*Betula* sp.), oak (*Quercus* sp.), linden (*Tilia* sp.) and willow (*Salix* sp.)	Improved total nitrogen and total phosphorus with enhancement in plant growth	Kasak et al. (2018)
Zhejiang, China	Stormwater laden with bisphenol A	Wood dust-derived biochar	Enhancement in bisphenol A removal and assistance in promoting the growth of *Phragmites australis* (Cav.) Trin. ex Steud., elevation in nutrients and increment in the removal of *E. coli*	Lu and Chen (2018)

influent C/N ratios. However, for the removal of PO_4^{3-} in CWs, biochars cannot be considered as a long-term solution as they only adsorb PO_4^{3-} for a short duration (Gao et al. 2018).

3.5.2 Metal Removal in Biochar-Amended Constructed Wetlands

Biochars derived from different types of biomass have shown potential for the removal of metals through processes such as adsorption, ion exchange, complexation, precipitation and electrostatic attraction (Figure 3.1). The driving mechanism for metal removal in biochar-amended CWs varies with the metal. For the removal of As, complexation and electrostatic attraction are the vital mechanisms, wherein the functional groups on the surface of biochar drive the sorption of As. A high amount of functional groups can be obtained by producing the biochar at low pyrolysis temperature. The Cr in wastewater exists in two forms: Cr(III) and Cr(VI). The sorption of Cr(III) by biochar is driven by three mechanisms: (i) cation exchange; (ii) electrostatic attraction between positively charged Cr(III) ions and negatively charged biochar; and (iii) complexation with oxygen-containing functional groups on the surface of biochar. However, for Cr(VI), the sorption takes place through (i) electrostatic attraction between positively charged Cr(VI) species and negatively charged biochar and (ii) reduction of Cr(VI) to Cr(III) primarily by oxygen-containing functional groups (such as carboxyl and hydroxyl groups), which subsequently leads to the complexation of Cr(III) with functional groups on the biochar. In a study conducted by Mohan et al. (2011), it is reported that the biochar derived from oak trees' bark and wood favoured Cr(VI) removal through adsorption.

The sorption of Cd and Pb by biochars is governed by cation exchange, complexation and precipitation, which depends upon the characteristics of biochar, which are affected by feedstock, pyrolysis temperature and the pH of the solution. Park et al. (2013) reported that Pb sorption by biochars (derived from chicken manure) was higher than the Cd sorption, which is attributed to the precipitation of Pb with various ions (such as carbonate, phosphate and sulphate) released from the biochar. Zhang et al. (2013) reported that the biochar derived from oil mallee (whole plants) or wheat chaff assisted in the immobilization of Cd. In the case of Hg, which exists as Hg^{2+} at $pH < 3.0$ and as $HgOH^+$ and $Hg(OH)_2$ at pH 3–7 in an aqueous solution, the principal mechanism for Hg sorption is complexation with carboxylic and phenolic hydroxyl groups (Dong et al. 2013). Xu et al. (2016) reported that the sorption of Hg^{2+} by biochar (derived from bagasse and hickory chips) is due to the formation of $(\text{–COO})_2Hg^{2+}$ and $(AO)_2Hg^{2+}$. The sorption of Hg^{2+} by hickory chips derived biochar primarily resulted from the π electrons of C=C and C=O induced Hg–π binding.

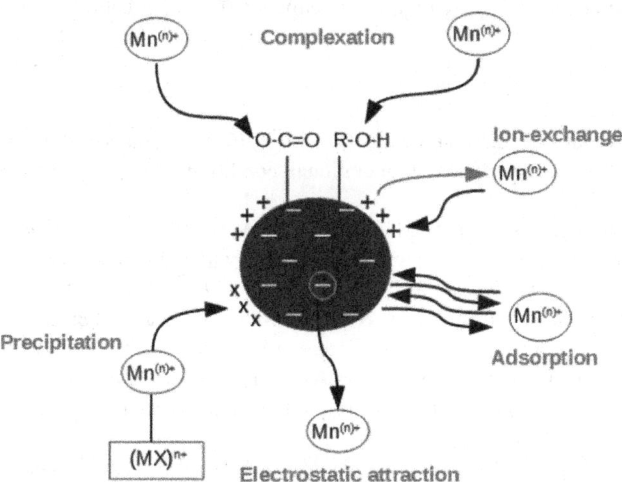

FIGURE 3.1 Metal removal mechanisms. (Adopted and modified from Inyang et al. 2016.)

3.5.3 Recent Advancements in Biochar-Amended Constructed Wetlands

As the metal sorption capacity of biochar is relatively lower than that of other biosorbents such as activated C, the recent research is being carried out to enhance its metal sorption potential. This can be achieved through (i) increasing the functional groups on the surface of biochar by treating it with minerals such as hematite, magnetite, zero-valent Fe, hydrous Mn oxide, calcium oxide and birnessite and (ii) increasing the surface area of biochar by activation or modification with an alkali solution such as NaOH and KOH.

Biochar-based nanocomposites enhanced the adsorption of heavy metals [such as As(III, V), Cd(II), Cr(VI), Cu(II), Hg(II) and Pb(II)] onto biochar surface due to the increase in the oxygen-containing functional groups (such as carboxylate, –COOH; hydroxyl, –OH; and carboxyl, –C=O) on the surface of the biochar after the introduction of nanomaterials (Gan et al. 2015). This increase in the oxygen-containing functional groups enhances the adsorption capacity and strengthens the interactions between the biochar and the heavy metals by forming surface complexes, cation–π bonding, electrostatic attraction and ion exchange. Also, biochar-based nanocomposites have shown a high affinity for organic contaminants such as dyes, phenols, naphthalene, ciprofloxacin and phenanthrene (Yang et al. 2020). As for the nano-metal oxide/hydroxide biochar composites and magnetic biochar composites, several mechanisms might be involved in the adsorption processes by the interactions between organic contaminants and functional groups of biochar, including π–π interactions, hydrogen bond, electrostatic attraction and hydrophobic interaction (Tan et al. 2016).

3.6 Conclusions

Urbanization and industrialization lead to the generation of domestic and industrial wastewater. In developing countries, a significant gap between sewage generation and treatment has been reported, which leads to the discharge of untreated or partially treated wastewaters into natural water bodies such as rivers, lakes and ponds. These wastewaters contain harmful inorganic and organic toxicants that may enter the aquatic and human food chain. To protect the environment and human health, alternative wastewater treatment technologies are required, which may help in the reduction in the load on existing wastewater treatment facilities. Constructed wetlands are engineered systems that are designed to utilize natural components such as soil, plants and microorganisms to degrade the inorganic/organic pollutants present in wastewater. CWs have shown satisfactory pollutant removal efficiency in the treatment of domestic and industrial wastewaters. Biochar amendment in CWs enhances nutrient uptake by plants, complexation and immobilization of metals, sorption and partitioning of organic pollutants, and climate change mitigation (low greenhouse gas emission and high carbon sequestration). Further advancements in improving the efficiency of biochar-amended CWs are required to enhance their applicability in wastewater treatment.

REFERENCES

Abdelhakeem, S.G., Aboulroos, S.A., and Kamel, M.M. 2016. Performance of a vertical subsurface flow constructed wetland under different operational conditions. *Journal of Advance Research* 7(5): 803–814.

Abou-Elela, S.I., Elekhnawy, M.A., Khalil, M.T., and Hellal, M.S. 2017. Factors affecting the performance of horizontal flow constructed treatment wetland vegetated with Cyperus papyrus for municipal wastewater treatment. *International Journal of Phytoremediation* 19(11): 1023–1028.

Agegnehu, G., Srivastava, A.K., and Bird, M.I. 2017. The role of biochar and biochar-compost in improving soil quality and crop performance: a review. *Applied Soil Ecology* 119: 156–170.

Akinbile, C.O., Ogunrinde, T.A., Man, H.C.B., and Aziz, H.A. 2016. Phytoremediation of domestic wastewaters in free water surface constructed wetlands using Azolla pinnata. *International Journal of Phytoremediation* 18(1): 54–61.

Ali, Z., Mohammad, A., Riaz, Y., Quraishi, U.M., and Malik, R.N. 2018. Treatment efficiency of a hybrid constructed wetland system for municipal wastewater and its suitability for crop irrigation. *International Journal of Phytoremediation* 20(11): 1152–1161.

Badejo, A.A., Omole, D.O., Ndambuki, J.M., and Kupolati, W.K. 2017. Municipal wastewater treatment using sequential activated sludge reactor and vegetated submerged bed constructed wetland planted with *Vetiveria zizanioides*. *Ecological Engineering* 99: 525–529.

Baldantoni, D., Alfani, A., Di Tommasi, P., Bartoli, G., and De Santo, A.V. 2004. Assessment of macro and microelement accumulation capability of two aquatic plants. *Environmental Pollution* 130(2): 149–156.

Batool, A., and Zeshan. 2017. Effect of chelators and substrates on phytoremediation of synthetic leachate for removal of trace elements. *Soil Sediment and Contamination* 26: 220–233.

Bohórquez, E., Paredes, D., and Arias, C.A. 2017. Vertical flow-constructed wetlands for domestic wastewater treatment under tropical conditions: effect of different design and operational parameters. *Environmental Technology* 38(2): 199–208.

Boonyapookana, B., Upatham, E.S., Kruatrachue, M., Pokethitiyook, P., and Singhakaew, S. 2002. Phytoaccumulation and phytotoxicity of cadmium and chromium in duckweed Wolffia globosa. *International Journal of Phytoremediation* 4(2): 87–100.

Burgos, V., Araya, F., Reyes-Contreras, C., Vera, I., and Vidal, G., 2017. Performance of ornamental plants in mesocosm subsurface constructed wetlands under different organic sewage loading. *Ecological Engineering* 99: 246–255.

Calheiros, C.S.C., Quitério, P.V.B., Silva, G., Crispim, L.F.C., Brix, H., Moura, S.C., and Castro, P.M.L. 2012. Use of constructed wetland systems with Arundo and Sarcocornia for polishing high salinity tannery wastewater. *Journal of Environmental Management* 95: 66–71.

Casierra-Martínez, H.A., Charris-Olmos, J.C., Caselles-Osorio, A., and Parody-Muñoz, A.E. 2017. Organic matter and nutrients removal in tropical constructed wetlands using *Cyperus ligularis* (Cyperaceae) and *Echinocloa colona* (Poaceae). *Water Air Soil Pollution Ion* 228(9): 338.

Caselles-Osorio, A., Vega, H., Lancheros, J.C., Casierra-Martínez, H.A. and Mosquera, J.E. 2017. Horizontal subsurface-flow constructed wetland removal efficiency using *Cyperus articulatus* L. *Ecological Engineering* 99: 479–485.

Chandanshive, V.V., Rane, N.R., Tamboli, A.S., Gholave, A.R., Khandare, R.V., and Govindwar S.P. 2017. Co-plantation of aquatic macrophytes *Typha angustifolia* and *Paspalum scrobiculatum* for effective treatment of textile industry effluent. *Journal of Hazardous Material* 338: 47–56.

Chen, J., Deng, W.J., Liu, Y.S., Hu, L.X., He, L.Y., Zhao, J.L., Wang, T.T., and Ying, G.G., 2019. Fate and removal of antibiotics and antibiotic resistance genes in hybrid constructed wetlands. *Environmental Pollution* 249: 894–903.

Chen, Y., Wen, Y., Tang, Z., Huang, J., Zhou, Q., and Vymazal, J. 2015. Effects of plant biomass on bacterial community structure in constructed wetlands used for tertiary wastewater treatment. *Ecological Engineering* 84: 38–45.

Chew, J.J., Soh, M., Sunarso, J., Yong, S.T., Doshi, V., and Bhattacharya, S. 2020. Gasification of torrefied oil palm biomass in a fixed-bed reactor: effects of gasifying agents on product characteristics. *Journal of the Energy Institute* 93(2): 711–722.

Christofilopoulos, S., Kaliakatsos, A., Triantafyllou, K., Gounaki, I., Venieri, D., and Kalogerakis, N. 2019. Evaluation of a constructed wetland for wastewater treatment: addressing emerging organic contaminants and antibiotic resistant bacteria. *New Biotechnology* 52: 94–103.

Comino, E., Riggio, V., and Rosso, M. 2011. Mountain cheese factory wastewater treatment with the use of a hybrid constructed wetland. *Ecological Engineering* 37: 1673–1680.

Davies, L.C., Carias, C.C., Novais, J.M., and Martins-Dias, S. 2005. Phytoremediation of textile effluents containing azo dye by using Phragmites australis in a vertical flow intermittent feeding constructed wetland. *Ecological Engineering* 25: 594–605.

De Jong, J. 1976. The purification of wastewater with the aid of rush or reed ponds. In: Tourbier, J., and Pierson, R.W. (eds.), *Biological Control of Water Pollution*, 133–139.

Dell'Osbel, N., Colares, G.S., Oliveira, G.A., Rodrigues, L.R., da Silva, F.P., Rodriguez, A.L., Lopez, D.A.R., Lutterback, C.A., Silveira, E.O., Kist, L.T., and Machado, Ê.L. 2020. Hybrid constructed wetlands for the treatment of urban wastewaters: increased nutrient removal and landscape potential. *Ecological Engineering* 158: 106072.

Dong, X., Ma, L.Q., Zhu, Y., Li, Y., and Gu, B. 2013. Mechanistic investigation of mercury sorption by Brazilian pepper biochars of different pyrolytic temperatures based on X-ray photoelectron spectroscopy and flow calorimetry. *Environmental Science & Technology* 47(21): 12156–12164.

Ekperusi, A.O., Sikoki, F.D., and Nwachukwu, E.O. 2019. Application of common duckweed (*Lemna minor*) in phytoremediation of chemicals in the environment: state and future perspective. *Chemosphere* 223: 285–309.

El Hanandeh, A., Albalasmeh, A.A., and Gharaibeh, M. 2017. Phosphorus removal from wastewater in biofilters with biochar augmented geomedium: effect of biochar particle size. *CLEAN Soil Air Water* 45(7): 1600123.

Flores, L., García, J., Pena, R., and Garfí, M. 2019. Constructed wetlands for winery wastewater treatment: a comparative life cycle assessment. *Science of the Total Environment* 659: 1567–1576.

Gan, C., Liu, Y., Tan, X., Wang, S., Zeng, G., Zheng, B., Li, T., Jiang, Z., and Liu, W. 2015. Effect of porous zinc–biochar nanocomposites on Cr (VI) adsorption from aqueous solution. *RSC Advances* 5(44): 35107–35115.

Gao, Y., Zhang, W., Gao, B., Jia, W., Miao, A., Xiao, L., and Yang, L. 2018. Highly efficient removal of nitrogen and phosphorus in an electrolysis-integrated horizontal subsurface-flow constructed wetland amended with biochar. *Water Research* 139: 301–310.

García, J. 2011. Advances in pollutant removal processes and fate in natural and constructed wetland. *Ecological Engineering* 37(5): 663–665.

Gong, Y., Zhao, D., and Wang, Q. 2018. An overview of field-scale studies on remediation of soil contaminated with heavy metals and metalloids: technical progress over the last decade. *Water Research* 147: 440–460.

Hamad, M.T. 2020. Comparative study on the performance of Typha latifolia and Cyperus *papyrus* on the removal of heavy metals and enteric bacteria from wastewater by surface constructed wetlands. *Chemosphere* 260: 127551.

Inyang, M.I., Gao, B., Yao, Y., Xue, Y., Zimmerman, A., Mosa, A., Pullammanappallil, P., Ok, Y.S., and Cao, X. 2016. A review of biochar as a low-cost adsorbent for aqueous heavy metal removal. *Critical Reviews in Environmental Science and Technology* 46(4): 406–433.

Jia, L., Wang, R., Feng, L., Zhou, X., Lv, J., and Wu, H. 2018. Intensified nitrogen removal in intermittently-aerated vertical flow constructed wetlands with agricultural biomass: effect of influent C/N ratios. *Chemical Engineering Journal* 345: 22–30.

Kasak, K., Truu, J., Ostonen, I., Sarjas, J., Oopkaup, K., Paiste, P., Koiv-Vainik, M., Mander, U., and Truu, M. 2018. Biochar enhances plant growth and nutrient removal in horizontal subsurface flow constructed wetlands. *Science of the Total Environment* 639: 67–74.

Khalifa M.E., Abou El-Reash Y.G., Ahmed M.I., and Rizk F.W. 2020. Effect of media variation on the removal efficiency of pollutants from domestic wastewater in constructed wetland systems. *Ecological Engineering* 143: 105668.

Kotti, I.P., Gikas, G.D., and Tsihrintzis, V.A. 2010. Effect of operational and design parameters on removal efficiency of pilot-scale FWS constructed wetlands and comparison with HSF systems. *Ecological Engineering* 36(7): 862–875.

Kumari, M., and Tripathi, B.D. 2015. Efficiency of *Phragmites australis* and *Typha latifolia* for heavy metal removal from wastewater. *Ecotoxicology and Environmental Safety* 112: 80–86.

Lu, L., and Chen, B. 2018. Enhanced bisphenol A removal from stormwater in biochar-amended biofilters: combined with batch sorption and fixed-bed column studies. *Environmental Pollution* 243: 1539–1549.

Marzec, M., Jóźwiakowski, K., Dębska, A., Gizińska-Górna, M., Pytka-Woszczyło, A., Kowalczyk-Juśko, A., and Listosz, A. 2018. The efficiency and reliability of pollutant removal in a hybrid constructed wetland with common reed, manna grass, and *Virginia mallow*. *Water* 10(10): 1445.

Mburu, C., Kipkemboi, J., and Kimwaga, R. 2019. Impact of substrate type, depth and retention time on organic matter removal in vertical subsurface flow constructed wetland mesocosms for treating slaughterhouse wastewater. *Physics and Chemistry of the Earth, Parts A/B/C* 114: 102792.

Mkandawire, M., and Tauert, B. 2004. Capacity of *Lemna gibba* L. (Duckweed) for uranium and arsenic phytoremediation in mine tailing waters. *International Journal of Phytoremediation* 6: 347–362.

Mohan, D., Rajput, S., Singh, V.K., Steele, P.H., and Pittman Jr, C.U. 2011. Modelling and evaluation of chromium remediation from water using low cost bio-char, a green adsorbent. *Journal of Hazardous Materials* 188(1–3): 319–333.

Mukherjee, B., Majumdar, M., Gangopadhayay, A., Chakraborty, S., and Debashish, C. 2015. Phytoremediation of parboiled rice mill wastewater using Water lettuce (*Pistia Stratiotes*). *International Journal of Phytoremediation* 17(7): 651–656.

Mustapha, H.I., van Bruggen, H.J.J.A., and Lens, P.N. 2018. Vertical subsurface flow constructed wetlands for the removal of petroleum contaminants from secondary refinery effluent at the Kaduna refining plant (Kaduna, Nigeria). *Environmental Science and Pollution Research* 25(30): 30451–30462.

Nussbaumer, T. 2003. Combustion and co-combustion of biomass: fundamentals, technologies, and primary measures for emission reduction. *Energy Fuel* 17(6): 1510–1521.

Papaevangelou, V.A., Gikas, G.D., Vryzas, Z., and Tsihrintzis, V.A. 2017. Treatment of agricultural equipment rinsing water containing a fungicide in pilot-scale horizontal subsurface flow constructed wetlands. *Ecological Engineering* 101: 193–200.

Park, J.H., Choppala, G., Lee, S.J., Bolan, N., Chung, J.W., and Edraki, M. 2013. Comparative sorption of Pb and Cd by biochars and its implication for metal immobilization in soils. *Water Air Soil Pollution* 224(12): 1711.

Prasad, M.N.V. 2006. Plants that accumulate and/or exclude toxic trace elements play an important role in phytoremediation. In: Prasad MNV, Sajwan KS, and Naidu, R. (eds.) *Trace Elements in the Environment: Biogeochemistry, Biotechnology and Bioremediation*. Taylor and Francis, Boca Raton, FL, 523–548..

Rana, V., and Maiti, S.K. 2018a. Metal accumulation strategies of emergent plants in natural wetland ecosystems contaminated with coke-oven effluent. *Bulletin of Environmental Contamination and Toxicology* 101(1): 55–60.

Rana, V., and Maiti, S.K. 2018b. Municipal wastewater treatment potential and metal accumulation strategies of *Colocasia esculenta* (L.) Schott and *Typha latifolia* L. in a constructed wetland. *Environmental Monitoring Assessment* 190(6): 328.

Saba, B., Jabeen, M., Khalid, A., Aziz, I., and Christy, AD. 2015. Effectiveness of rice agricultural waste, microbes and wetland plants in the removal of reactive black-5 azo dye in microcosm constructed wetlands. *International Journal of Phytoremediation* 17(11): 1060–1067.

Sarkhot, D.V., Ghezzehei, T.A., and Berhe, A.A. 2013. Effectiveness of biochar for sorption of ammonium and phosphate from dairy effluent. *Journal of Environmental Quality* 42(5): 1545–1554.

Seidel, K. 1955. Die flechtbinse scirpus lacustris. In: *Okologie, Morphologie und Entwicklung, ihre Stellung bei den Volkern und ihre wirtschaftliche Bedeutung*. Schweizerbartısche Verlagsbuchnadlung, Germany, Stuttgart. 37–52.

Seidel, K. 1976. Macrophytes and water purification. In: Tourbier, J., Pierson, R.W. (Eds.), *Biological Control of Water Pollution*, University of Pennsylvania Press, Philadelphia, 109–122.

Sharif, F., Westerhoff, P., and Herckes, P. 2014. Impact of hydraulic and carbon loading rates of constructed wetlands on contaminants of emerging concern (CECs) removal. *Environmental Pollution* 185: 107–115.

Shi, Y., Zhang, G., Liu, J., Zhu, Y., and Xu, J. 2011. Performance of a constructed wetland in treating brackish wastewater from commercial recirculating and super-intensive shrimp growout systems. *Bioresource Technology* 102: 9416–9424.

Shrikhande, A.N., Nema, P., and Mhaisalkar, V.A. 2014. Performance of free water surface constructed wetland using *Typha latifolia* and Canna lilies for the treatment of domestic wastewater. *Journal of Environmental Science and Engineering* 56(1): 93–104.

Tan, X.F., Liu, Y.G., Gu, Y.L., Xu, Y., Zeng, G.M., Hu, X.J., Liu, S.B., Wang, X., Liu, S.M., and Li, J. 2016. Biochar-based nano-composites for the decontamination of wastewater: a review. *Bioresource Technology* 212: 318–333.

Tang, X., Yang, Y., Tao, R., Chen, P., Dai, Y., Jin, C., and Feng, X. 2016. Fate of mixed pesticides in an integrated recirculating constructed wetland (IRCW). *Science of the Total Environment* 571: 935–942.

Thullner, M., Stefanakis, A.I., and Dehestani, S. 2018. *Constructed Wetlands for Industrial Wastewater Treatment*. Wiley-Blackwell, Chichester, UK.

Tripathi, M., Sahu, J.N., and Ganesan, P. 2016. Effect of process parameters on production of biochar from biomass waste through pyrolysis: a review. *Renewable Sustainable Energy Reviews* 55: 467–481.

Vymazal, J., and Březinová, T. 2016. Accumulation of heavy metals in aboveground biomass of *P. australis* in horizontal flow constructed wetlands for wastewater treatment: a review. *Chemical Engineering Journal* 290: 232–242.

Wang, B., Liu, S.Y., Li, F.Y., and Fan, Z.P. 2017. Removal of nitrate from constructed wetland in winter in high-latitude areas with modified hydrophyte biochars. *Korean Journal of Chemical Engineering* 34(3): 717–722.

Wu, H., Lai, C., Zeng, G., Liang, J., Chen, J., Xu, J., Dai, J., Li, X., Liu, J., Chen, M., and Lu, L. 2017. The interactions of composting and biochar and their implications for soil amendment and pollution remediation: a review. *Critical Reviews of Biotechnology* 37(6): 754–764.

Xu, X., Schierz, A., Xu, N., and Cao, X. 2016. Comparison of the characteristics and mechanisms of Hg (II) sorption by biochars and activated carbon. *Journal of Colloid Interface Science* 463: 55–60.

Yalcuk, A., and Ugurlu, A. 2009. Comparison of horizontal and vertical constructed wetland systems for landfill leachate treatment. *Bioresource Technology* 100(9): 2521–2526.

Yang, Z., Xing, R., Zhou, W., and Zhu, L. 2020. Adsorption characteristics of ciprofloxacin onto g-MoS_2 coated biochar nanocomposites. *Frontiers in Environmental Science and Engineering* 14(3): 1–10.

Zhang, S., Zhou, Q., Xu, D., He, F., Cheng, S., Liang, W., Du, C., and Wu, Z. 2010. Vertical-flow constructed wetlands applied in a recirculating aquaculture system for channel catfish culture: effects on water quality and zooplankton. *Polish Journal of Environmental Studies* 19: 1063–1070.

Zhang, X., Zhao, F.J., Huang, Q., Williams, P.N., Sun, G.X., and Zhu, Y.G. 2009. Arsenic uptake and speciation in the rootless duckweed Wolffia globosa. *New Phytology* 182: 421–428.

Zhang, Z., Solaiman, Z.M., Meney, K., Murphy, D.V., and Rengel, Z. 2013. Biochars immobilize soil cadmium, but do not improve growth of emergent wetland species *Juncus subsecundus* in cadmium-contaminated soil. *Journal of Soils and Sediments* 13(1): 140–151.

Zheng, Y., Wang, X., Dzakpasu, M., Ge, Y., Xiong, J., and Zhao, Y. 2016. Feasibility study on using constructed wetlands for remediation of a highly polluted urban river in a semi-arid region of China. *Journal of Water Sustainability* 6(4): 139.

Zhou, W., Zhu, D., Tan, L., Liao, S., Hu, H., and David, H. 2007. Extraction and retrieval of potassium from water hyacinth (*Eichhornia crassipes*). *Bioresource Technology* 98: 226–231.

Zhou, X., Wang, R., Liu, H., Wu, S. and Wu, H., 2019. Nitrogen removal responses to biochar addition in intermittent-aerated subsurface flow constructed wetland microcosms: Enhancing role and mechanism. *Ecological Engineering* 128: 57-65.

Zhou, X., Wu, S., Wang, R., and Wu, H. 2019. Nitrogen removal in response to the varying C/N ratios in subsurface flow constructed wetland microcosms with biochar addition. *Environmental Science and Pollution Research* 26(4): 3382–3391.

Zhu, X., Chen, B., Zhu, L., and Xing, B. 2017. Effects and mechanisms of biochar-microbe interactions in soil improvement and pollution remediation: a review. *Environmental Pollution* 227: 98–115.

Zhu, Y.F., and Haynes, R.J. 2010. Sorption of heavy metals by inorganic and organic components of solid wastes: significance to use of wastes as low-cost adsorbents and immobilizing agents. *Critical Reviews of Environmental Science and Technology* 40(11): 909–977.

4

Treatment of Metalworking Effluent: Chemical Precipitation, Advanced Oxidative Processes and Biological Treatments

Daniel Delgado Queissada, Jesiel Alves da Silva, and Vanessa Cruz dos Santos
UniAGES - University Center

Iraí Tadeu Ferreira de Resende
Instituto Federal de Sergipe

Débora da Silva Vilar
Tiradentes University (UNIT)

Ram Naresh Bharagava
Babasaheb Bhimrao Ambedkar University (A Central University)

Luiz Fernando Romanholo Ferreira
Tiradentes University (UNIT)

CONTENTS

4.1 Introduction

Oily effluents have a high environmental impact, once, if discarded out of physicochemical parameters established by the legislation may cause a number of harmful effects to aquatic organisms, for example, asphyxia in birds and fish, abrupt reduction of photosynthesis by algae, besides the animals and plants contamination living on the water bodies shore. These effects can be amplified or minimized according to some factors, whether they are environmental or related to effluent involved characteristics. The expected impacts are also associated with the type and volume of oil, its behaviour in the

DOI: 10.1201/9781003181224-4

environment and, also the contaminated environment type, either by the product persistence or by the sensitivity of biota (Borsa 2014).

Among oily effluents, those produced by the metal-mechanical industry are the most abstruse and are difficult to treat and reuse due to high variety and complexity of compounds present. The main component of metal-mechanical effluents is cutting fluids or cutting oils, which are used to cool and lubricate metal parts in the machining process (Curia 2010). Chromium, iron and zinc ions, as well as a variety of other compounds, such as sulphuric acid, hydrochloric acid, potassium hydroxide, synthetic refrigerants and surfactants, are also found in the metal-mechanical effluents (Monteiro 2006). All these compounds give the characteristics of a potential polluter to the metal-mechanical effluents, among oily effluents in general. Even so, few studies are directed to this type of effluent. Generally, the same treatments have been applied for all types of oily effluents. This lack of specificity often makes the treatment ineffective. Therefore, a complete characterization of the effluent is necessary before treatment, whether physical, chemical or biological, in order to provide a more directed treatment for each type of effluent and its constituents, in order to make it more effective.

In addition to seeking the best treatment for each type of effluent, it is of utmost importance that studies are carried out aiming at producing new technologies and alternatives to conventional treatment processes. Among oily effluent treatment systems, biological methods provide the greatest economic flexibility in the search for alternatives that result in effective and economically viable solutions for their treatments (Samson et al. 2018). These oily effluents can contain several types of microorganisms, and their presence and type depend on the metal-mechanical effluent constitution where they are found. Bacteria, yeasts and filamentous fungi have been cited in the literature as effective transforming agents, because of their ability to degrade a wide variety of organic substances commonly found in oily effluents (Silva 2017). However, sometimes, the lack of specificity in biological treatments also makes this type of process ineffective. As a solution to this problem, the most indicated is the bioaugmentation accomplishment of the microorganisms that can most effectively degrade the pollutants present in the effluent to be treated. Therefore, it is of great importance to accomplish biological treatment with microorganisms isolated from the effluent to be treated, since the treatment process performed with these microorganisms (bioaugmentation) tends to be more effective, as they are already adapted to the existing pollutants (Queissada 2013). This isolation may also lead to the discovery of new microorganisms that can degrade the recalcitrant compounds present in the metal-mechanical effluents, which are not yet mentioned in the literature for this type of industrial effluent treatment processes.

While increasing the specificity of biological treatments of metal-mechanical effluents, generally only these techniques are not enough to make the treated effluents meet the standards of disposal required by legislation. It is necessary to use complementary treatments to the adequate manner disposal of final effluent in the receiving body (Souza 2010). The search for new technologies to complement the techniques used for treating the effluents contaminated with recalcitrant compounds gives advanced oxidative processes (AOPs), due to their effectiveness and treatment speed. The AOPs are processes in which the main oxidative agent corresponds to the hydroxyl radical (•OH). These radicals are not selective and promote the degradation of all organic compounds, reacting 106–1,012 times faster than oxidants such as ozone (Oturan and Aaron 2014). Generally, when there is an integrated treatment use, constituted of advanced and biological oxidative processes, these processes integration occurs in the same order and is based on the premise that oxidative process technologies generally enable biodegradation resistant compounds elimination or transformation (refractories) in products with higher biodegradability (Esguerra et al. 2017).

4.2 Metal-Mechanical Effluents

4.2.1 Characteristics of Metal-Mechanical Effluents

The study of effluent characteristics is extremely important for the effectiveness of treatment processes. Besides physicochemical characteristics, it is very important to determine the microbiological characteristics (microorganisms isolation) of the effluent, since this is a crucial factor for an eventual biological treatment, through bioaugmentation, to be accomplished effectively (Muszynski and Lebkowska 2005).

4.2.2 Cutting Fluids

Cutting fluids are used in the machining process in order to provide lubrication and cooling in the cutting region (Figure 4.1).

Cooling promoted by the cutting fluids removes the heat generated during the cutting operation. This provides longer tool life and ensures the dimensional accuracy of the parts by reducing thermal distortion. Lubrication is important for improving the surface finish of the part produced, because small saliences in the cutting tool can collide with existing saliences on the machining chips, resulting in its faster wear (Rodrigues et al. 2016).

The cutting fluids' constitution depends on each typology; however, there are compounds that are common to all types of fluids. Among them is mineral oil, which is considered the cutting fluids' basic component. It can be used in pure state or with additives (Carvalho 2018). The paraffinic mineral oils are the most used because of their high viscosity index, which allows them to better resist the variations in viscosity that occur due to temperature variations, higher oxidation resistance and lighter colour, resulting in better cutting machine operation products. These oily compounds after leaving the industrial line proceed to the treatment system and shall be removed to the maximum extent before the biological treatment process, because in large concentrations they may inhibit microbial development and occasionally disrupt the aerators' operation (Byers 2006). The additives are the other compounds, besides the mineral oil, found in the cutting fluids. They are incorporated into the fluids to obtain properties that improve their use and effectiveness. Among the most commonly used additives are biocides such as H_2O_2; anticorrosive agents such as sodium nitrite ($NaNO_2$); antifoams (silicone oils); and emulsifiers such as fatty acid soaps (Assenhaimer 2015).

Other important substances present in the effluents containing the cutting fluids are the phenolic compounds. These compounds represent the largest class of organic pollutants from various industrial activities, including metal mechanics, which are generally recalcitrant to biodegradation and toxic to most microorganisms. Phenolic compounds are also found in these types of effluents as intermediates during the biodegradation of compounds containing aromatic rings, such as benzene, toluene and xylene.

4.2.3 Machining Chips

Other constituents of metal-mechanical effluents are the machining chips. The machining chip is the metallic material portion removed from the original part with the cutting fluid aid. From the final effluent of a machining production system, the machining chips are withdrawn by a physical treatment, generally sieving. But pieces of these parts remain in the effluent and may be in suspension. For this

FIGURE 4.1 Cutting fluid application. (From CIMM, 2009.)

reason, effluents from the metal-mechanical industry are generally composed of high concentrations of metals (Kurniawan et al. 2006). Among the metals mostly found in the metal-mechanical effluents are iron, chromium, zinc and aluminium. However, other metals, such as manganese, copper and arsenic, can be found depending on the nature of the machining process. It is important to emphasize that other metals can still be incorporated into the metal-mechanical effluent, originating from other sources such as the wear of mechanical components of the cutting machine and the formulations of some additives (Machado et al. 2011). The removal of these metals, up to the limits required by the legislation, is usually carried out using the chemical precipitation treatment (Kurniawan et al. 2006).

4.2.4 Microorganisms

The microorganisms present in the metal-mechanical effluents are normally incorporated into the same working environment, since the concentrated cutting fluids are practically free of them. This is due to the fact that the fluid's osmotic pressure without dilution is high, besides the presence of biocides found in them to avoid their biodegradation. However, for their use, cutting fluids are diluted, and with this dilution, a favourable environment is formed for the development of microorganisms, which, occasionally, resist the biocides present in them (Piubeli et al. 2004). It is worth mentioning that some microorganisms present in the metal-mechanical effluents are also incorporated into them, thus initiating the treatment line. This is because most industrial effluent treatment systems operate in open environments, which provides the absorption of microorganisms from other localities, such as plants and the wind.

Some genera of bacteria such as *Pseudomonas, Bacillus, Acinetobacter, Achromobacter, Mycococcus, Flavobacterium* and *Geobacillus* are commonly found in environments with crude oil and its by-products such as metal-mechanical effluents (Iwashita et al. 2004). On the other hand, the fungi found in these oily environments are of the genus *Cladosporium, Aspergillus* and *Penicillium*, besides yeasts such as *Candida* and *Rhodotorula* (Ollivier and Magot 2005). However, the microorganism's survival and predominance in these environments depends on factors such as pH, temperature and oily concentration. Most filamentous fungi and yeasts are adapted to acid environments with pH values close to 5, as opposed to bacteria that predominate in slightly neutral environments, with pH values between 6.5 and 7.5.

Regarding the environment temperature, most bacteria survive at temperatures around 37°C. On the other hand, filamentous fungi and yeasts develop best in temperatures around 28°C (Ollivier and Magot 2005). In oily effluents, the oil concentration may also be an important factor for the predominance of a type of microorganism as it has already been observed in several samples that oil concentration alters the population of predominant microorganisms.

4.3 Effluent Treatment Processes

The treatment processes for effluents generated by the metal-mechanical industry are divided into physicochemical and biological treatments (Marques 2018). The objectives of these treatments are both the disposal of the treated effluent in recipient bodies and their reuse. The choice depends on the economic necessity and availability of the industry, since, normally, the treated effluent reuse is initially more expensive than effluent disposal. However, in the long run, reuse becomes more economically viable (Queissada 2013). Generally, the treated industrial effluent, when reused, is directed to gardening and cleaning of toilets and patios. Therefore, the water quality characterization is fundamental, assuring the right water quality for each type of reuse is chosen (Monteiro 2006).

4.3.1 Physical and Chemical Treatments

Physical and chemical processes are considered as primary treatments in a metal-mechanical effluent treatment system, being responsible for the removal of solids and metal ions from the effluents, and, generally, also oxidizing several compounds increasing their biodegradability. This transformation of complex compounds into simpler molecules facilitates further biological treatment (Moura et al. 2011).

4.3.1.1 Chemical Precipitation

The main objective of chemical precipitation in the treatment systems of metal-mechanical industries is to remove metallic ions in the form of machining chips (Kurniawan et al. 2006). The most common metals in this type of effluent are zinc, cadmium and manganese (Nascimento et al. 2006). However, the concentration each specific metal can vary greatly from industry to industry, depending on the machining process used (Pereira Neto et al. 2008).

The chemical precipitation process is a simple system to execute and is relatively inexpensive. According to Queissada et al. (2013), this process consists in adjusting the effluent pH until the environment is in right conditions to decrease the solubility of the existing metals. With low solubility, these metals tend to precipitate, which are generally removed and sent for co-processing. However, chemical precipitation treatment processes can become more expensive if the metals present in the effluent precipitate at very different pH values. Other increasingly used methods to treat the oily effluents from metal mechanics industries are the AOPs.

4.3.1.2 Advanced Oxidation Processes (AOPs)

Advanced oxidative processes (AOPs) are alternative treatment systems for treating effluents contaminated with compounds resistant to conventional treatments. Such systems are, by definition, processes in which the main oxidizing agent corresponds to the hydroxyl radical (•OH), which is not selective and, therefore, promotes the degradation, on a greater or lesser scale, of a large part of organic compounds, reacting from 10^6 to 10^{12} times faster than oxidants such as ozone (Kanakaraju et al. 2018). Table 4.1 shows the oxidation potential of the most important oxidizing agents used.

Due to its high reactivity, the hydroxyl radical must be generated in the reaction environment itself. With this objective, it is possible to resort to heterogeneous or homogeneous processes, assisted (or not) by ultraviolet radiation. In this way, several systems can be used, among which the best known are shown in Table 4.2.

4.3.1.2.1 Ultrasonic Treatment

Ultrasonic treatment is a technique applied in several fields of research, such as AOPs. It is applied in processes such as ultrasonic imaging, cell rupture, polymerization processes, degassing, nanotechnology, food preservation and sonar detection. This sonochemical process used for the oxidation of organic compounds consists of OH radical synthesis by water decomposition, under ultrasonic action, and the formation of hydrogen peroxide, a secondary oxidant, occurs due to the recombination of OH radicals, aiding in the degradation process (Steter et al. 2014). According to Eren (2012), there are three ultrasonic frequency bands: low frequency, ranging from 20 to 100 kHz; medium frequency, ranging from 300 to 1,000 kHz; and high frequency, ranging from 2 to 10 MHz. Among these, low and medium frequencies are more commonly used for the degradation of organic compounds (Eren 2012).

TABLE 4.1

Oxidation Potential of Various Oxidants

Oxidant Species	Oxidation Potential (V)
Fluorine	3.03
Hydroxyl radical (•OH)	2.80
Ozone	2.07
Hydrogen peroxide	1.78
Potassium permanganate	1.68
Chlorine dioxide	1.57
Hypochlorous acid	1.49
Chlorine	1.36
Bromine	1.09
Iodine	0.54

Source: Adapted from Legrini et al. (1993).

TABLE 4.2

Advanced Oxidative Processes – Typical Systems

Homogeneous Systems		**Heterogeneous Systems**	
With irradiation	Without irradiation	With irradiation	Without irradiation
UV/H_2O_2	H_2O_2/O_3	UV/TiO_2	Electro-Fenton
UV/O_3	Fe^{2+}/H_2O_2(Fenton)	$UV/TiO_2/H_2O_2$	
Electron beam			
US (ultrasound)			
US/H_2O_2			
US/UV			

Source: Adapted from Morais, J.L., *Estudo da potencialidade de processos oxidativos avançados, isolados e integrados com processos biológicos tradicionais, para tratamento de chorume de aterro sanitário*, Universidade Federal do Paraná, Curitiba, 2005.

The sonochemical oxidation techniques are improved by ultrasonic radiation use in order to create an oxidative environment and, under ultrasonic radiation, the hydroxyl radicals are formed in the presence of different gases also in combination with other processes such as O_3/US, H_2O_2/US and photocatalysis/US (Sathishkumar et al. 2016). In the compression cycle, a positive pressure is exerted on the liquid, bringing the molecules closer, whereas in the rarefaction cycle, the pressure exerted is negative, pushing them away. When a sufficiently high negative pressure is applied, the average distance between the molecules exceeds the critical molecular distance necessary to leave the liquid intact and then voids and cavities arise and thus the cavitation bubbles are formed. The bubbles then undergo an expansion and contraction movement as the compression/expansion cycles alternate until they reach a critical size and collapse resulting in extreme temperature and pressure conditions (Pang et al. 2011).

The free radicals formation and/or excited states from the water, vapours and other substrates dissociation are a result of high temperatures and post-collapse pressures, which provide the activation energy required for the homolytic cleavage. The water molecules dissociation and dissolved oxygen convert them into radicals such as hydroxyl (•OH), hydrogen atoms (•H) and hydroperoxyl radicals (•HO_2). These radicals, as demonstrated by Equations 4.1–4.5, where the symbol')))' represents the ultrasonic irradiation, can act by oxidizing the organic pollutants or, in the absence of solutes, recombination (Pang et al. 2011).

$$H_2O+))) \rightarrow \bullet OH + \bullet H \tag{4.1}$$

$$O_2+))) \rightarrow 2O\bullet \tag{4.2}$$

$$\bullet OH + \bullet O \rightarrow \bullet OOH \tag{4.3}$$

$$\bullet O + H_2O \rightarrow 2HO\bullet \tag{4.4}$$

$$\bullet H + O_2 \rightarrow \bullet OOH \tag{4.5}$$

A combination of oxidative processes can be a good alternative, so when the combination of ultrasound with ultraviolet (US/UV) or ultrasound with hydrogen peroxide (H_2O_2/US) is employed, there is an increased rate of free radical generation (Chakinala et al. 2009).

4.3.1.2.2 Ultrasonic/UV Treatment

Ultrasonic (US) irradiation mediated by catalysts (sonocatalysis) has received particular attention as an environmentally friendly technique for the removal of organic pollutants from wastewater. However, the degradation is slow compared to other established methods. Studies in order to increase the sonocatalysis efficiency by combination with other AOP techniques are underway in many countries. An alternative that has been studied is the coupling US with ultraviolet (UV) irradiation, which shows an efficiency increase in synergistic pollutants semiconductor-mediated degradation (Jyothi et al. 2014).

4.3.1.2.3 Ultrasonic/H_2O_2 Treatment

In order to accelerate the ultrasonic degradation of organic compounds, external oxidants such as hydrogen peroxide (H_2O_2) can be added. Under hotspot conditions (generation under extremely high temperature and pressure conditions) in ultrasonic cavitation, H_2O_2 easily suffers from sonolysis and decomposes into •OH, since the O–O bond dissociation energy for H_2O_2 is only 213 kJ/mol, which is lower than that of O–H bond in H_2O (418 kJ/mol). It is also facilitated by the low concentration of H_2O_2 inside the cavitation bubbles as a result of its low volatility and high solubility in water (Pang et al. 2011).

Following the same principle, the amount of H_2O_2 that can be produced by the ultrasound itself is too small to dissociate into large amounts of •OH. Therefore, an additional amount of H_2O_2 is generally required to significantly accelerate the degradation process. H_2O_2 has the function of forming •OH by reducing H_2O_2 in the conduction band or by self-decomposition due to ultrasound, as shown in Equations 4.6 and 4.7.

$$\bullet H + H_2O_2 \rightarrow HO\bullet + H_2O \tag{4.6}$$

$$H_2O_2 +))) \rightarrow 2HO\bullet \tag{4.7}$$

Cavitation is the origin of the sonic effects, which result in microbubbles that end in a violent collapse and act as a localized microreactor that causes high temperatures and pressures. Organic pollutants can be degraded by this process in aqueous solutions by sonochemistry, pyrolysis or reactions with free radicals. The simplicity of the process is one of its advantages, but degradation rates are still low for practical uses, so some techniques such the addition of solid particles, addition of H_2O_2 and combination with UV can be adopted to improve the degradation efficiency (Behnajady et al. 2009).

4.3.1.2.4 UV/H_2O_2 Treatment

An excellent alternative for the treatment of effluents from the metal-mechanical industry is the UV/H_2O_2 system, because H_2O_2 has excellent thermal stability, commercial availability, infinite solubility in water and high oxidation potential (Vandevivere et al. 1998).

The mechanism to obtain •OH from H_2O_2 is the homolytic cleavage of the hydrogen peroxide molecule (H_2O_2), involving a high energy sigma bond breakage (O–O/48.5 kcal/mol), resulting in the formation of two hydroxyl radicals (Equation 4.8). The required cleavage energy corresponds to ultraviolet irradiation with wavelength in the region of 254 nm (Leahy and Shreve 2000).

$$H_2O_2 + hv(254\,\text{nm}) \rightarrow 2HO\bullet \tag{4.8}$$

According to Vandevivere et al. (1998), the hydroxyl radical generation process efficiency is dependent on the substrate pH and concentration to be treated. In general, excess hydrogen peroxide favours radical–radical recombination, again generating H_2O_2 (Equation 4.9) and consequently reducing the efficiency of the reaction. Therefore, the H_2O_2 concentration must be optimized to avoid oxidation reactions.

$$HO\bullet + HO\bullet \rightarrow H_2O_2 \tag{4.9}$$

Several studies enhanced the efficiency of oily effluents treatment through UV/H_2O_2 system (Adams and Kuzhikannil 2000). However, for the system be effective some parameters need to be considered, as they directly influence the oxidation of several compounds present in them. Among these parameters are, in addition to the H_2O_2 concentration, the temperature and pH. The treatment temperature is important since its increase can reduce the H_2O_2 concentration. Santos et al. (2007) observed that temperatures close to 70°C can reduce up to 25% H_2O_2 in the treatment environment. However, pH may have a strong influence on the oxidative H_2O_2 efficacy, since an acidic treatment environment may increase the concentration of chloride ions (Cl^-), usually present in oily effluents, in the solution. These environmental reactions react with •OH and produce inorganic ionic radicals with lower oxidation potential, which consequently interfere with the degradation efficiency of the compounds (Behnajady et al. 2004).

Studies with UV/H_2O_2 showed degradation and mineralization of several recalcitrant and toxic compounds; however, their use in pilot and industrial scale to treat effluents still has limitations due to the high cost of treatment, mainly due to the high consumption of energy and oxidizing agents used. Therefore, this technique should be applied in cases of substances that are refractory to biological treatments, which are more economical (Britto and Rangel 2008). However, some alternatives can be used to minimize this problem, such as the solar radiation use (Bandala et al. 2004).

4.3.1.2.5 UV/O_3 Treatment

To improve the efficiency of the process, UV radiation can be used with a strong oxidizing agent such as ozone (O_3). When combined with UV radiation, the oxidizing power of ozone increases due to increased hydroxyl radical generation. The radiation treatment of a solution with O_3 (UV of $\lambda = 254\,\text{nm}$) induces the photolysis of ozone to H_2O_2 (Equation 4.10), which, in turn, reacts with UV radiation (Equation 4.11) generating hydroxyl and also peroxyl radicals: the O_3/UV system becomes more efficient than the isolated O_3 treatment (Verlicchi et al. 2015).

$$O_3 + H_2O \rightarrow H_2O_2 + O_2 \tag{4.10}$$

$$H_2O_2 \rightarrow 2HO\bullet \tag{4.11}$$

According to Chang et al. (2015), O_3/UV process involves the same behavioural processes as the UV/H_2O_2 and O_3/H_2O_2 processes. However, O_3/UV provides the highest yield of hydroxyl radicals by oxidants. One of the process obstacles is that the environment must not contain suspended solids because they will reduce the UV transmission, which is fundamental for the formation of radicals.

4.3.1.2.6 UV/TiO_2/H_2O_2 Treatment

The catalytic process UV/TiO_2/H_2O_2 is another alternative that is generally recommended due to some parameters such as very high photocatalytic oxidation potential, relatively low cost, operation under environmental conditions, commercial availability and photochemical stability (Affam et al. 2018). The H_2O_2 in the UV/TiO_2/H_2O_2 system according to Wahyuni et al. (2016) acts a free radical delivery agent. Thus, increasing its concentration increases the number of OH radicals, which highly promote photodegradation. However, the photodegradation of a given effluent can decrease as the H_2O_2 concentration increases, too, because the excess H_2O_2 reacts with OH radicals, forming water and oxygen, according to Equations 4.12 and 4.13.

$$H_2O_2 + OH \rightarrow HOO + H_2O \tag{4.12}$$

$$HOO + OH \rightarrow H_2O + O_2 \tag{4.13}$$

The presence of TiO_2 at suitable concentrations may provide a greater number of OH radicals. However, an excessive dose creates more turbulence, which can inhibit the penetration of light (Wahyuni et al. 2016).

4.3.1.2.7 H_2O_2/O_3 Treatment

H_2O_2 and O_3 are among the main oxidants and can be used both in direct oxidation and in conjunction with other techniques aiming at the generation of hydroxyl radicals (•OH), which are the basis of AOPs (Cardoso et al. 2016). H_2O_2 is one of the most powerful and harmless oxidants. Many organisms can degrade it into water and oxygen. Its isolated application in effluent treatment is not as efficient as its association with other oxidants, with iron species (Fenton) or other techniques, such as ultraviolet irradiation (Ghaly et al. 2014).

The homogeneous irradiation process UV/H_2O_2/O_3 has been reported to show high efficiencies at a relatively short exposure time. Other processes using H_2O_2 and O_3 have great applicability similar to H_2O_2/O_3, O_3/UV and H_2O_2/UV. However, it has already been verified that the process H_2O_2/UV needs more time to remove some parameters such as colour than the process UV/O_3 (Moreno et al. 2015).

4.3.1.2.8 TiO_2/UV Treatment

Among AOPs, heterogeneous photocatalysis with titanium dioxide (TiO_2/UV) has some advantages due to its non-selective action, high efficiency oxidant, and its ability to treat complex mixtures of contaminants effectively. In addition, it offers the possibility to use solar radiation as a primary source of energy, giving a significant environmental and sustainable characteristic to the process (Chong et al. 2015). The TiO_2 is the most photoactive semiconductor employed ($300 < \lambda < 390$ nm), due to its high activity (Tafurt-García et al. 2018). In addition, the degradation of some compounds using TiO_2/UV follows first-order reaction kinetics, which significantly increases the effectiveness of the process (Cortazar et al. 2012).

The TiO_2 is notable for its existence in three polymorphic forms: anatase (tetragonal), rutile (tetragonal) and brookite (orthorhombic), among which the tetragonal crystallographic structures, anatase and rutile, are the most used. The rutile phase has a high refractive index and better chemical and thermal stability. The anatase phase has a relatively low refractive index. Finally, brookite, which belongs to a group of minerals called hydrothermal (formed by processes of dissolution and precipitation related to magma solidification in pre-existing rocks bodies), has aroused great interest due to its applications, mainly in waste water treatment. One of its most important characteristics is its photosensitivity to any type of light, since anatase only reacts with ultraviolet light (González et al. 2018).

4.3.1.2.9 Fe^{2+}/H_2O_2 (Fenton) Treatment

The Fenton process has a great prominence as an efficient technique for the degradation of several types of pollutants. The process Fe^{2+}/H_2O_2 generates hydroxyl radicals through electron transfer between the hydrogen peroxide and the ferrous ions, which act as catalysts (Trovó et al. 2013). Pignatello et al. (2006) described that in the Fenton process, several reactions can occur (Equations 4.14–4.23).

$$Fe^{2+} + H_2O_2 \rightarrow Fe^{3+} + HO^- + HO\bullet \tag{4.14}$$

$$Fe^{3+} + H_2O_2 \rightarrow Fe^{2+} + H^+ + HO_2\bullet \tag{4.15}$$

$$H_2O_2 + HO\bullet \rightarrow H_2O + HO_2\bullet \tag{4.16}$$

$$Fe^{2+} + HO\bullet \rightarrow Fe^{3+} + HO^- \tag{4.17}$$

$$Fe^{3+} + HO_2\bullet \rightarrow Fe^{2+} + O_2 + H^+ \tag{4.18}$$

$$Fe^{2+} + HO_2\bullet + H^+ \rightarrow Fe^{3+} + H_2O_2 \tag{4.19}$$

$$HO_2\bullet + H_2O\bullet \rightarrow O_2 + H_2O_2 \tag{4.20}$$

$$HO_2\bullet + H_2O_2 \rightarrow HO\bullet + H_2O_2 + O_2 \tag{4.21}$$

$$HO_2\bullet + HO\bullet \rightarrow O_2 + H_2O \tag{4.22}$$

$$2HO\bullet \rightarrow O_2 + H_2O_2 \tag{4.23}$$

The oxidation of ferrous ions to ferric ions initiates and catalyses the decomposition reaction of the hydrogen peroxide molecules, resulting in fast hydroxyl radical generation (Equation 4.14), known as the Fenton reaction. The efficiency of Fenton techniques and the degradation of pollutants depend entirely on several parameters, with temperature, pH, ratio [H_2O_2]/[Fe^{2+}] and incidence of UV radiation being the most relevant. It should be noted that most of the tests carried out for the oily effluents treatment by

Fenton technique take place at ambient temperature, without the need for reaction environment heating or cooling, since at 20°C–40°C, the process is very efficient and temperatures of approximately 50°C accelerate the hydrogen peroxide decomposition into water and oxygen, disfavouring the hydroxyl radical generation process (Babuponnusami and Muthukumar 2014).

The optimum pH for efficient compound degradation is generally less than 4, because at higher values, ferrous ions will easily be converted to ferric ions, which, in the presence of hydroxide, will produce complexes capable of coagulating as well as precipitating organic matter; that is, the organic compounds reaction must always occur in acidic environment and the reaction environment must be adjusted to basic pH at the end of the treatment in order to interrupt the oxidative process (Ghanbari and Moradi 2015). In relation to $[H_2O_2]/[Fe^{2+}]$, an excess of iron ions restricts the amount of photo-energy that would be transferred to the reaction environment, because the iron particles present in the reaction hinder the light passage. When the iron amount added is much higher than the amount of hydrogen peroxide, the treatment tends to show a coagulant effect and cause the formation of sludge (Nogueira et al. 2017).

One of the most important factors in the UV radiation incidence on the photo-Fenton process is the radiation source position. With an increase in UV radiation intensity, the number of photons that react with the iron ions increases, generating a higher amount of hydroxyl radicals and, consequently, degrading the compounds. However, the costs associated with the generation of this energy must always be evaluated (Vaishnave et al. 2014).

4.3.1.2.10 Electron-Fenton (EF) Treatment

Hydrogen peroxide is an oxidant applied in various types of processes, for degrading most diverse compounds and producing non-toxic by-products such as O_2 and water. However, its individual use is impracticable due to its low oxidation power compared to other oxidants. Thus, the H_2O_2 can be activated in an acid environment, with Fe^{2+} (as a Fenton catalyst), producing hydroxyl radicals as a more potent oxidizing agent (Equation 4.14) (Borràs et al. 2013). Yet, there is an adverse factor, which is the slight consumption of Fenton reagent in the course of mineralization of contaminants, but this disadvantage can be overcome by the electro-Fenton process, in which the peroxide is electrochemically continuously produced by the cathode through the reduction of O_2 in an aqueous environment (Xu and Wang 2012).

The following advantages of the electro-Fenton process can be highlighted when compared to the traditional Fenton method, such as the progressive recovery of Fe^{2+} at the cathode; prevention of transport problems, hydrogen peroxide handling and storage, favourable to its production *in loco*; and the possibility of total mineralization of pollutants, with relatively low costs (Gao et al. 2015). Yet, the process application is only possible in situations in which the solution under treatment is acidic with pH between 2 and 4, because in this pH range, the maximum concentration of active Fe^{2+} species occurs with lowest decomposition of H_2O_2. Thus, depending on the contaminated matrix, it is necessary to use extensive amounts of reagents to acidify the same and then also to carry out their respective neutralization (Harichandran and Prasad 2016).

The reaction represented by Equation 4.14 is driven by the catalytic behaviour of the set Fe^{2+}/Fe^{3+}. The Fe^{2+} is progressively regenerated, by the simple reduction of Fe^{3+} in cathode according to Equation 4.24, through hydrogen peroxide, as shown in Equation 4.15; for $HO_2\bullet$ in Equation 4.18 or by the radical organic intermediates as demonstrated by Equation 4.25.

$$Fe^{3+} + e^- \rightarrow Fe^{2+} \tag{4.24}$$

$$Fe^{3+} + R\bullet \rightarrow Fe^{2+} + R^+ \tag{4.25}$$

In the case of only one distribution electrolytic cells, this organic matter present in the contaminated effluent can undergo conjugated oxidations with heterogeneous •OH radicals produced on the surface of the anode by reaction with water (Equation 4.26) (Oturan et al. 2012).

$$H_2O \rightarrow \bullet OH + H^+ + e^- \tag{4.26}$$

4.3.1.2.11 Electron Beam Treatment

The effluents treatment technique that uses ionizing radiation focuses on containing the desired sample to the electron beam, respecting the water layer thickness according to the equipment energy that will be used. Ionizing radiation has been applied for degrading different types of effluents, such as liquids and gases, being domestic or industrial, in order to promote the biodegradability (Boiani 2016).

The ionizing radiation used, generated from an electron beam accelerator, has been developed as an advanced alternative treatment that can be applied to treatment of both surface water and effluent. The process occurs by accelerating subatomic particles from very low values up to several million and several billion electron volts (eV) and high kinetic energies, by the electric and magnetic fields combination. The unit electron volt corresponds to the change in the electron energy passing through a potential difference of 1 V in vacuum. This high voltage potential is established between the cathode and the anode, and it is precisely this potential difference that will be responsible for the acceleration of particles (Romanelli 2004).

4.3.1.3 Biological Treatments

Biological processes are considered as secondary treatments in a metal-mechanical effluent treatment system. The treatment techniques that apply biological processes are called bioremediation, which, by definition, are systems that use microorganisms with metabolic potentials to remove pollutants from effluents or soils (Pereira and Lemos 2004). These types of treatments have several advantages, mainly low cost. However, they are processes that are used in a unitary manner and hardly provide the complete mineralization of pollutants, but when combined with other methods, such as the physical and chemical processes, provide better treatment efficiency (Kambourova et al. 2003).

As mentioned above, it is of great importance that biological treatments of industrial effluents be carried out with autochthonous microorganisms, since these are already adapted to the pollutants existing in the environment, which increases the process efficacy. Generally, these treatments are performed with microorganisms that develop in the effluent itself; however, it is important to increase the microorganisms concentration, which when tested, have a higher compounds effluent interest degradation rate. This technique is called bioaugmentation and, unfortunately, is rarely used in effluent treatment systems in Brazilian industries. They generally resist this technique use because it initially entails an additional effluent treatment line cost. However, when the bioaugmentation process occurs effectively, it reduces the time for the biological treatment of effluents, thus reducing process costs (Queissada et al. 2003).

Generally, biological treatment using bioaugmentation is optimized when there is an integrated chemical treatment. Most of the times, chemical treatment is performed before the biological process, because it increases the biodegradability of compounds, besides degrading compounds that would be toxic to the microorganisms used in the biological treatment. However, it can also be used as a tertiary or polishing treatment, degrading recalcitrant compounds from biological treatment (Neyens and Baeyens 2003). One of the most commonly used chemical treatments in this integration is AOP. However, this integration occurs mainly in bench and pilot studies and is still little used on an industrial scale (Gogate and Pandit 2004).

These integrated treatment systems are normally more effective when the biological process comes after than the AOP, which oxidizes complex molecules into smaller molecules, making them more biodegradable (Neyens and Baeyens 2003). It is important to emphasize that the integrated biological and chemical treatment effectiveness, it is necessary a non-production of toxic compounds, when the pretreatment occurs, or microorganisms inhibitors to the later biological treatment. When these phenolic compounds reach the biological treatment system, they may inhibit the various microorganisms present in the treatment system (Alnaizy and Akgerman 2000). Once again, the importance of carrying out the biological treatment with autochthonous microorganisms is clear, since they are adapted to the pollutants and the phenolic compound. Yet, some microorganisms may not be inhibited by these compounds, but at the same time, they do not degrade them.

4.3.1.3.1 Aerobic Biological Treatment

In aerobic biological treatment, microorganisms have the function of degrading organic substances through bio-oxidative processes, metabolizing the nutrients and using them as an energy source. The aerobic systems of aerated lagoons and biological filters, for example, demonstrate good results, although activated sludge systems such as flocculent microbial mass, which is formed when sewage and other biodegradable effluents are submitted to aeration, are the most applied and have a greater efficiency in the removal of the load, reaching up to 98% efficiency in the removal of BOD under prolonged aeration in some cases (Kispergher 2013). Bacteria, the most aerobic and facultative heterotroph, are mainly responsible for the organic matter elimination process and act with a greater efficiency in appropriate temperature conditions (20°C–30°C), pH (6.0–8.0) and dissolved oxygen (DO, 1–4 ppm) (Cyprowski et al. 2014).

The use of aerobic processes integrated with AOPs for oily effluents treatment has benefited communities and industries. Another advantage is the reduction in the risk of odour emissions with a greater capacity of refractory substances absorption. Yet, it is necessary to standardize the mass relation with biological oxygen demand (BOD), carbon (C), nitrogen (N), phosphor (P) = 100:5:1, which vary with the biota formed in each treatment season. Another important point that needs to be highlighted is that if the process occurs in the pH range between 3.0 and 5.0, there will be development of fungal cells and poor sedimentation of the sludge. However, in the pH range between 8.0 and 10.0, another important factor to be highlighted is if the process occurs with pH between 3.0 and 5.0, will be development of fungal cells and poor sludge sedimentation. However, in the pH range between 8.0 and 10.0, the water clarity will be compromised with yellow-brownish sludge and lack of nutrients (N and P) will cause dispersed flakes formation and excessive growth of filamentous bacteria, reducing the treatment action (Esguerra et al. 2017).

4.3.1.3.2 Anaerobic Biological Treatment

In anaerobic effluent treatment, the bacteria are mainly responsible for the contaminants degradation. The process is operated with the aim of converting the biodegradable and soluble particulate matter into methane and carbon dioxide. The formation of methane is a difficult process that can be divided into four stages of degradation defined as hydrolysis, acidogenesis, acetogenesis and methanogenesis (Deublein and Steinhauser 2015). The most known anaerobic treatment processes are anaerobic lagoons, septic tanks, anaerobic filters and high-rate reactors, capable of receiving higher amounts of organic load per unit volume such as UASB (upflow anaerobic sludge blanket). Compared with aerobic processes, anaerobic treatment offers advantages such as lower consumption of non-renewable energy, as it does not require aeration; the process requires a lower nutrient supplementation and produces less sludge; and most of the reactors takes up less space, making it a more economical and environmentally sustainable process (Almeida and Grossi 2014).

According to Sawatdeenarunat et al. (2015), in anaerobic digestion, several processes occur that together result in the matter involving bacterial fermentation of organic residues in the absence of free oxygen decomposition. Fermentation is responsible for the decomposition of complex, biodegradable organic matter in four stages:

- **Hydrolysis:** This is the stage in which some anaerobic species, such as those belonging to families *Streptococcaceae, Enterobacteriaceae, Bacteroides, Clostridium, Butyrivibrio, Eubacterium, Bifidobacterium* and *Lactobacillus*, hydrolyse the soluble products in a mixture of organic acids, hydrogen and carbon dioxide (Kamali et al. 2016).
- **Acidogenesis:** This is the stage where other groups of anaerobic bacteria metabolize the organic acids into acetate, H_2 and CO_2. In this stage, methanogenic bacteria rapidly fix H_2 and maintain the partial gas pressure very low, which results in a thermodynamically favourable condition for the hydrogen-producing bacteria to degrade the organic compounds to acids, including formic and acetic acids, H_2 and CO_2 (Abbasi et al. 2012).
- **Acetogenesis:** In this third step, the products obtained in the previous step are used as substrates. This process can be either autotrophic, using CO_2 as a carbon source for cellular

synthesis, or heterotrophic, with organic substrates such as methanol and formic acid being carbon sources (Abbasi et al. 2012).

- **Methanogenesis:** This is the fourth and last stage. Its main characteristic is methane production, presenting exclusively anaerobic character, and can occur by three main routes, ethanogenesis (CO_2 reduction); hydrogenotrophic, acetotrophic or acetoclastic methanogenesis; and methylotrophic methanogenesis. Another important factor when using anaerobic digestion is the existence of inhibitors of the process observation. The most important inhibitors that should always be determined are oxygen, sulphur and light and heavy metal ions (Almeida and Grossi 2014).

4.4 Conclusions

Oily effluents have a high polluting potential, especially those from the metal-mechanical industry, due to their high oils, greases and metallic ions concentration. However, one of the most effective treatments for the removal of these ions is the chemical precipitation. For the reduction of other parameters such as COD (chemical oxygen demand), oils, greases and colour, the biological treatments have a relative efficacy, having lower cost, in comparison with other types of treatments. However, it is necessary to fully characterize the matrix to be treated and to decide whether after the treatment it will be reused or discarded into the environment, to choose the best type of biotreatment, aerobic or anaerobic, always considering that the use of native microorganisms considerably increases the treatment success, because they are already adapted to the existing pollutants. Advanced oxidative processes (AOPs) are an excellent alternative to traditional treatments. However, some treatments have a higher cost (reagents and/or energy), but also have high treatment effectiveness, especially in the recalcitrant compounds removal, which treatments like biological cannot. In most treatment processes, the integrated treatment procedure is used instead of isolated techniques. Such integration generally intensifies the pollutants removal. However, any treatment prior to the biological one needs to be done with more care, because in certain situations, it can promote the synthesis of compounds that inhibit the biological process.

REFERENCES

Abbasi T et al (2012) Anaerobic digestion for global warming control and energy generation-an overview. *Renewable and Sustainable Energy Reviews*, v. 16, n. 5, pp. 3228–3242.

Adams CD, Kuzhikannil JJ (2000) Effects of UV/H2O2 peroxidation on the aerobic biodegradability of quaternary amine surfactants. *Water Research*, v. 34, pp. 668–672.

Affam AC et al (2018) Comparison of Five Advanced Oxidation Processes for Degradation of Pesticide in Aqueous Solution. *Bulletin of Chemical Reaction Engineering & Catalysis*, v. 13, n. 1, pp. 179–186.

Almeida EJM, Grossi LJ (2014) *Estudo do processo de tratamento de agua da industria de laticinio. Trabalho de conclusão de curso de graduação em Engenharia Química*, Universidade Federal de Alfenas - UNIFAL, 30 p.

Alnaizy R, Akgerman A (2000) Advanced oxidation of phenolic compounds. *Advances in Environmental Research*, v. 4, pp. 233–244.

Assenhaimer C (2015) *Evaluation of Emulsion Destabilization by Light Scattering Applied to Metalworking Fluids. 2015.* 130 p. Tese (Doutorado)-Escola Politécnica, Universidade de São Paulo, São Paulo.

Babuponnusami A, Muthukumar K (2014) A review on Fenton and improvements to the Fenton process for wastewater treatment. *Journal of Environmental Chemical Engineering*, v. 2, n. 1, pp. 557–572.

Bandala ER et al (2004) Solar photoreactors comparison based on oxalic acid photocatalytic degradation. *Solar Energy*, v. 77, pp. 503–512.

Behnajady MA et al (2004) Photo destruction of acid orange 7 (AO7) in aqueous solutions by UV/H2O2: influence of operational parameters. *Chemosphere*, v. 55, pp. 129–134.

Behnajady MA et al (2009) Evaluation of electrical energy per order (EEO) with kinetic modeling on the removal of Malachite Green by US/UV/H2O2 process. *Desalination*, v. 249, n. 1, pp. 99–103.

Boiani NF (2016) *Removal of Toxicity the Pharmaceutical Propranolol and Your Mixture with Fluoxetine Hydrochloride in Aqueous Solution Using Radiation with Electron Beam.* Tese (Doutordo) - Instituto de Pesquisas Energeticas e Nucleares (IPEN/CNEN-SP), Sao Paulo, SP, Brazil.

Borràs N et al (2013) Anodic oxidation, electro-Fenton and photoelectro-Fenton degradation of cyanazine using a boron-doped diamond anode and an oxygen-diffusion cathode. *Journal of Electroanalytical Chemistry*, v. 689, pp.158–167.

Borsa MBNO (2014) *Remoção de grafite de um efluente oleoso através de técnicas eletroquímicas, 2014.* 133 p. Tese (Doutorado em Engenharia). Programa de pós-graduação em engenharia de Minas, Metalúrgica e de Materiais; Universidade Federal do Rio Grande do sul. Porto Alegre.

Britto JM, Rangel MC (2008) Processos avançados de oxidação de compostos fenólicos em efluentes industriais. *Química Nova*, v. 31, pp. 114–122.

Byers JP (2006) *Metalworking Fluids.* 2nd ed. CRC Press, Boca Raton, 450 p.

Cardoso JC et al (2016) Análise crítica dos processos empregados no tratamento de efluentes têxteis. In: Zanoni MVB, Yamanaka H. (Orgs.). *Corantes: caracterização química, toxicológica, métodos de detecção e tratamento*, Cultura Acadêmica, São Paulo, pp. 215–239.

Carvalho DOA (2018) *Comparação do desempenho de fluidos de corte de base vegetal e mineral no torneamento do aço ABNT 1050. 2017.* 169 f. Tese (Doutorado em Engenharia Mecânica) - Universidade Federal de Uberlândia, Uberlândia.

Chakinala AG et al (2009) Industrial wastewater treatment using hydrodynamic cavitation and heterogeneous advanced Fenton processing. *Chemical Engineering Journal*, v. 152, n. 2–3, pp. 498–502.

Chang J et al (2015) What have we learned from worldwide experiences on the management and treatment of hospital effluent? — An overview and a discussion on perspectives. Science of the Total Environment, v. 514, pp. 467–491.

Chong MN et al (2015) Evaluation of Titanium dioxide photocatalytic technology for the treatment of reactive Black 5 dye in synthetic and real greywater effluents. *Journal of Cleaner Production*, v. 89, pp. 196–202.

CIMM (2009) *Centro de Informação Metal Mecânica.* Disponível em: http://www.cimm.con.br - Acesso em: 19/07/2009.

Cortazar A et al. (2012) *Biotechnology Applied to the Degradation of Textile Industry Dyes*, Univ. y Cienc, v. 28, pp. 187–199.

Curia AC (2010) *Banhados construídos como sistema terciário para reúso da água industrial em uma empresa metal-mecânica.* 237 p. Tese (Doutorado em Engenharia). Programa de pós graduação em engenharia de Minas, Metalúrgica e de Materiais; Universidade Federal do Rio Grande do sul, Porto Alegre.

Cyprowski M et al (2014) Peptidoglycans in cutting fluids-a good indicator of bacterial contamination? *Annals of Agricultural and Environmental Medicine*, v. 21, n. 2, pp. 256–258.

Deublein D, Steinhauser A (2015) *Biogas from Waste and Renewable Resources: An Introduction.* 2nd ed. Wiley-VCH, Weinhein. 2011. Engineering Journal, v. 276, pp. 97–105.

Eren Z (2012) Ultrasound as a basic and auxiliary process for dye remediation: a review. *Journal of Environmental Management*, v. 104, pp. 127–141, abr.

Esguerra KVN et al (2017) Catalytic aerobic oxidation of halogenated phenols. *Inorganica Chimica Acta*, v. 17, pp.31021–31026.

Gao G et al (2015) Carbon nanotube membrane stack for flow-through sequential regenerative electro-Fenton. *Environmental Science & Technology*, v. 49, n. 4, pp. 2375–2383.

Ghaly AE et al (2014) Production, characterization and treatment of textile effluents: a critical review. *Journal of Chemical Engineering and Process Technology*, v. 5, pp. 1–18.

Ghanbari F, Moradi M (2015) A comparative study of electrocoagulation, electrochemical Fenton, electro-Fenton and peroxi-coagulation for decolorization of real textile wastewater: electrical energy consumption and biodegradability improvement. *Journal of Environmental Chemical Engineering*, v. 3, n. 1, pp. 499–506.

Gogate PR, Pandit AB (2004) A review of imperative technologies for wastewater treatment II: hybrid methods. *Advances in Environmental Research*, v. 8, pp. 553–597.

González LG et al. (2018) Synthesis and characterization of nanostructured TiO2 and TiO2/W thin films deposited by co-sputtering. *Matéria (Rio de Janeiro)*, v. 23, n. 2.

Harichandran G, Prasad S (2016) Sono Fenton degradation of an azo dye, Direct Red. *Ultrasonic Chemistry*, v. 29, pp. 178–185.

Iwashita S, Callahan TP, Haydu J, Wood TK (2004) Mesophilic aerobic degradation of metal lubricant by a biological consortium. *Applied Microbiology Biotechnol*ogy, v. 65, pp. 620–626.

Jyothi KP et al (2014) Ultrasound (US), Ultraviolet light (UV) and combination (US+ UV) assisted semiconductor catalysed degradation of organic pollutants in water: oscillation in the concentration of hydrogen peroxide formed in situ. *Ultrasonics Sonochemistry*, v. 21, n. 5, pp. 1787–1796.

Kamali M et al (2016) Anaerobic digestion of pulp and paper mill wastes–An overview of the developments and improvement opportunities. *Chemical Engineering Journal*, v. 298, pp. 162–182.

Kambourova M et al. (2003) Purification and properties of thermostable lipase from a thermophilic *Bacillus stearothermophilus* MC7. *Journal of Molecular Catalysis B: Enzymatic*, v. 22, pp. 307–313.

Kanakaraju D et al (2018) Advanced oxidation process-mediated removal of pharmaceuticals from water: a review. *Journal of Environmental Management*, v. 219, pp. 189–207.

Kispergher EM (2013) *Digestão Anaeróbia de Efluentes da Industria de Alimentos*. Dissertação de Mestrado em Engenharia de Alimentos. Universidade Federal do Paraná – UFPR, p. 102.

Kurniawan TA et al (2006) Physico–chemical treatment techniques for wastewater laden with heavy metals. *Chemical Engineering Journal*, v. 118, pp. 83–98.

Leahy JG, Shreve GS (2000) The effect of organic carbon on the sequential reductive dehalogenation of tetrachloroethylene in landfill leachates. *Water Research*, v. 34, pp. 2390–2396.

Legrini O et al (1993) Photochemical processes for water treatment. *Chemical Reviews*, v. 93, pp. 671–698.

Machado ÁR et al (2011) *Teoria da Usinagem dos Materiais*. 2nd ed. Blucher, São Paulo.

Marques MG (2018) *Avaliação técnico-econômica do reuso de efluentes em indústria metal-mecânica*. 69 f. Dissertação (Mestrado em Ciência e Tecnologia Ambiental), Universidade do Sagrado Coração, Bauru.

Monteiro MI (2006) *Tratamento de efluentes oleosos provenientes da indústria metal-mecânica e seu reuso*. 148 p. Tese (Doutorado em Biotecnologia Industrial). Departamento de Biotecnologia Industrial, Universidade de São Paulo, Lorena.

Morais JL (2005) *Estudo da potencialidade de processos oxidativos avançados, isolados e integrados com processos biológicos tradicionais, para tratamento de chorume de aterro sanitário*. 207 p. Tese (Doutorado em Química). Setor de Ciências Exatas; Universidade Federal do Paraná, Curitiba.

Moreno FLA et al (2015) Propuesta de diseño para un sistema de tratamiento y reutilización de efluentes textiles combinando tecnologíasconvencionalesconelproceso de oxidación avanzada (O3/H2O2/UV). *INVENTUM*, v. 10, n. 18, pp. 54–62.

Moura FN et al (2011) Desempenho de sistema para tratamento e aproveitamento de esgoto doméstico em áreas rurais do semiárido brasileiro. *Engenharia Ambiental: Pesquisa e Tecnologia*, v. 8, n. 3, pp. 264–276.

Muszynski A, Lebkowska M (2005) Biodegradation of used metalworking fluids in wastewater treatment. *Polish Journal of Environmental Studies*, v. 14, pp. 73–79.

Nascimento SC et al (2006) Disponibilidade de metais pesados em aterro de indústria siderúrgica. *Engenharia Sanitária Ambiental*, v. 11, pp. 196–202.

Neyens E, Baeyens JA (2003) Review of classic Fenton's peroxidation as an advanced oxidation technique. *Journal of Hazardous Materials*, v. 98, pp. 33–50.

Nogueira AA et al (2017) Ferrioxalate complexes as strategy to drive a photo-FENTON reaction at mild pH conditions: a case study on levofloxacin oxidation. *Journal of Photochemistry and Photobiology A: Chemistry*, v. 345, pp. 109–123.

Ollivier B, Magot M (2005) *Petroleum Microbiology*, Ed. ASM Press, Washington DC, p. 365.

Oturan N et al (2012) Unprecedented total mineralization of atrazine and cyanuric acid by anodic oxidation and electro-Fenton with a boron-doped diamond anode. *Environmental Chemistry Letters*, v. 10, n. 2, pp. 165–170.

Oturan MA, Aaron JJ (2014) Processos Oxidacionais Avançados em Água/Tratamento de Efluentes: Princípios e Aplicações: Uma Revisão. *Critical Reviews in Environmental Science and Technology*, v. 44, pp. 2577–2641.

Pang YL, Abdullah AZ, Bhatia S (2011) Review on sonochemical methods in the presence of catalysts and chemical additives for treatment of organic pollutants in wastewater. *Desalination*, v. 277, n. 1–3, pp. 1–14.

Pereira Neto A et al. (2008) Alternativas para o tratamento de efluentes da indústria galvânica. *Engenharia Sanitária Ambiental*, v. 13, pp. 263–270.

Pereira LTC, Lemos JLS (2004) *Degradação de Hidrocarbonetos de Petróleo por Aspergillus niger e Penicillium corylophilum*, Centro de Tecnologia Mineral – CETEM, Rio de Janeiro, pp. 1–11. Artigo Técnico.

Pignatello JJ et al (2006) Advanced oxidation processes for organic contaminant destruction based on the Fenton reaction and related chemistry. *Critical Reviews in Environmental Science and Technology*, v. 36, n. 1, pp. 1–84.

Piubeli FA et al (2004) *Análise microbiológica de fluido de corte em operação de usinagem*. Editorial do Centro de Informação Metal-Mecânica, Junho, pp. 25–26.

Queissada DD et al (2003) Biotransformação de agroquímicos presentes em água de lavagem de cana por microrganismos nativos. In: *5º Encontro Nacional De Biólogos, agosto 3–6*, Conselho Federal de Biologia, Natal, p. 131.

Queissada DD et al (2013) *Epicoccum nigrum* and *Cladosporium* sp. for the treatment of oily effluent in an air-lift reactor. *Brazilian Journal of Microbiology*, v. 44, n. 2, pp. 607–612.

Rodrigues AC et al (2016) Effect of grinding parameters on surface finishing of advanced ceramics. *Matéria. Rio de Janeiro*, v. 21, n. 4, pp. 1012–1020.

Romanelli MF (2004) *Avaliação da toxicidade aguda e crônica dos surfactantes DSS e LAS submetidos a irradiação com feixes de elétrons*. Dissertação (Mestrado) – Instituto de Pesquisas Energéticas e Nucleares - Universidade de São Paulo, São Paulo.

Samson MEFG et al (2018) Biological treatments in giant cell arteritis & Takayasu arteritis. *European Journal of Internal Medicine*, v. 50, pp. 12–19.

Santos A, Yustos P, Rodriguez S, Simon E, Garcia-Ochoa F (2007) Abatement of phenolic mixtures by catalytic wet oxidation enhanced by fenton's pretreatment: Effect of H_2O_2 dosage and temperature. *Journal of Hazardous Materials*, v. 146, pp. 595–601.

Sathishkumar P, Sambandam A (2016) Review on the recent improvements in sonochemical and combined sonochemical oxidation processes – A powerful tool for destruction of environmental contaminants. *Renewable and Sustainable Energy Reviews*, v. 55, pp. 426–454.

Sawatdeenarunat C et al (2015) Anaerobic digestion of lignocellulosic biomass: challenges and opportunities. *Bioresource Technology*, v. 178, pp. 178–186.

Silva TC (2017) *Tratamento de efluente oleoso de biodiesel por extrato bruto de lipase fúngica*. 106 p. Dissertação de Mestrado. Universidade Federal de Pernambuco.

Souza MB (2010) *Avaliação de processos oxidativos avançados acoplados com carvão ativado granular com biofilme para reuso de efluentes de refinaria de petróleo*. 179 p. Dissertação de Mestrado, PEQ/COPPE/UFRJ, Rio de Janeiro.

Steter JR et al (2014) Electrochemical and sonoelectrochemical processes applied to amaranth dye degradation. *Chemosphere*, v. 117, pp. 200–207.

Tafurt-García G et al (2018) Decolorization of reactive black 5 dye by heterogeneous photocatalysis with TiO_2/UV. *Revista Colombiana de Química*, v. 47, n. 2, pp. 36–44.

Trovó AG et al (2013) Degradation of caffeine by photo-Fenton process: optimization of treatment conditions using experimental design. *Chemosphere*, v. 90, n. 2, pp. 170–175.

Vaishnave P et al (2014) Photo oxidative degradation of azure-B by sono-photo-Fenton and photo-Fenton reagents. *Arabian Journal of Chemistry*, v. 7, n. 6, pp. 981–985.

Vandevivere PC et al (1998) Treatment and reuse of wastewater from the textile wet-processing industry. Review of emerging technologies. *Journal of Chemistry & Biotechnology*, v. 72, pp. 289–302.

Verlicchi P et al (2015) What have we learned from worldwide experiences on the management and treatment of hospital effluent? - An overview and a discussion on perspectives. *Science of the Total Environment*, v. 514, pp. 467–491.

Wahyuni ET et al (2016) Photodegradation of detergent anionic surfactant in wastewater using $UV/TiO_2/H_2O_2$ and $UV/Fe+2/H_2O_2$ processes. *American Journal of Applied Chemistry*, v. 4, n. 5, pp. 174–180.

Xu L, Wang J (2012) Fenton-like degradation of 2, 4-dichlorophenol using Fe_3O_4 magnetic nanoparticles. *Applied Catalysis B: Environmental*, v. 123, pp. 117–126.

5

Reducing Heavy Metal Toxicity using Biochar as a Naturally Derived Adsorbent

Iqbal Azad and Tahmeena Khan
Integral University

Bhanu Pratap
Babasaheb Bhimrao Ambedkar University

Abdul Rahman Khan
Integral University

CONTENTS

DOI: 10.1201/9781003181224-5

5.1 Introduction

Heavy metals are highly hazardous, non-biodegradable elements which bioaccumulate in food webs. Their occurrence in waters or soils may cause threat to environmental and human well-being even at relatively low concentrations (Agrafioti et al. 2014). Growing industrialization and extensive agricultural activities have led to severe heavy metals incidences in recent years. Pb and Cd are non-essential elements for living organisms, which are carcinogenic even in low quantity, and are usually sourced in industrial wastewater from mining, smelting, electroplating and petrochemical production (Rai et al. 2019). The prescribed drinking water guidelines by the World Health Organization (WHO) and the American Water Works Association (AWWA) are 0.05 mg/L for Pb and 0.005 mg/L for Cd, respectively. Dumping of heavy metals directly or indirectly into soils and waters causes momentous risks to environmental sustainability.

Various methods have been adopted for the treatment of contaminated water, including electroplating, ion exchange, reverse osmosis, precipitation and membrane processes (Inyang et al. 2016). Adsorption is a commonly used technique for heavy metal remediation from wastewater. Biosorption using retro-applied organic materials is an ecologically friendly and cost-effective method, as biosorbents are environmentally safe and easily available in large quantities. Agricultural by-products with suitable changes may prove handy as active adsorbents for the removal of heavy metals from wastewater.

The accumulation of heavy metals is increasing, which affects water and soil quality and has emerged as a food safety risk. Several potent approaches have been suggested for heavy metal remediation from water and soil, which includes (i) phytoremediation, (ii) physical remediation, (iii) chemical remediation and (iv) bioremediation (Wang et al. 2015a). The residue obtained from biomass has been used for heavy metal remediation and has received worldwide attention because of its environment-friendly nature, low cost and easy availability. Biochar is a material with abundant prospective applications, but the properties of biochar can be altered by many factors.

5.2 Heavy Metals

Heavy metals are well defined as naturally occurring elements that are characterized by relatively high density compared to water (5 g/cm^3 or five times greater than water) and high atomic mass. Heavy metals are also broadly classified as the elements that have an atomic number greater than 20 and atomic mass higher than 23. They are usually metallic elements found in the earth's crust and present in different environmental matrices in trace amount and thus are also called trace elements. Some physical phenomena such as temperature, phase association, adsorption and impounding influence the bioavailability of heavy metals.

Various chemical and biological factors such as thermodynamic equilibrium, complication mechanisms, lipid solubility, octanol/water partition coefficients, species characteristics, trophic interactions and biochemical/physiological variation also influence their bioavailability. The United States Environmental Protection Agency (USEPA) and the International Agency for Research on Cancer (IARC) also categorize them as possible human carcinogenesis promoter on the basis of epidemiological and tentative studies that have displayed a link between cancer and heavy metals exposure in humans and animals. Various heavy metals such as Co, Se or Zn are essential trace elements and have an imperative role in the human and plant metabolism. They are components of numerous significant enzymes and play a vital role in several redox reactions (Stern 2010). Co is one of most common examples of essential trace elements present as an essential cofactor in some oxidative stress-related enzymes such as catalase, superoxide dismutase, peroxidase and cytochrome oxidases (Al-Fartusie and Mohssan 2017).

Similarly, other essential trace elements are also mandatory for different biological metabolisms; on the other hand, a higher concentration of metals is also involved in the cellular and tissue injury, leading to various diseases. Some other heavy metals do not show any biological significance and include non-essential heavy metals such as aluminium (Al), antinomy (Sb), arsenic (As) and barium (Ba). Many heavy metals also possess very well hi-tech implications, such as iron (Fe), zinc (Zn), tin (Sn), lead (Pb), copper (Cu) and tungsten (W). Some are also a part of bioinorganic catalysis and play an important role in novel chemical transformations (Chellan and Sadler 2015). Additionally, some heavy metals are valuable and expensive, such as gold (Au), silver (Ag), iridium (Ir), rhodium (Rh) and platinum (Pt) (Seehra and Bristow 2018). Besides these, many such as mercury (Hg), cadmium (Cd), arsenic (As), chromium (Cr), thallium (Tl) and lead (Pb) are recognized as the dark side of chemistry. They show high toxicity even at lower concentration (Koller and Saleh 2018).

5.2.1 Sources of Heavy Metals

Heavy metals are widely spread contaminants found in terrestrial and freshwater ecosystems. They usually occur naturally in the earth's crust. Natural and anthropogenic activities are two probable routes that can derive heavy metals, including different environmental components such as air, water, soil and their interface. Unlike organic pollutants which ultimately break down to carbon dioxide and water, heavy metals persist in environment and assimilate in different environmental components such as air, water and soil. They can also be transported from one component to another. The concentration of heavy metals in environment is assessed by its source, adsorption or precipitation. The degree of adsorption depends upon the various physicochemical characteristics of the environment, such as water pH, hardness and redox potential as well as the presence of other such metals.

5.2.1.1 *Natural Routes*

Many natural sources of heavy metals have been recognized by different studies. Under the diverse environmental circumstances, natural emissions of heavy metals take place by rock weathering, volcanic eruptions, biogenic sources, forest fires, wind-borne soil particles and sea-salt sprays, out of which weathering is a well-known source of metals responsible for their widespread distribution to diverse environment components. Heavy metals exist in diverse forms such as oxides, sulphides, hydroxides, phosphates, sulphates, silicates and also as components of various organic compounds.

5.2.1.2 *Anthropogenic Routes*

Anthropogenic sources are associated with industrialization as atmospheric deposition, waste disposal, waste incineration, urban effluents, traffic emissions, fertilizer application and long-term application of wastewater in agricultural lands. Metals discharged in wind-blown dirt are frequently reported from industrial regions. Various key anthropogenic sources that considerably participate to heavy metal pollution of various environment sections include automobile emissions, which mainly emit Pb, smelting and insecticides processes, which mainly emit As, Cu and Zn, and combustion of fossil fuels, which emits Ni, V, Hg, Se and Sn. Many human activities also add significantly to environmental deterioration (He et al. 2005) (Figure 5.1).

5.2.2 Effects of Heavy Metals

Heavy metals can also cause cellular and tissue damage in various biological systems, mainly affecting nucleus, mitochondria, cell membrane, endoplasmic reticulum, lysosome and several enzymes

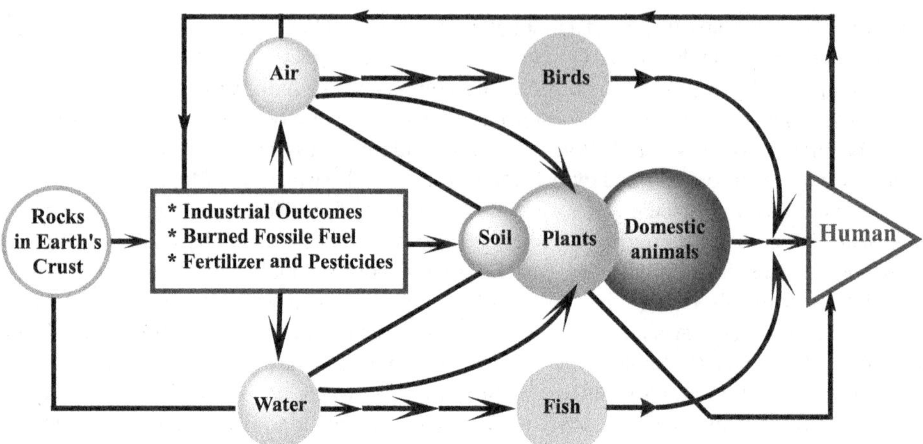

FIGURE 5.1 Sources and general environmental cycle of heavy metals.

responsible for detoxification, metabolism and damage restoration (Yedjou and Tchounwou, 2007). Metallic ions are believed to interact with DNA and nuclear proteins and lead to DNA damage and change the cell cycle inflection, carcinogenesis or apoptosis (Tchounwou et al. 2004). A number of studies have confirmed that the production of reactive oxygen species (ROS) as well as oxidative stress (OS) plays a vital role in the toxicity and carcinogenicity of heavy metals such as As, Cd, Cr, Pb and Hg. These five metals show high toxicity in low concentration; thus, EPA has categorized them as the most toxic heavy metals or priority pollutants (Figure 5.2).

5.2.2.1 Heavy Metal Pollution of Water

In water sources, heavy metals can be found in limited amounts, due to very high toxicity; still, they cause severe health problems in living organisms (Lawrence and Khan 2020). All heavy metals have their specific toxicity concentration that depends on the entities which are exposed to it. Water systems are polluted by various contaminants, but urbanization and industrialization are leading causes of the acceleration of heavy metal water pollution. Heavy metals are mainly transported through the run-off from urban, municipal and industrial areas. Surface and ground waters are mainly polluted by the direct release of untreated industrial wastes into the water system. Heavy metal contamination of sediments is a globally significant concern for aquatic animals as well as human health. In the aquatic system, sediments are the core site of metals and represent the extent of water pollution. In sediments, many physicochemical factors such as temperature, particle size, redox state, organic content, hydrodynamic state, microorganism and salinity affect the adsorption, desorption and consequent concentrations of heavy metals, and their distribution is also influenced by the chemical composition of the sediments, such as particle size and total organic content (TOC) (Ali et al. 2019). pH is another significant factor for the determination of bioavailability of heavy metals in sediments. Lower pH indicates higher risk of pollution.

5.2.3 Heavy Metals Toxicity

As, Cd, Pb, Cu, Hg, Cr and Zn are all are included in the list of priority toxic pollutants and toxic trace metals, and these are the highest toxic heavy metals in the environment. They cause high toxicity in various environmental constituents, including water, air, soil, plants, animals and wildlife. Due to natural and anthropogenic activities, heavy metals are released and cause pollution in water system, sediments and soils. Various factors such as physicochemical and climatic properties affect the biogeochemical cycling and global dynamics of heavy metals in the environmental system. In toxicological investigation, the general mode of uptake of heavy metals is found to be inhalation, ingestion and dermal absorption.

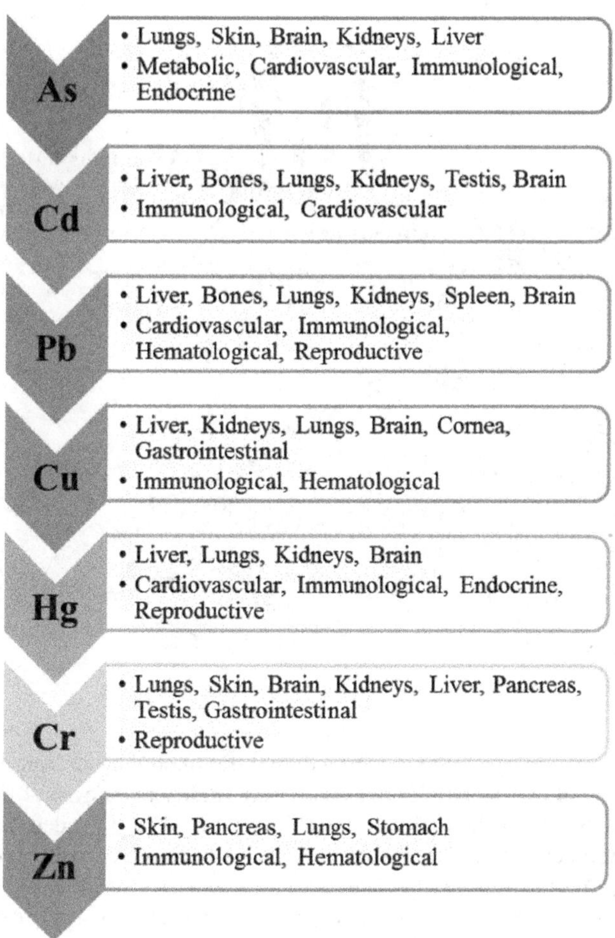

FIGURE 5.2 Impact of heavy metals on the biological system.

In the living organism, absorption occurs through active or passive diffusion (Tchounwou et al. 2012). Numerous factors such as gender, age, method of exposure, different susceptibilities and time interval of contact are used to determine the health risks posed by these heavy metals. The contaminated water affects every living being. Human beings are more susceptible to severe health hazards because of the greater uptake of heavy metals in the food chain (Jaishankar et al. 2014).

In general, source route of heavy metals in the body of an organism is directly from the environmental system including soil, sediments and water or enter through its food/prey, the entry route in the body of organism from water or sediments through the gills/skin and by alimentary canal. In the food chain, the concentration of heavy metals may vary with consecutive trophic levels. In the body of an entity, the accumulation of heavy metals relies on a lot of reasons such as the concentration of metals, physiological mechanisms involved in homeostasis, detoxification and regulation of the heavy metals.

5.3 Toxicity Mechanism of Heavy Metals

The exact mechanism through which heavy metals exert hazardous effects is not known; still, it has been found that metals cause oxidative stress, inflammation and changes in the immune response. In the earth, some heavy metals are known as essential heavy metals due to their noteworthy part

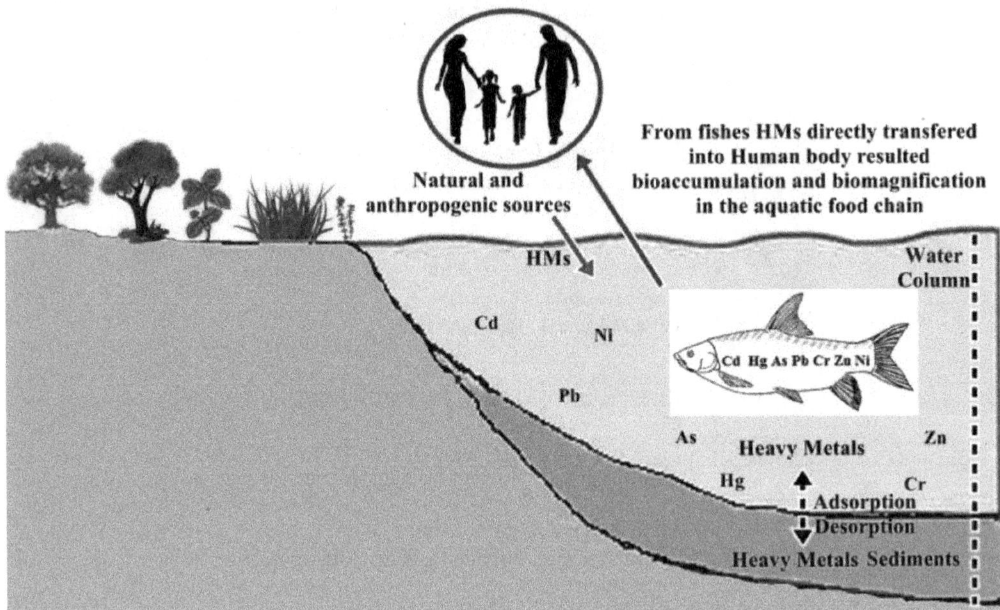

FIGURE 5.3 Transfer of heavy metals from freshwater fish to humans.

in the ecological systems, and the toxicity is based on duration and dose. Non-essential heavy metals such as Pb, Hg and Cd and metalloids such as As are highly toxic even at low exposure; on the contrary, in the development of the body, essential heavy metals are needed in trace amounts, but when their concentration crosses the threshold limits, they become toxic. For some heavy metals, a marginal difference is found between the essentiality and toxicity. Many heavy metals are tumorigenic, carcinogenic and mutagenic due to the formation of ROS that lead to the development of induced oxidative stress. Furthermost, they cause several diseases and abnormalities. Heavy metals are well known as metabolic poisons. The toxicity is mainly attributed to the primary interaction with sulphhydryl (SH) enzyme systems and its inhibition. Glutathione (GSH) is a significant antioxidant in the body and heavy metals interact with SH and replace H atoms on two adjacent GHS molecules that lead to the formation of a strong bond and thus GSH and heavy metals form an active complex (Figure 5.3).

$$2\text{GSH} + M^{2+}(\text{metal ion}) \rightarrow M(\text{glutathione})_2 + 2\text{H}^+$$

5.4 Biochar: Composition and Sources

Pollutants can be removed effectively by adsorption, which is an effective physical phenomenon. Silica gel, activated carbon and aluminium oxide are some of the common adsorbents. Biochar is a black substance containing carbon. It is porous in nature, and because of the porosity and presence of several functional groups such as carboxyl, hydroxyl and phenolic groups, biochar has a robust binding potential for inorganic and organic compounds present in the environment (Park et al. 2011). Biochar is usually produced from the thermal degradation of cellulose-rich biomass waste from agriculture and forestry, such as plant residues, wood waste, peat, cattle manure, which are carbon rich. The steady carbon framework of the biochar supplements it with fascinating properties and capacity to trap water, air, metals and organic moieties. Temperature plays a pivotal role in the characterization of biochar. Biochars obtained by pyrolysis at high temperatures have a high surface area and microporosity and decreased molar H/C and O/C ratios required for the sorption of organic contaminants. On the other hand, biochars obtained at low temperatures are energy-saving and suitable for the remediation of inorganic contaminants as they are rich in oxygen-containing functional groups (Zhang et al. 2015). Biochar has a tendency to act as a sorbent for metals such as Cd, Pb, Cr and Hg from soil as well as water through exchangeable ions or by complexation with functional groups. The metals can also be leached through physical adsorption and surface precipitation (Li et al. 2017). Wood biochar is a potent adsorbent because of its efficient porosity (from 10 to 3,000 μm) and specific area (from 5 to 600 m^2/g). Several factors including organic groups, inorganic materials and cations of biochar influence the metal sorption. Physical characteristics such as surface area are also responsible for the sorption of metals (Khan et al. 2016). As adsorbent, activated carbon has been used frequently, but biochar is a relatively cheap alternative for heavy metal remediation from the environment.

5.4.1 Synthesis of Biochar

Biochars are mainly produced by pyrolysis. During the process, some vital trace elements found in plants become part of the carbon structure; therefore, they are not leached and become accessible to plants via root exudates and microbial symbiosis. Because of the ability of biochars to serve as a carrier for plant nutrients, they may be used to synthesize organic fertilizers by mixing with molasses, ash, slurry, etc. Various organic chemicals are also formed during pyrolysis. Some of them get trapped in the pores and surfaces and may stimulate a plant's immune system, thereby increasing the resistance to pathogens. Some of the biochars have been obtained from plant roots. *Eichhornia crassipes* roots have proven as a source of biochars, which have been obtained by the pyrolysis of root powder in a muffle furnace under O_2-limited conditions at 200°C–500°C (Li et al. 2018).

Pyrolysis depends upon the time, temperature of the pyrolytic material, pressure, size of adsorbent, heating rate and methods such as burning of fuel, electrical heating or by microwaves. In slow pyrolysis, oxygen-free feedstock biomass is transferred to a pre-heated kiln or furnace. Fast pyrolysis depends on fast heat transfer, at less than 650°C with a high heating rate (100–1,000°C/s). The characteristics of biochars are influenced by the extent of pyrolysis (pyrolytic temperature and residence pressure) and by biomass size and kiln or furnace residence time (Meyer et al. 2011). Residence time determines the rate at which volatile gases are removed from the furnace. Chances of secondary reactions on the biochar surface increase with the increase in residence time. Scots pine (*Pinus sylvestris L.*) among coniferous trees and silver birch (*Betula pendula*) among deciduous trees have been selected for their local availability and cost-effectiveness, and the prevalence has been obtained from potentially clean areas as wood biomass materials for biochar production. The tree selection had been done on the basis of similarity in age, and the materials are collected from low pollution concentration areas (Komkiene and Baltrenaite 2016).

Four types of biochars were produced upon pyrolysis on different temperatures, viz. 450°C±5°C and 700°C±5°C, for varying time periods. The four different biochars were: (i) Scots pine biochar produced at slow pyrolysis; (ii) Scots pine biochar produced at fast pyrolysis; (iii) silver birch biochar produced at slow pyrolysis; and (iv) silver birch biochar produced at fast pyrolysis. Some new biochars from the leaves of *Tectona* and *Lagerstroemia speciosa* have been obtained at 800°C, which have been used for the remediation of As(III) and As(V) from water solution. The biochar material was crystalline in nature and was identified by using scanning electron microscopy (SEM), energy-dispersive X-ray (EDX) spectroscopy, Brunauer–Emmet–Teller (BET) method, Fourier transform infrared (FTIR) spectroscopy, X-ray diffraction (XRD), zeta potential, particle size and X-ray photoelectron spectroscopy (XPS) (Verma and Singh 2019). Soil stabilization is one of the most efficient methods for the leaching of heavy metals from the soil. Soil stabilization is considered as a sustainable remediation approach (O'Connor et al. 2018) (Figure 5.4).

Usage of only biochar cannot compulsorily decrease heavy metal bioavailability as dependent on the soil texture and nature of contaminants due to limited adsorption mechanisms. Some modifications have been made to improve the adsorption capacity, including chemical, physical and magnetic modifications and impregnation with mineral sorbents (Rajapaksha et al. 2016). A MgO-coated corncob biochar was produced and used in a lead-laden soil. Magnesium oxide (MgO), having a high potential for soil stabilization (Shen et al. 2019), was anticipated to potentially increase the immobilization of

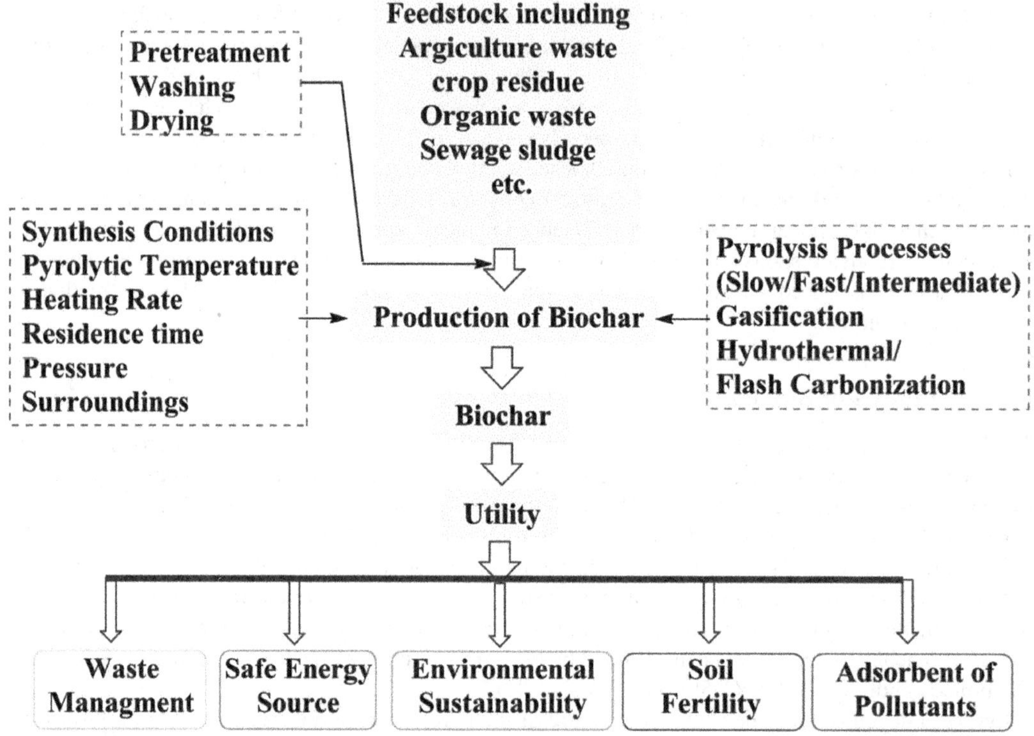

FIGURE 5.4 Production outline of biochar.

heavy metals by corncob biochar. Therefore, it was synthesized and characterized, and its performance in lead immobilization in soil was studied. The investigation revealed that the MgO-coated biochar adsorbed lead onto its surface through cation–π interaction and amplified surface adsorption owing to the greater surface area, and then the MgO coated on MCB's surface additionally boosted the adsorption via precipitation. The synergistic roles of biochar mineral composites make them favourable for soil remediation. The characteristics of biochar may vary upon changing pyrolytic conditions. It has been found that the biochar obtained by fast or slow pyrolysis had the same fused aromatic ring cluster, while the biochar obtained from gasification at high temperature had 17 rings per cluster and was much more condensed.

Fewer aromatic C-H functional groups were obtained from slow pyrolysis than from fast pyrolysis. The final product is usually more porous when obtained upon slow pyrolysis. The biomass and pyrolysis temperature are the vital factors that influence the characteristics of the biochar.

5.4.1.1 The Starting Biomass

Lignocellulosic biomass contains cellulose, hemicellulose and lignin. The percentage of these components in the starting biomass results in varying properties of the resultant biochar (Xiong et al. 2017). The pyrolysis of a high-lignin biomass obtained from pine wood and spruce wood produces a carbon-rich biochar (Antal and Gronli 2003).

5.4.1.2 Inorganic Species

The carbon content of the biochar varies from 45% to 60% by weight. A substantial amount of hydrogen and oxygen is found in biochar. Inorganic elements such as K, Na, Ca, Mg, Si, Al and Fe also occur frequently with biochar. The biochar derived from wood-based biomass has a lower inorganic content than the biochar derived from herbaceous biomass (Zhao et al. 2013). Inorganic species present in the biochar influence the catalytic applications of the biochar, such as tar cracking, bio-oil upgrading and methane decomposition (Dong et al. 2015). The inorganic portion in the biochar endows it with a catalytic activity towards reactions such as methane decomposition and tar cracking.

5.4.1.3 Functional Groups

Usually, biochar has functional groups bound at the surface. These functional groups are important for the characterization and functionalization of the biochar (Lua et al. 2004). To enhance the performance of metal catalysis by biochar, some functional groups are added to the biochar network, which can bind to the metal ions (Liu et al. 2018). Some surface functional groups may boost the biochar-based catalysts for several processes. It has been found that biochar having sulphonic groups as surface functional groups can efficiently catalyse the hydrolysis of cellohexaose because the phenolic, –OH and –COOH groups in the biochar act as adsorption *sites* (Cha et al. 2016).

5.4.1.4 The Biochar Framework

In some cases, the microstructure of the biomass is maintained after the pyrolysis as well. The biochar derived from wood possessed vertically aligned microchannels and fibrous ridged surfaces as in raw wood. The biochar obtained from the sisal leaves also retained the natural quality of the starting biomass having associated porous frameworks (Zhao et al. 2018). These examples suggested the localized formation of biochar via biomass pyrolysis. The biochar framework is mainly amorphous with some crystalline conjugated aromatic sheets. These sheets are cross-linked in an unsystematic fashion. The crystalline nature increases with the increase in the temperature of pyrolysis, and the structure attains more order (Li et al. 2015). Heteroatoms such as N, P and S may be introduced into the biochar matrix. The difference in the electronegativity between these atoms and the aromatic C leads to chemical heterogeneity in biochar, which has an imperative role in catalytic applications (Muradov et al. 2012).

5.4.1.5 Temperature and Its Effect

The properties of biochar may be greatly affected by pyrolysis temperature. The effect of temperature is more significant than that of the heating rate and residence time (Jeffery et al. 2015). A positive correlation has been obtained between the pyrolysis temperature and pH, carbon content, biochar stability and ash content, whereas a negative correlation is obtained with biochar yield, O and H contents and the surface functional groups. The porosity is also affected by the pyrolysis temperature; as the temperature increases, volatile substances evaporate from the biomass surface, which increases the porosity and surface area (Smernik et al. 2002). A rise in temperature has led to smaller average pore sizes, while a previous study (Haefele et al. 2011) quoted that a high pyrolysis temperature facilitated the fusion of pores with expanded pore sizes. The chemical behaviour of biochar is also somewhat related to the pyrolysis temperature.

5.4.2 Physical and Chemical Characterization Methods

The characteristics of biochar depend upon the pyrolytic conditions and the type of feedstock used, such as crop residues, industrial organic by-products or bioenergy crops, thereby causing diversity in its composition. Biochars synthesized by wood having a high lignin content possess a greater carbon content than those obtained from herbaceous feedstocks (Yu et al. 2011), but they lack N. Still, these chars can be useful for soils favouring the growth of microbes. A synthesized biochar is often characterized through the variations in the elemental concentrations of C, H, O, S and N (Lee et al. 2010). A fixed amount of carbon is left behind once the sample is carbonized after the expulsion of volatile matter (Lu et al. 2012). It is used to quantify the carbonaceous substances yielded from a solid sample. H/C and O/C ratios are used to assess the degree of aromaticity and maturation, as often described in van Krevelen diagrams

(Swiatkowski et al. 2004). O/C, O/H and C/H ratios have been found to provide a consistent measure of both the degree of pyrolysis and level of oxidative adjustment of biochar in the soil and solution systems and are fairly easy to be determined.

Biochars can be characterized by Fourier transform infrared (FTIR) spectroscopy technique. The technique helps in the identification of the main functional groups such as carbohydrates, lignin, cellulose, lipids and proteins. Recently, the characterization of biochars by FTIR has become a very common practice. The FTIR spectra usually change with the change in temperature. However, biochars produced by fast pyrolysis at high temperatures cannot be characterized by FTIR due to the carbon formation that results in weak IR signals. Structural changes due to high-temperature pyrolysis mainly occur between 1,800 and 1,200 cm^{-1}. Feedstocks rich in protein, cellulose, lignin and fatty acids mainly show bands due to amide I (1,642–1,638 cm^{-1}) and amide II (1,515–1,536 cm^{-1}) groups, respectively. C=O stretching band due to ketonic group is obtained at 1,642 cm^{-1}, and C=C aromatic skeletal vibrations in lignin structure occur around 1,515 cm^{-1}. In cellulose, CH_2 scissoring vibration occurs at 1,434 cm^{-1} and O–H bending vibrations in saturated fatty acids occur between 1,500 and 1,240 cm^{-1} (Zhang et al. 2013). It has been found that biochars obtained by fast pyrolysis show characteristic bonds as in cellulose. Baseline correction of FTIR spectra showed that with rise in temperature, there is a distinct decrease in frequencies associated with O–H (3,600–3,100 cm^{-1}), C=C and C=O stretching (1,740–1,600 cm^{-1}) and aromatic C=C and C–H deformation modes of alkenes (1,500–1,100 cm^{-1}), and the C–O–C symmetric stretching (1,097 cm^{-1}) characteristics of cellulose and hemicelluloses.

Biochar derived from sludge when combined with Pb^{2+} has been shown to have abundant carboxyl and hydroxyl groups (Hale et al. 2011). The band for the complexed carboxyl (–COOMe) showed no shifting in its position even after being replaced by Pb at 1,400–1,500 cm^{-1}. The noticeable shift of the band at 3,404–3,406 cm^{-1} (pH=2.0), 3420 cm^{-1} (pH=3.0), 3,429 cm^{-1} (pH=4) and 3,422 cm^{-1} (pH=5.0) supported the proposed complexation between Pb^{2+} and hydroxyl as well as carboxyl groups. The identification of acidic and basic groups on biochars may be done by Boehm titration (Hammes et al. 2008), where the biochar is equilibrated in the presence of strong bases such as HCO_3^-, CO_3^{2-} or OH^- or strong acids such as H_2SO_4, HCl and HNO_3, followed by titration of the extract with a strong acid or base to estimate the reacting fraction. The amount of acid or base determines the quantification of lactonic, phenolic and carboxylic groups (Li et al. 2013). Surface charge on the biochar can be measured by cation exchange capacity (CEC), which increases as the biochar ages and has been described to grow in some oxygenated functional groups on the surface of the biochar (Bagreev et al. 2001).

Scanning electron microscopy is a microscopic technique through which the physical morphology and microporosity of a solid substance are measured (Chen et al. 2014). The pore size of the biochar is dependent upon the inherent structure of the feedstock and is an important factor for water-retaining and adsorption capacities in different phases (Chen et al. 2014). The biochars produced at different pyrolytic temperatures have different structures. The biochars possess a high Brunauer, Emmett and Teller (BET) area, which usually increases with an increase in carbon burn-off, irrespective of the pyrolytic temperature (Yuan et al. 2019), indicating that the carbon burn-off has a most important consequence on the growth of the surface area.

Functional group identification in rice straw and bran biochars at 100°C–800°C using ^{13}C NMR spectroscopy showed dehydroxylation/dehydrogenation and aromatization. It was observed that with the increase in temperature, the formation of O-alkylated groups and anomeric O-C-O carbons occurred before aromatization. At the temperature below 300°C, aliphatic O-alkylated carbons were more prominent, whereas at higher temperatures, aromatic structure predominated. A relatively new correlative technique based on mid-infrared (MIR) spectroscopy has evolved for the quantification of biochars. Earlier, the technique was used for the estimation of carbon content in soil and solutions. Through UV oxidation method, several algorithms have been proposed, which can relate MIR response spectrum to biochar (Al-Wabel et al. 2013) (Figure 5.5).

5.4.3 Properties/Characteristics of Biochar

The physicochemical properties of biochars significantly affect their sorption efficacy. Following are the important parameters.

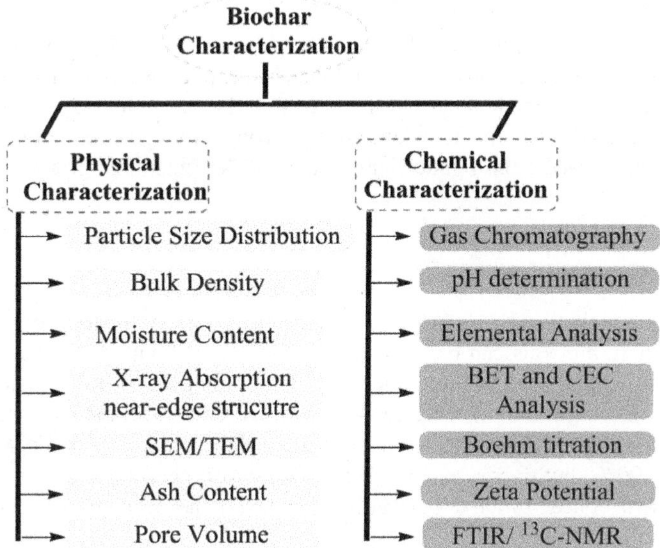

FIGURE 5.5 Characterization methods of biochar.

5.4.3.1 Surface Area and Porosity of the Biochar

The sorption efficiency of biochar is dependent on the surface area and porosity. Porosity is developed because of water or dehydration upon pyrolysis of biomass (Chen et al. 2014). Biochar pore size varies from nano- (<0.9 nm), micro- (<2 nm) and macropores (>50 nm). The biochars having small pores are unable to trap large particles irrespective of their charge or polarity. Pyrolysis conditions lead to variation in pore size. It has been found that a high temperature leads to the formation of large pore size and large surface area. With increasing temperature from 500°C to 900°C, the porosity of biosolids biochar increased from 0.056 to 0.099 cm^3/g, while the surface area increased from 25.4 to 67.6 m^2/g (Hossain et al. 2011). However, in some cases, at high temperatures the biochar porous structure gets disturbed by the formation of tar, leading to a lower surface area. The biochar composition also depends upon the material or feedstock used for the preparation of biochar. Generally, biomass rich in lignin (e.g. bamboo and coconut shell) develops a macroporous-structured biochar, while biomass rich in cellulose (e.g. husks) yields a predominantly microporous-structured biochar.

5.4.3.2 Biochar pH and Charge Over the Surface

The pH of biochar also varies with pyrolysis temperature and feed material. Usually, biochar is alkaline. Biochars obtained by pyrolysis at low temperature are generally acidic, and the biochar produced from oak wood at 300°C and 600°C was acidic with its pH lying between 4.84 and 4.91 (Subedi et al. 2016). With increasing temperature, the pH usually increases. A positive correlation was observed between the biochar pH and pyrolysis temperature as in wheat, corn and maize residues (Subedi et al. 2016). This may be due to the formation of ash at high temperature. Metal sorption is also influenced by its surface charge. The point of zero charge refers to the pH at which the surface charge becomes zero. When solution pH is > pHPZC, the biochar is negatively charged and binds to metal cations such as Cd^{2+}, Pb^{2+} and Hg^{2+}, whereas at pH < pHPZC, the biochar is positively charged and binds to metal anions such as $HAsO_4^{2-}$ and $HCrO_4^-$. With increasing temperature from 500°C to 900°C, the pHPZC of biosolids biochar increased from 8.58 to 10.2 (Mohan et al. 2014). At elevated temperature, the anionic functional groups on biochars (e.g. $-COO^-$, –COH and –OH) have been found to be fewer in number, which leads to high pHPZC and relatively less negative charge on the surface.

5.4.3.3 Presence of Different Functional Groups

For effective metal sorption, the presence of carboxylic, amino and phenolic groups is highly required. The presence of these functional groups can be controlled by the pyrolysis conditions and surface charge. Generally, the availability of functional groups reduces with the rise in temperature owing to high carbonization. At high temperatures, the abundance of hydroxyl, carboxylic and amino groups increases. In the case of biochar derived from lignocellulose, most of the functional groups are lost at elevated temperature.

5.4.3.4 Mineral Composition

Mineral concentration including K, Ca, Mg and P is also accountable for the sorption of metals from aqueous solution. These minerals can be exchanged or precipitated with heavy metals, reducing their availability. Pyrolysis temperature and feedstock both determine the mineral content in the biochar. It has been found that the concentration of K, Ca, Mg and P increases with the increase in pyrolysis temperature in the biochar obtained from biosolids. The water-soluble amount of K, Ca, Mg and P increases when heated up to 200°C, but decreases at higher temperatures. Feedstock is also a significant factor upon which the mineral concentration of biochar is dependent. It has been observed that the P content in oak wood biochar is much lower than `.

5.4.4 Applications of Biochar as an Adsorbent

Biochar is a promising and reasonable adsorbent used for metal removal from water. Metal sorption capacities of biochar are 2.4–147, 19.2–33.4, 0.3–39.1 and 3.0–123 mg/g for Pb, Ni, Cd and Cr, respectively (Wang et al. 2010). Suitable modifications have been suggested in the biochar framework, which improve the metal binding potency usually by infusing the parent biochar with minerals, oxygen-rich functional groups, reducing agents and nanoparticles or by activation of biochar with alkali solution.

5.4.4.1 Sorption of Organic Materials

The sorption of organic pollutants on biochar is dependent upon the pyrolysis conditions as well as the feedstock. It has been observed that biochar has a high adsorptive potential for pesticides and organic pollutants such as polycyclic aromatic hydrocarbons (PAHs), polychlorinated biphenyls (PCBs), chlorinated compounds and dyes (Yao et al. 2012). The sorption mechanism was assisted by π-electrons in case of wood and manure, and the reaction propagated through the pore-filling action. Biochars obtained from carbonaceous materials such as peat, soya bean stalk and coke remediate organic substances and reduce their bioavailability by repartitioning organic pollutants to carbonaceous adsorbents. Biochars derived from bamboo when mixed in soil amended the leachability and bioavailability of pentachlorophenol. Biochars derived from pepper wood and sugarcane bagasse were found useful in the removal of sulphamethoxazole from water (Dong et al. 2011).

5.4.4.2 Application of Biochar in Water Treatment

Biochars have a high surface area, fine porous structure and several functional groups on their surface, and these properties make them very useful for water purification. According to previous studies, the biochars may be applied with the removal efficiency of nearly 45% for heavy metals, 40% for organic contaminants, 13% for nanoparticles and 2% for other pollutants. Heavy metal contamination in water has become an enormous issue. Thus, their remediation has grown as one of the major research areas using biochar. Adsorption isotherm emphasizes on optimizing adsorbents, which describes the interaction between the adsorbates and adsorbents. Arsenic is a metalloid and is carcinogenic in nature. In natural water, As(V) and As(III) are dominant species with As(III) being more toxic. The arsenic sorption usually occurs through complexation and electrostatic interactions (Cao et al. 2009).

The biochar obtained from pinewood pyrolysis at 600°C sorbed As(V) up to 0.3 mg/g at pH 7. As mainly occurred as $HAsO_4^{2-}$ with an abundance of positively charged functional groups at its surface. Some functional groups were protonated when the solution pH was $<pH_{PZC}$. As(V) interacted with positively charged functional groups. Cr is another metal of environmental concern. Cr(VI) species have a high solubility in aqueous medium and of great significance because of the carcinogenic, mutagenic and teratogenic nature of chromium in +6 oxidation state. Cr(III) is comparatively 300 times less toxic than Cr(VI). $HCrO_4^-$, CrO_4^{2-} and $Cr_2O_7^{2-}$ are the dominant species in +6 oxidation state. For Cr(VI) sorption, two mechanisms have been proposed, *viz.* electrostatic attraction between Cr(VI) and biochar and reduction of Cr(VI) to Cr(III) in the presence of oxygenated groups such carboxyl and hydroxyl groups. The reduction is subsequently followed by Cr(III) complexation. The biochar obtained from sugar beet tailing has been an effective Cr(VI) adsorbent with the highest sorption of 123 mg/g at pH 2.0 (Kong et al. 2011). The mechanism involves reduction of Cr from +6 to +3 oxidation state followed by complexation by hydroxyl and hydroxyl groups. Oak wood and oak bark biochars having a high quantity of lignin, cellulose and hemicellulose have a high capacity for Cr(VI) sorption. The biochar obtained from oak wood has been found to contain considerable percentage of oxygen (8%–12%). Cr(III) sorption mainly occurs through (i) complexation with oxygen-containing functional groups, (ii) ion exchange and (iii) electrostatic attraction between positively charged Cr(III) ions and negatively charged biochar. Biochars obtained from crop straws have been studied for Cr(III) sorption, and it was found that peanut-derived biochar had a sorption capacity of 0.48 mmol/kg. Cr(III) complexation with functional groups is important for its sorption by biochar (Singh et al. 2010). Pb in aquatic environment is present as Pb^{+2} and $Pb(OH)^+$ at $pH<5.5$ and as $Pb(OH)_2$ at $pH>12.5$ (Liu and Zhang 2009).

Biochars derived biosolids, dairy manure, oak wood, oak bark and bagasse have been studied for their role in lead sorption from water. Pb sorption mainly occurs through cation exchange, complexation and precipitation. Ca and Mg are the ions that are exchanged, contributing to 40%–52% of Pb sorption at pH 2–5, while exchange with Na and K contributed to 4.8%–8.5% Pb sorption. Cation exchange is also responsible for Pb sorption by biochar derived from oak wood and bark (Chen et al. 2011). Divalent Hg is a very common species in the environment. It exists as Hg^{2+} at $pH<3.0$ and as $HgOH^+$ and $Hg(OH)_2$ at pH 3.0–7.0. Complexation of Hg usually occurs with carboxylic and phenolic groups. In Brazilian pepper biochar, the sorption capacity decreased at higher temperatures as it was found that there was a reduction in these functional groups as the temperature rose above 600°C (Regmi et al. 2012). Chemical reduction is another route for Hg sorption as investigated in a soya bean stalk biochar. Hg^{2+} was reduced to Hg_2Cl_2 in the presence of Cl^-, which was precipitated on the biochar surface.

5.4.5 Mechanism of Interaction between Biochar and Heavy Metals

Heavy metals can be removed from the aqueous phase as well as soils using biochar. The removal can be done through various mechanisms including precipitation, complexation, ion exchange, electrostatic interaction and physical sorption. Biochars have a large surface area and porosity; therefore, they have a strong affinity for metal cations which can be physically sorbed on the surface and get trapped in pores (Yuan et al. 2011). Heavy metal absorption mechanism can be studied by various methods including adsorption isotherms and kinetic models, desorption studies and industrial analyses such as XRD, FTIR and SEM. Sorption may be due to electrostatic attraction and formation of inner sphere complex with the available carboxyl, alcoholic hydroxyl or phenolic hydroxyl groups on biochar surface (Ahmad et al. 2014) as well as co-precipitation. Carboxyl (R-COOH) and alcoholic or phenolic hydroxyl groups (R-OH) present on the sorbent surface are considered as the main groups that coordinate with the heavy metals. The presence of mineral impurities such as ash and basic nitrogen groups present in heterocyclic bases such as pyridine serve as other adsorption sites of the carbon-containing substance. Thermodynamic aspects have shown that metal sorption is an endothermic process (Li et al. 2019), which takes place by the electrostatic interaction between the positively charged metal cations and π-electrons of C=O ligand or C=C of a shared electron cloud on the aromatic part of the biochar.

To improve the sorption capacity, methods such as magnetization, incorporation of nanoparticles, modifications in coating and alkaline treatment have been used. Alkali activation process is an efficient method to enhance the porosity and sorption properties. The metal sorption ability of the biochar varies between

1 and 3 from 1 to 200 mg/g. The pH of the solution also influences the metal sorption and specific charge of the biochar. An alteration in pH influences the complexation behaviour of the functional groups such as carboxyl, hydroxyl and amino groups. The carboxyl group ionization is maximum at pH 3–4. Based on the previous studies in the literature, five different mechanisms have been proposed for metal sorption: (i) electrostatic interactions between metals and biochar, (ii) cation exchange between metals and other metal ions such as alkali and alkaline earths and H^+ ions on the biochar surface, (iii) metal complexation with functional groups and π-electrons present in the aromatic framework of biochar, (iv) precipitation of metal ions and (v) reduction of metal ions and their subsequent sorption. The mechanistic action and the potential of sorption vary for different biochars and heavy metals. Many researchers have studied and devised numerous methods for the synthesis of biochars and for determining their metal binding capacity, which include thermodynamics and kinetic factors from water (Batool et al. 2018).

5.4.5.1 Chemical Sorption

Upon exposure to the environment, oxygenation occurs, resulting in the abundance of oxygenated functional groups on the internal surface area of the biochar (Fraga et al. 2019). These functional groups including hydroxyl and carbonyl groups induce a negative charge on the biochar surface. This happens at low temperature, and as the pyrolysis temperature increases, the abundance of anionic functional groups decreases, which reduces the surface charge on the biochar. Upon surface sorption, H^+ ions are released from the biochar directly (Fraga et al. 2019), but at the same time, many alkali and alkaline earth ions are passed into the solution, which shows metal exchange with other positively charged ions. The preferential sorption of a single metal cation from an aqueous solution is more prominent than the sorption of multiple metals because of the competition for binding sites on the biochar surface (Nartey and Zhao 2014). Ligands which have P and S donor atoms show a greater affinity for metal ions such as Pb and Hg that have strong likeness towards sulphates and phosphates (Verheijen et al. 2010).

5.4.5.2 Precipitation

The mineral fraction in biochar, such as CO_3^{2-}, PO_4^{3-}, SiO_4^{4-}, Cl^-, SO_4^{2-} and OH^-, reacts with heavy metal ions to form water-insoluble compounds such as metal oxides, phosphates and carbonates, leading to adsorption and immobilization of heavy metals. Biochar source material consists of varying fraction of minerals, which result in non-organic ash formation. The mineral fraction found in ash fraction is usually in oxidized form, and its concentration increases with the increase in pyrolysis temperature (Hassan and Carr 2018). Pb minerals have very low solubility, and their formation results in an increase in Pb-absorbing capacity as compared to other bivalent cations. Cu, Cd and Zn can also be precipitated as insoluble carbonates and phosphates at high pH.

5.4.5.3 Physical Sorption

In physical sorption, the biochar adsorbs heavy metals and other contaminants on its surface. The large surface area and porosity are responsible for physical adsorption. The pollutants usually diffuse through the fine pores present in the biochar framework. The adsorption can be either single- or multi-layer, which is usually fitted by Langmuir and Freundlich model (Dong et al. 2017).

Thermodynamic parameters show that sorption is an endothermic process (Kołodynska et al. 2017). Sorption occurs as a result of electrostatic interaction between metal cation and π-electrons associated with C=O and C=C present in biochar. Because of the presence of vacant orbitals on metal, pi-electrons associated with the biochar are slightly shifted towards the metal ions, causing a polarity in the bond and creating a week interaction between them. An increase in pyrolysis temperature enhances the aromatic character and decreases the oxygenated functional groups (Zhao et al. 2017). The low-temperature pyrolysis results in metal immobilization, which is for a small term owing to the synthesis of inner and outer sphere complexes with oxygen-rich functional groups. Pyrolysis at high temperature results in a negative surface charge that lasts for a longer time period, but that makes the biochar a week adsorbent for metal ions. Over a certain time period, owing to oxidation on the biochar surface, negatively charged functional groups are created.

5.4.5.4 Ion Exchange

In the process of ion exchange, ions having like charges are mutually exchanged with each other. Carboxyl, carbonyl and hydroxyl groups present on the surface of biochar possess replaceable H^+ ions, which can be exchanged with alkali and alkaline earth metal ions or other organic contaminants possessing positive charge (Sizmur et al. 2017).

5.4.5.5 Electrostatic Interaction

Electrostatic interaction occurs between the negative charge on the surface of biochar and the positively charged metal ions, and that usually occurs when the pH of the solution is more than the charge point of the biochar. The phenomenon is also termed as electrostatic adsorption. The interaction mainly occurs due to the presence of oxygen-containing functional groups such as carboxyl, carbonyl and hydroxyl groups (Dong et al. 2017).

5.5 Modification of Biochar to Improve Sorption

Several modifications have been made in the structure of biochar to improve sorption properties. Physical and chemical methods have been adopted to achieve the modified and enhanced activity. The modifications in biochar can be brought about by varying starting biomass, pyrolysis temperature, reaction time, etc. The alteration in the framework can be brought about prior to pyrolysis by using an appropriate catalyst, usually NaOH or KOH, or by using metal salts that undergo reduction to metal or metal oxide nanoparticles. Modifications can also be made by chemical or physical methods such as enrichment with different functional groups on the surface of biochar by different methods such as oxidation, sulphonation, amidation or reaction of monomers (oligomers) to form composite materials. Some of the methods used for the improvement of sorption capacity of the biochar are discussed below.

5.5.1 Chemical Oxidation

In chemical oxidation, hydrophilicity is enhanced by oxidation on the biochar surface, which yields oxygen-containing functional groups such as –OH and –COOH. Oxidation also leads to an increase in the pore size, thereby improving the adsorption capacity for the polar substances. HCl, HNO_3, H_2O_2 and H_3PO_4 are commonly used oxidants for the purpose (Ho et al. 2014). It has been found that biochar modified by HNO_3 has more acidic oxygen-rich functional groups and a greater sorption potential for NH_3–N. Biochar modified by H_3PO_4 is more productive in the remediation of Pb. The increased specific surface area and pore volume, as well as the role of phosphate precipitation, increase the biochar affinity for Pb (Zhao et al. 2017).

5.5.2 Chemical Reduction

Non-polarity is increased by the reduction of functional groups. Chemical alterations also improve pore formation and surface area of biochar. The common reducing agents used for the purpose are NaOH, KOH, NH_4OH, etc. (Sizmur et al. 2017). Different reducing agents have varying effects. KOH and NaOH activate biochar through different mechanisms. It has been found that potassium atoms which are released during potassium hydroxide activation intercalate between adjacent carbon crystallite layers, although the phenomenon is not observed in case of interaction of sodium with carbon (Figure 5.6).

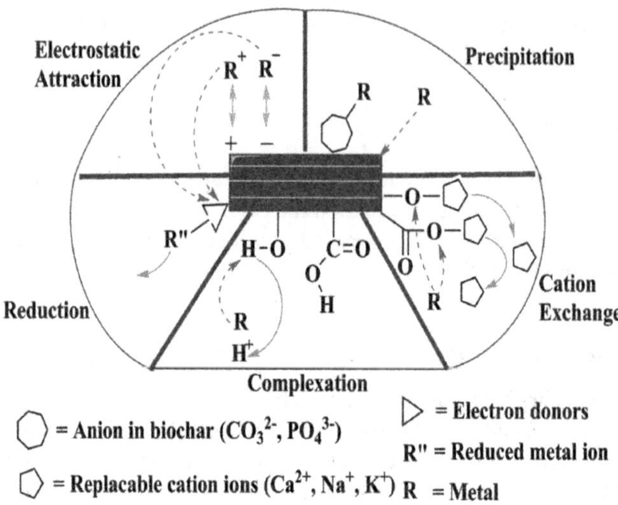

FIGURE 5.6 Mechanistic approaches for the modification of biochar.

5.5.3 Impregnation of Metal

Some of the heteroatoms are adsorbed onto the biochar surface during metal impregnation. Metal ions combine with the adsorbate to improve the adsorption efficiency. Fe, Mg, Ag and Zn (Angin et al. 2013) are some of the ions that improve the biochar sorption upon impregnation. Biochar material (CMC–FeS@biochar) through association with carboxymethyl cellulose (CMC) and iron sulphide (FeS) showed active sorbing capacity of CMC–FeS@biochar composite for Cr(VI).

5.6 Future Prospects of Biochar

Biochar has lesser ability to act as an adsorbent than activated carbon. Studies and researches have been concentrated to enhance the surface area, porosity and functional groups of the biochar. These modifications have been done by adding minerals, organic functional groups and other nanoparticles. Haematite (γ-Fe_2O_3), magnetite, Fe, hydrated MnO and CaO have been used to enrich the biochar during or after pyrolysis (Wang et al., 2015b). The adsorption capacity of a biochar loaded with Fe oxides for As(V) increased because of the electrostatic attraction between the two. Biochar sorption is also modified by the presence of exogenous functional groups such as amino, carboxylic and hydroxyl groups. Amino groups were incorporated into biochars obtained from husk and saw dust via polyethylenimine modification and through nitration and reduction. The sorption capacity of such a modified biochar for Cr(VI) increases ten times from 30 to 120 mg/g (Zhang et al. 2015). The surface area can also be modified and enhanced for sorption by incorporating nanoparticles. Magnetic biochars infused with ZnS nanocrystals induced magnetic behaviour in the biochar. The biochar thus produced showed a better sorption capacity for Pb, which was up to 368 mg/g (Yan et al. 2015). Biochars incorporated with Zn and obtained from sugarcane bagasse showed an increased Cr(VI) absorption. A graphene-coated biochar resulted in π–π interaction, and an increase in surface area was obtained from 4.5 to 17.3 m^2/g for acidic functional groups and the sorption capacity for Hg increased from 0.77 to 0.85 mg/g (Tang et al. 2015). Other than nanoparticles, the surface area of biochar has been seen to increase with the solution of bases such as sodium and potassium hydroxide, leading to improved sorption of metals.

Thermodynamic studies of biochar are very scarce, which makes uncertain whether a particular biochar can be used for the sorption of each heavy metal or not. Mathematical models should be implied to explore the exact mechanism of action of biochar and how the interaction occurs. Pre-nourishment of biochar can be useful for wastewater management. Although studies have been focused on the removal of heavy metals, very less work has been done to understand and modify the interaction with metalloids such as As. Although the biochar has conventionally been prepared by pyrolysis, the properties of biochar are greatly affected by pyrolysis temperature. Hydrothermal carbonization (HTC) is usually performed at preferably low temperatures (180°C–250°C) in the presence of water. The technique has a distinct advantage of producing high-moisture biomass and structure in the presence of inorganics and surface functional groups made up of highly conjugated aromatic sheets. Strategies should be adopted so that the modified biochar can act as a better catalyst. The biochar should also be used in small amounts so that the wastewater treatment is cost-effective.

Studies have reported the quantification of extractable toxic elements in the biochar itself, and recommendations and guidelines have been implemented to minimize the potential risk of toxicity. The International Biochar Initiative (IBI) has established standards (Standardized Product Definition and Product Testing Guidelines for Biochar that Is Used in Soil) to characterize crucial qualities and biochar material characteristics in soils. Similar guidelines are needed for water treatment as well, where biochar is used as a remedial agent. Very few reports in the literature are available that describe the usage of biochar for the treatment of contaminated sediments, which needs to be explored because of the proven remedial properties of biochar in overcoming heavy metal contamination in water.

5.7 Conclusions

This chapter has summarized the heave metal contamination in environment with an emphasis to heavy metal pollution in water. Several diseases are associated with consumption of polluted water and to address the issue numerous remedial measures have been explored, of which biochars have emerged as promising adsorbents. This chapter has taken a closer look at different aspects of biochar including its synthesis, characterization, properties applications and mechanistic action. Chemically, biochar mostly contains carbon, hydrogen, oxygen, nitrogen, phosphorus and sulphur. The physicochemical features of biochar depend upon the feedstock used and temperature of pyrolysis. Potential applications of biochars include their use as an energy source, as an additive in soil amendment and as a sorbent in waste management. Biochar as an adsorbent has received significant attention due to its advantageous physical and chemical characteristics including cation exchange capacity and porous network. However, large-scale studies have been scare and experiments using biochars derived from different sources have been done on a laboratory scale. Usually, the studies have been focused on the removal of one single metal. In natural water bodies, several metals coexist with other contaminants, creating a competition for the binding site on the biochar surface. Very few reports are available on the remediation of heavy metals from water. Simulated water should be replaced with natural water to ascertain the actual removal ability and functionality of biochar in a particular environment. This can be achieved by using various physicochemical conditions. Studies are also needed to understand the actual mechanism and efficiency of heavy metal removal, which is mostly dependent on the application rate, dosage of the biochar and recovery approaches, and the regeneration and disposal approaches of the adsorbed biochar.

Acknowledgement

The authors are thankful to the faculty of the Department of Chemistry, Integral University, Lucknow for their whole-hearted support.

REFERENCES

Agrafioti. E., Kalderis, D., and and Diamadopoulos, E. 2014. Arsenic and chromium removal from water using biochars derived from rice husk, organic solid wastes and sewage sludge. *Journal of Environmental Management* 133: 309–314. doi: 10.1016/j.jenvman.2013.12.007.

Ahmad, M., Rajapaksha, A.U., and Lim, J.E. 2014. Biochar as a sorbent for contaminant management in soil and water: A Review. *Chemosphere* 99: 19–33.

Al-Fartusie, F.S., and Mohssan, S.N. 2017. Essential trace elements and their vital roles in human body. *Industrial Journal of Advances in Chemical Science* 5(3): 127–136. doi: 10.22607/IJACS.2017.503003.

Ali, H., Khan, E., and Ilahi, I. 2019. Environmental chemistry and ecotoxicology of hazardous heavy metals: environmental persistence, toxicity, and bioaccumulation. *Journal of Chemistry* 2019: 1–14. doi: 10.1155/2019/6730305.

Al-Wabel, M.I., Al-Omran, A., and El-Naggar, A.H. 2013. Pyrolysis temperature induced changes in characteristics and chemical composition of biochar produced from conocarpus wastes. Bioresource Technology 131: 374–379. doi: 10.1016/j.biortech.2012.12.165.

Angin, D. 2013. Effect of pyrolysis temperature and heating rate on biochar obtained from pyrolysis of safflower seed press cake. Bioresource Technology 128: 593–597. doi: 10.1016/j.biortech.2012.10.150.

Angin, D., Altintig, E., and Kose, T.E. 2013. Influence of process parameters on the surface and chemical properties of activated carbon obtained from biochar by chemical activation. Bioresource Technology 148: 542–549. doi: 10.1016/j.biortech.2013.08.164.

Antal, M.J., and Gronli, M. 2003. The art, science, and technology of charcoal production. *Industrial & Engineering Chemistry Research* 42(8): 1619–1640.

Bagreev, A., Bandosz, T.J., and Locke, D.C. 2001. Pore structure and surface chemistry of adsorbents obtained by pyrolysis of sewage sludge-derived fertilizer. *Carbon* 39(13): 1971–1979. doi: 10.1016/S0008-6223(01)00026-4.

Batool, F., Akbar, J., Iqbal, S., et al. 2018. Study of isothermal, kinetic, and thermodynamic parameters for adsorption of cadmium: an overview of linear and nonlinear approach and error analysis. *Bioinorganic Chemistry and Applications* 2018: 1–11. doi: 10.1155/2018/3463724.

Boehm, H.P. 1994. Some aspects of the surface chemistry of carbon blacks and other carbons. *Carbon* 32(5): 759–769.

Bordoloi, N., Goswami, R., Kumar, M., et al. 2017. Biosorption of Co (II) from aqueous solution using algal biochar: kinetics and isotherm studies. *Bioresource Technology* 244: 1465–1469. doi: 10.1016/j.biortech.2017.05.139.

Cao, X., Ma, L., and Gao, B. 2009. Dairy-manure derived biochar effectively sorbs lead and atrazine. *Environmental Science and Technology* 43(9): 3285–3291. doi: 10.1021/es803092k.

Cha, J.S., Park, S.H., Jung, S.C., et al. 2016. Production and utilization of biochar: a review. *Journal of Industrial Engineering and Chemistry* 40: 1–15.

Chellan, P., and Sadler, P.J. 2015. The elements of life and medicines. *Philos Transaction A Math Physics Engineering and Science* 373(2037). doi: 10.1098/rsta.2014.0182.

Chen, B., Chen, Z., and Lv, S., 2011. A novel magnetic biochar efficiently sorbs organic pollutants and phosphate. *Bioresource Technology* 102(2): 716–723. doi: 10.1016/j.biortech.2010.08.067.

Chen, T., Zhang, Y., Wang, H. et al. 2014. Influence of pyrolysis temperature on characteristics and heavy metal adsorptive performance of biochar derived from municipal sewage sludge. *Bioresource Technology* 164: 47–54. doi: 10.1016/j.biortech.2014.04.048.

Dong, H., Deng, J., Xie, Y., et al. 2017. Stabilization of nanoscale zero-valent iron (nZVI) with modified biochar for Cr(VI) removal from aqueous solution. *Journal of Hazardous Materials* 332: 79–86. doi: 10.1016/j.jhazmat.2017.03.002.

Dong, T., Gao, D., Miao, C., et al. 2015. Two-step microalgal biodiesel production using acidic catalyst generated from pyrolysis-derived bio-char. *Energy Conversion and Management* 105: 1389–1396. doi: 10.1016/j.enconman.2015.06.072.

Dong, X., Ma, L.Q., and Li, Y. 2011. Characteristics and mechanisms of hexavalent chromium removal by biochar from sugar beet tailing. *Journal of Hazardous Materials* 190(1–3): 909–915. doi: 10.1016/j.jhazmat.2011.04.008.

Fraga, T.J.M., Carvalho, M.N., Ghislandi, M.G., et al. 2019. Functionalized graphene-based materials as innovative adsorbents of organic pollutants: a concise overview. *Brazilian Journal of Chemistry Engineering* 36(1): 1–31. doi: 10.1590/0104-6632.20190361s20180283.

Haefele, S.M., Konboon, Y., Wongboon, W., et al. 2011. Effects and fate of biochar from rice residues in rice-based systems. *Field Crops Research* 121(3): 430–440. doi: 10.1016/j.fcr.2011.01.014.

Hale, S., Hanley, K., Lehmann, J., et al. 2011. Effects of Chemical, Biological, and Physical Aging As Well As Soil Addition on the Sorption of Pyrene to Activated Carbon and Biochar. *Environmental Science and Technology* 45(24): 10445–10453. doi: 10.1021/es202970x.

Hammes, K., Smernik, R.J., Skjemstad, J.O., et al. 2008. Characterisation and evaluation of reference materials for black carbon analysis using elemental composition, colour, BET surface area and ^{13}C NMR spectroscopy. *Applied Geochemistry* 23(8): 2113–2122. doi: 10.1016/j.apgeochem.2008.04.023.

Hassan, M.M., and Carr, C.M. 2018. A critical review on recent advancements of the removal of reactive dyes from dye house effluent by ion-exchange adsorbents. *Chemosphere* 209: 201–219.

He, Z.L., Yang, X.E., and Stoffella, P.J. 2005. Trace elements in agroecosystems and impacts on the environment. *Journal of Trace Elements in Medicine and Biology* 19(2–3): 125–140.

Ho, P.H., Lee, S.Y., Lee, D., et al. 2014 Selective adsorption of tert-butylmercaptan and tetrahydrothiophene on modified activated carbons for fuel processing in fuel cell applications. *International Journal of Hydrogen Energy* 39(12): 6737–6745. doi: 10.1016/j.ijhydene.2014.02.011.

Hossain, M.K., Strezov, V., Chan, K.Y., et al. 2011. Influence of pyrolysis temperature on production and nutrient properties of wastewater sludge biochar. *Journal of Environmental Management* 92(1): 223–228. doi: 10.1016/j.jenvman.2010.09.008.

Inyang, M.I., Gao, B., Yao, Y. et al. 2016 A review of biochar as a low-cost adsorbent for aqueous heavy metal removal. *Critical Reviews in Environmental Science and Technology* 46(4): 406–433.

Jaishankar, M., Tseten, T., Anbalagan, N., et al. 2014. Toxicity, mechanism and health effects of some heavy metals. *Interdisciplinary Toxicology* 7(2): 60–72.

Jeffery, S., Bezemer, T.M., Cornelissen, G., et al. 2015. The way forward in biochar research: targeting trade-offs between the potential wins. *GCB Bioenergy* 7(1): 1–13. doi: 10.1111/gcbb.12132.

Khan, N., Clark, I., Sanchez-Monedero, M.A., et al. 2016. Physical and chemical properties of biochars co-composted with biowastes and incubated with a chicken litter compost. *Chemosphere* 142: 14–23. doi: 10.1016/j.chemosphere.2015.05.065.

Koller, M. and Saleh, H.M. 2018. *Introductory Chapter: Introducing Heavy Metals.* Heavy Metals, Hosam El-Din M. Saleh and Refaat F. Aglan, IntechOpen, DOI: 10.5772/intechopen.74783. Available from: https://www.intechopen.com/chapters/59857

Kołodynska, D., Krukowska, J., and Thomas, P. 2017. Comparison of sorption and desorption studies of heavy metal ions from biochar and commercial active carbon. *Chemical Engineering Journal* 307: 353–363. doi: 10.1016/j.cej.2016.08.088.

Komkiene, J., and Baltrenaite, E. 2016. Biochar as adsorbent for removal of heavy metal ions [Cadmium(II), Copper(II), Lead(II), Zinc(II)] from aqueous phase. *International Journal of Environmental Science and Technology* 13(2): 471–482. doi: 10.1007/s13762-015-0873-3.

Kong, H., He, J., Gao, Y., et al. 2011. Cosorption of Phenanthrene and Mercury(II) from Aqueous Solution by Soybean Stalk-Based Biochar. *Journal of Agriculture and Food Chemistry* 59(22): 12116–12123. doi: 10.1021/jf202924a.

Lawrence, A.J., Khan, T. 2020. Quantification of Airborne Particulate and Associated Toxic Heavy Metals in Urban Indoor Environment and Allied Health Effects. In: Gupta T., Singh S., Rajput P., Agarwal A. (eds) *Measurement, Analysis and Remediation of Environment, and Sustainability.* Springer, Singapore. doi: 10.1007/978-981-15-0540-9_2.

Lee, J.W., Kidder, M., Evans, B.R. et al. 2010. Characterization of Biochars Produced from Cornstovers for Soil Amendment. *Environmental Science and Technology* 44(20): 7970–7974. doi: 10.1021/es101337x.

Li, H., Dong, X., and da Silva, E.B. 2017. Mechanisms of metal sorption by biochars: biochar characteristics and modifications. *Chemosphere* 178: 466–478.

Li, N., Rao, F., and He, L. 2019. Evaluation of biochar properties exposing to solar radiation: a promotion on surface activities. *Chemical Engineering Journal* 123353. doi: 10.1016/j.cej.2019.123353.

Li, X., Shen, Q., Zhang, D., et al. 2013. Functional Groups Determine Biochar Properties (pH and EC) as Studied by Two-Dimensional^{13}C NMR Correlation Spectroscopy. Motta, A. (ed.). *PLoS One* 8(6): e65949. doi: 10.1371/journal.pone.0065949.

Li, Q., Tang, L., Hu, J., et al. 2018. Removal of toxic metals from aqueous solution by biochars derived from long-root *Eichhornia crassipes. Royal Society Open Science* 5(10): 180966. doi: 10.1098/rsos.180966.

Li, Y., Zhang, Q., Zhang, J., et al. 2015. A top-down approach for fabricating free-standing bio-carbon supercapacitor electrodes with a hierarchical structure. *Scientific Reports* 5: 14155. doi: 10.1038/srep14155.

Liu, S., Wang, M., Sun, X., et al. 2018. Facilitated oxygen chemisorption in heteroatom-doped carbon for improved oxygen reaction activity in all-solid-state zinc-air batteries. *Advanced Materials* 30(4): 1704898. doi: 10.1002/adma.201704898.

Liu Z and Zhang FS 2009. Removal of lead from water using biochars prepared from hydrothermal liquefaction of biomass. *Journal of Hazardous Materials* 167(1–3): 933–939. doi: 10.1016/j.jhazmat.2009.01.085.

Lou, L., Luo, L., Yang, Q., et al. 2012. Release of pentachlorophenol from black carbon-inclusive sediments under different environmental conditions. *Chemosphere* 88(5): 598–604. doi: 10.1016/j.Chemosphere.2012.03.039.

Lu, H., Zhang, W., Yang, Y., et al. 2012. Relative distribution of Pb^{2+} sorption mechanisms by sludge-derived biochar. *Water Research* 46(3): 854–862. doi: 10.1016/j.watres.2011.11.058.

Lua, A.C., Yang, T., and Guo, J. 2004. Effects of pyrolysis conditions on the properties of activated carbons prepared from pistachio-nut shells. *Journal of Analytical and Applied Pyrolysis* 72(2): 279–287. doi: 10.1016/j.jaap.2004.08.001.

Meyer, S., Glaser, B., and Quicker, P. 2011. Technical, economical, and climate-related aspects of biochar production technologies: a literature review. *Environmental Science and Technology* 45(22): 9473–9483.

Mohan, D., Sarswat, A., Ok, Y.S., et al. 2014. Organic and inorganic contaminants removal from water with biochar, a renewable, low cost and sustainable adsorbent - a critical review. *Bioresource Technology* 160: 191–202. doi: 10.1016/j.biortech.2014.01.120.

Muradov N, Fidalgo B, Gujar AC, et al. 2012. Production and characterization of Lemna minor bio-char and its catalytic application for biogas reforming. *Biomass and Bioenergy* 42: 123–131. doi: 10.1016/j.biombioe.2012.03.003.

Nartey, O.D., and Zhao, B. 2014. Biochar preparation, characterization, and adsorptive capacity and its effect on bioavailability of contaminants: an overview. *Advances in Materials Science and Engineering* 1–12. doi: 10.1155/2014/715398.

O'Connor, D., Peng, T., Zhang, J., et al. 2018. Biochar application for the remediation of heavy metal polluted land: a review of in situ field trials. *Science of the Total Environment* 619–620: 815–826. doi: 10.1016/j.scitotenv.2017.11.132.

Park, J.H., Choppala, G.K., Bolan, N.S., et al. 2011. Biochar reduces the bioavailability and phytotoxicity of heavy metals. *Plant and Soil* 348(1–2): 439–451. doi: 10.1007/s11104-011-0948-y.

Pellera, F.M., Giannis, A., Kalderis, D., et al. 2012. Adsorption of Cu(II) ions from aqueous solutions on biochars prepared from agricultural by-products. *Journal of Environmental Management* 96(1): 35–42. doi: 10.1016/j.jenvman.2011.10.010.

Rai, P.K., Lee, S.S., Zhang, M., et al. 2019. Heavy metals in food crops: health risks, fate, mechanisms, and management. *Environment International* 125: 365–385. doi: 10.1016/j.envint.2019.01.067.

Rajapaksha, A.U., Chen, S.S., Tsang, D.C.W. et al. 2016. Engineered/designer biochar for contaminant removal/immobilization from soil and water: potential and implication of biochar modification. *Chemosphere* 148: 276–291. doi: 10.1016/j.Chemosphere.2016.01.043.

Regmi, P., Garcia Moscoso, J.L., Kumar, S., et al. 2012. Removal of copper and cadmium from aqueous solution using switchgrass biochar produced via hydrothermal carbonization process. *Journal of Environmental Management* 109: 61–69. doi: 10.1016/j.jenvman.2012.04.047.

Seehra, M.S., and Bristow, A.D. 2018. Introductory chapter: overview of the properties and applications of noble and precious metals. *Noble and Precious Metals - Properties, Nanoscale Effects and Applications*. InTech Open.

Shen, Z., Zhang, J., Hou, D., et al. 2019. Synthesis of MgO-coated corncob biochar and its application in lead stabilization in a soil washing residue. *Environmental International* 122: 357–362. doi: 10.1016/j.envint.2018.11.045.

Singh, B., Singh, B.P., and Cowie, A.L. 2010. Characterization and evaluation of biochars for their application as a soil amendment. *Soil Research* 48(7): 516. doi: 10.1071/SR10058.

Sizmur, T., Fresno, T., Akgul, G., et al. 2017. Biochar modification to enhance sorption of inorganics from water. *Bioresource Technology* 246: 34–47.

Smernik, R.J., Baldock, J.A., Oades, J.M., et al. 2002. Determination of T1ρH relaxation rates in charred and uncharred wood and consequences for NMR quantitation. *Solid State Nuclear Magnetic Resonance* 22(1): 50–70. doi: 10.1006/snmr.2002.0064.

Stern, B.R. 2010. Essentiality and toxicity in copper health risk assessment: overview, update and regulatory considerations. *Journal of Toxicology and Environmental Health, Part A* 73(2): 114–127. doi: 10.1080/15287390903337100.

Subedi, R., Taupe, N., Pelissetti, S., et al. 2016. Greenhouse gas emissions and soil properties following amendment with manure-derived biochars: influence of pyrolysis temperature and feedstock type. *Journal of Environmental Management* 166: 73–83. doi: 10.1016/j.jenvman.2015.10.007.

Swiatkowski, A., Pakula, M., Biniak, S., et al. 2004. Influence of the surface chemistry of modified activated carbon on its electrochemical behaviour in the presence of lead(II) ions. *Carbon* 42(15): 3057–3069. doi: 10.1016/j.carbon.2004.06.043.

Tang, J., Lv, H., Gong, Y., et al. 2015. Preparation and characterization of a novel graphene/biochar composite for aqueous phenanthrene and mercury removal. *Bioresource Technology* 196: 355–363. doi: 10.1016/j.biortech.2015.07.047.

Tchounwou, P.B., Yedjou, C.G., Foxx, D.N., et al. 2004. Lead-induced cytotoxicity and transcriptional activation of stress genes in human liver carcinoma (HepG2) cells. *Molecular and Cellular Biochemistry* 255(1–2): 161–170. doi: 10.1023/b:mcbi.0000007272.46923.12.

Tchounwou, P.B., Yedjou, C.G., Patlolla, A.K., et al. 2012. Heavy metal toxicity and the environment. *EXS* 101: 133–164.

Verheijen, F., Jeffery, S., Bastos, A.C., et al. 2010. *Biochar Application to Soils A Critical Scientific Review of Effects on Soil Properties, Processes and Functions*. Office for the Official Publications of the European Communities ISBN: 9789279142932. doi: 10.2788/472.

Verma, L., and Singh, J. 2019. Synthesis of novel biochar from waste plant litter biomass for the removal of Arsenic (III and V) from aqueous solution: a mechanism characterization, kinetics and thermodynamics. *Journal of Environmental Management* 248: 109235. doi: 1016/j.jenvman.2019.07.006.

Wang, H., Gao, B., Wang, S., et al. 2015a. Removal of Pb(II), Cu(II), and Cd(II) from aqueous solutions by biochar derived from KMnO4 treated hickory wood. *Bioresource Technology* 197: 356–362. doi: 10.1016/j.biortech.2015.08.132.

Wang, S., Tang, Y., Chen, C., et al. 2015b. Regeneration of magnetic biochar derived from eucalyptus leaf residue for lead(II) removal. *Bioresource Technology* 186: 360–364. doi: 10.1016/j.biortech.2015.03.139.

Wang, F.Y., Wang, H., and Ma, J.W. 2010. Adsorption of cadmium (II) ions from aqueous solution by a new low-cost adsorbent-Bamboo charcoal. *Journal of Hazardous Materials* 177(1–3): 300–306. doi: 10.1016/j.jhazmat.2009.12.032.

Xiong, X., Yu, I.K.M., Cao, L., et al. 2017. A review of biochar-based catalysts for chemical synthesis, biofuel production, and pollution control. *Bioresource Technology* 246: 254–270.

Yan, L., Kong, L., Qu, Z., et al. 2015. Magnetic Biochar Decorated with ZnS Nanocrytals for Pb (II) Removal. *ACS Sustainable Chemistry and Engineering* 3(1): 125–132. doi: 10.1021/sc500619r.

Yao, Y., Gao, B., Chen, H., et al. 2012. Adsorption of sulfamethoxazole on biochar and its impact on reclaimed water irrigation. *Journal of Hazardous Materials* 209–210: 408–413. doi: 10.1016/j.jhazmat.2012.01.046.

Yedjou, C.G. and Tchounwou, P.B. 2007. N-acetyl-l-cysteine affords protection against lead-induced cytotoxicity and oxidative stress in human liver carcinoma (HepG2) cells. *International Journal of Environmental Research and Public Health* 4(2):132–137. doi: 10.3390/ijerph2007040007.Yu, J.T., Dehkhoda, A.M., and Ellis, N. 2011. Development of biochar-based catalyst for transesterification of canola oil. *Energy and Fuels* 25(1): 337–344. doi: 10.1021/ef100977d.

Yuan, J.H., Xu, R.K., and Zhang, H. 2011. The forms of alkalis in the biochar produced from crop residues at different temperatures. *Bioresource Technology* 102(3): 3488–3497. doi: 10.1016/j.biortech.2010.11.018.

Yuan, P., Wang, J., Pan, Y., et al. 2019. Review of biochar for the management of contaminated soil: preparation, application and prospect. *Science of the Total Environment* 659: 473–490.

Zhang, M., Gao, B., Varnoosfaderani, S., et al. 2013. Preparation and characterization of a novel magnetic biochar for arsenic removal. *Bioresource Technology* 130: 457–462. doi: 10.1016/j.biortech.2012.11.132.

Zhang, M.M., Liu, Y.G., Li, T.T., et al. 2015. Chitosan modification of magnetic biochar produced from Eichhornia crassipes for enhanced sorption of Cr(vi) from aqueous solution. *RSC Advances* 5(58): 46955–46964. doi: 10.1039/c5ra02388b.

Zhao, B., O'Connor, D., Zhang, J., et al. 2018. Effect of pyrolysis temperature, heating rate, and residence time on rapeseed stem derived biochar. *Journal of Cleaner Production* 174: 977–987 doi: 10.1016/j.jclepro.2017.11.013.

Zhao, L., Cao, X., Masek, O., et al. 2013. Heterogeneity of biochar properties as a function of feedstock sources and production temperatures. *Journal of Hazardous Materials* 256–257: 1–9. doi: 10.1016/j.jhazmat.2013.04.015.

Zhao, L., Zheng, W., Masek, O., et al. 2017. Roles of phosphoric acid in biochar formation: synchronously improving carbon retention and sorption capacity. *Journal of Environmental Quality* 46(2): 393–401. doi: 10.2134/jeq2016.09.0344.

6

Bioremediation of Potentially Toxic Metals by Microorganisms and Biomolecules

Luciana Maria Saran, Bárbara Bonfá Buzzo, Cinara Ramos Sales, and Lucia Maria Carareto Alves
São Paulo State University (UNESP)

Renan Lieto Alves Ribeiro
Instituto Chico Mendes de Conservação da Biodiversidade (ICMBio/CENAP)

CONTENTS

6.1 Introduction

Accelerated population growth coupled with technological advances in recent years has increasingly been generating a high volume of waste with the potential to negatively impact the environment. Generally, such waste is disposed of in the environment without any treatment, and this has resulted in the pollution of the soil, water resources and the atmosphere. Large amounts of organic and inorganic waste are

DOI: 10.1201/9781003181224-6

generated daily by the urban population, and these types of waste are responsible for much of the damage to the environment and public health because of their improper disposal (Kaza et al. 2018).

One of the major problems encountered in these improper disposal areas is the high content of certain potentially toxic metals (PTMs) that are commonly referred to as 'heavy metals'. Thus, elements such as arsenic (As), nickel (Ni), copper (Cu), zinc (Zn), cadmium (Cd), lead (Pb), mercury (Hg) and chromium (Cr) and/or compounds containing them are present in various types of waste, such as fluorescent lamps, batteries, waste paint and cans, that are deposited in landfills (Milhome et al. 2018).

Technological advances in the new antropocene era have also propelled the industrial sector and have raised concerns regarding the release of pollutants, particularly those containing PTMs and/or their compounds that may negatively impact water and soil quality. In this sense, mining activities related to current high consumption may also exert several negative environmental impacts.

Thus, considering the interference of modern man in nature that is caused primarily by technological developments and unrestrained consumption and by inappropriate disposal of products and accidents with tailings, this chapter aims to describe the remediation techniques that can assist or minimize the anthropogenic impacts on the soil or water that result from the disposal of PTMs by using organisms and microorganisms that are naturally found in these environments.

6.2 Mining Accidents as a Source of Potentially Toxic Metals

Despite the outstanding reuse capacity of the waste metal material produced today, waste from the mining process does not always receive adequate treatment and/or disposal. Extractive operations of mineral resources are a major source of waste in many diverse locations throughout the world. These wastes involve materials that are removed while attempting to obtain mineral resources such as topsoil and rocks and are also present in remnant tailings after the ore has been mined. Some waste is inert and therefore does not pose a significant threat of pollution to the environment, except for the possible collapse of rivers that can occur when this waste is stored in large quantities. Additionally, waste generated by the non-ferrous mining industry may contain large amounts of PTMs.

It is noteworthy that the management of mining tailings is an inherently risky activity that typically involves the use of waste processing chemicals and high amounts of metals. In many cases, tailings are stored in mounds or in large lagoons where they are held by dams. The collapse of dams or mounds can result in serious impacts on the environment and on human health and safety. Examples of these types of incidents include accidents in Aberfan (Wales, 1966), Stava (Italy, 1985), Aznalcóllar (Spain, 1998), Baia Mare, Baia Borsa (Romania, 2000) and, most recently, Itabirito in Brazil (2014), Mariana (2015) and Brumadinho (2019) in the state of Minas Gerais.

All these accidents resulted in significant human, social and environmental damage. In particular, the most recent accident that occurred in Brumadinho was considered one of the largest disasters involving ore dam disruption in the world during the last 34 years, and this accident resulted in more than 250 dead or missing individuals and affected almost 40,000 people. The Brumadinho accident released 12 million m^3 of tailings that polluted at least 300 km of the river (mainly Paraopeba River) and caused a significant impact on its margins that resulted in the burying of animals, destruction of civil construction and soil degradation (Figure 6.1).

In Brazil, according to the National Department of Mineral Production, mining dams (also known as tailings dams) are built to store the tailings from ore extraction and beneficiation processes (DNPM 2017). When a dam begins to lose structural integrity, severe environmental impacts can occur that include hydrological disturbances, socio-economic problems, contamination of the physical and biotic environment, and deaths and compromised health in populations living in proximity to the dam (Queiroz et al. 2018).

The mining industry utilizes processes such as drilling, extraction and processing techniques for the extraction of minerals for commercial use. Waste is generated during all stages of the mining. Mining industry waste can be solid, aqueous or paste waste that remains deposited on the ground and can be a major source of contamination and environmental damage. Soils contaminated by mining waste can leach large amounts of heavy metals that are considered one of the worst environmental contaminants,

FIGURE 6.1 Satellite images of Brumadinho region, Minas Gerais (Brazil), prior to the dam rupture (a) and after the dam rupture (b). (From Earth Observatory images by Lauren Dauphin, using Landsat data from the U.S. Geological Survey, 2019.)

as they can trigger hazardous metal contamination of large amounts of land at long distances from the source of contamination.

One of the most important problems with PTMs is that they are not biodegradable and can enter the food chain to ultimately accumulate in certain tissues where they are toxic to many living organisms, even at low concentrations. This toxicity depends on the concentration, the chemical form and their persistence within tissues, as in trace concentrations some of these metals may be indispensable for living beings. The metal ions coexist with other cations and anions in acidic aqueous solutions; however, they are highly insoluble in an alkaline medium.

Mercury (Hg), lead (Pb) and cadmium (Cd) are the primary metals produced by mining activities, while arsenic (As) is the main semi-metal produced during these activities. These wastes must be subjected to adequate remediation technologies to avoid becoming a major problem concerning public health (Khalid et al. 2017). Remediation tactics such as chemical and physical manipulations are often not satisfactory in mitigating pollution problems due to the successive threats of new recalcitrant pollutants produced by anthropogenic activities. Thus, bioremediation using microorganisms provides an ecological, economical and socially acceptable alternative to reduce environmental pollution (Dangi et al. 2019).

6.3 What Does Bioremediation Mean?

Bioremediation involves the use of organisms and microorganisms that can remove or decrease the concentration of pollutants in the environment. It can be considered an ecologically and economically viable

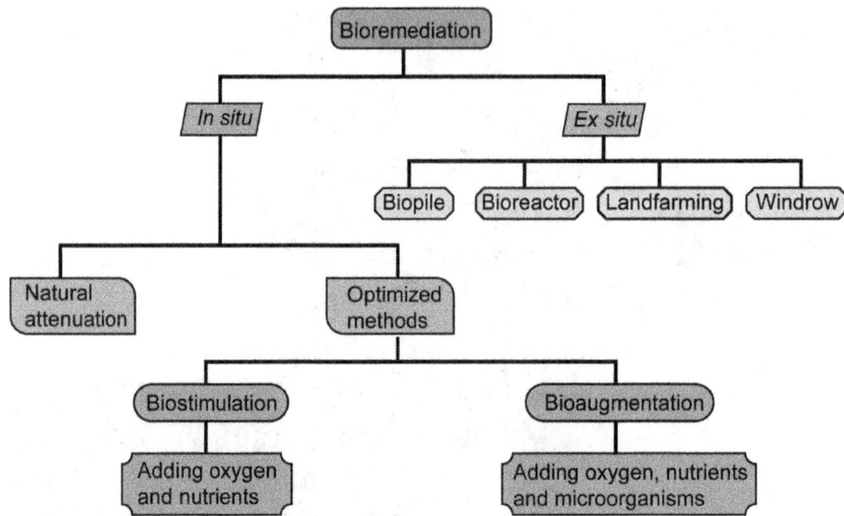

FIGURE 6.2 Schematic diagram of bioremediation types. (Adapted from Azubuike, C.C., Chikere, C.B., and Okpokwasili, G.C., *World J. Microbiol. Biotechnol.*, 32, 180, 2016.)

solution for decontamination and/or recovery of polluted environments. Various organisms can be used as bioremediation agents, including bacteria, fungi, plants and algae. The efficiency of such organisms varies according to the structure of the chemical species to be degraded and the presence of enzymes capable of degrading these chemical species. Organisms and biomolecules can be used as bioremediation agents. Bacteria are the most widely used at approximately 57%, and this is followed by enzymes (19%), fungi (13%), algae (6%), plants (4%) and protozoa (Quintella et al. 2019).

Bioremediation (Figure 6.2) can be performed at the contamination site, and this is termed *in situ* bioremediation and is used when the area affected by the contamination is very extensive. Conversely, in *ex situ* bioremediation, the contaminated material is transferred to a proper location to achieve decontamination (Angelucci and Tomei 2016).

Three basic methods are used during the *in situ* technique, and these include natural attenuation, biostimulation and bioaugmentation. The first method makes use of microorganisms that are native to the contaminated environment to allow for the degradation of contaminants at the specific contaminant location, thus avoiding damaging the habitat and allowing the ecosystem to revert to its original condition. To increase and accelerate the efficiency of *in situ* bioremediation, biostimulation and bioaugmentation methods are used.

The *ex situ* method is more efficient than the *in situ* method, as it allows for more efficient removal of contaminants and greater control of physical–chemical parameters, thus causing recovery time to become shorter. During the process, the contaminated material is extracted and transferred to a specific location where the remediation of the contaminants will be performed. These locations include wetland systems where liquid materials can be decontaminated; mud bioreactors where semi-solid or solid residues can be decontaminated; and composts or biopiles where solid contaminants themselves can be biodegraded (Angelucci and Tomei 2016).

6.4 Bioremediation Agents

Several organisms (living or dead) are used in bioremediation processes, including bacteria, fungi, algae, plants, and genetically manipulated microorganisms. These microorganisms are cosmopolitan in their distribution and possess the potential to tolerate, alter and participate in various environmental conditions (Ma et al. 2016). Under natural conditions, microorganisms participate in biogeochemical cycles in nature such as the cycles of different metals, where microorganisms that participate in these cycles are already resistant and adapted to environments with high concentrations of PTMs.

Fungi and bacteria are important microorganisms used for the recovery, immobilization or detoxification of metallic and semi-metallic pollutants and radionuclides. Such organisms can sense and respond quickly to environmental changes, and they can develop resistance to stress caused by pollutants or adapt physiologically to new conditions. The primary organisms used in bioremediation are described below, and the species that stand out most in these processes are highlighted.

6.4.1 Fungi

Fungi are microorganisms capable of living in a variety of habitats, and these organisms spread through the dispersion of their spores in the air and often are associated with processes of biogeochemical cycles in litter and soil. These characteristics, combined with their metabolism that is rich in lignocellulolytic enzymes, make them good candidates for the bioremediation of several sites.

Fungi play an important role in the bioremediation of various pollutants, including textile dyes, petroleum hydrocarbons, effluents from the pulp and paper industry, effluents from the leather tanning industries, pesticides, pharmaceuticals and PTMs. These microorganisms play an important role in the bioremediation of PTMs, as they can transform metallic contaminants into soluble and insoluble forms using their existing biological mechanisms (Gupta et al. 2017). The main species of fungi that are efficient in processes involved in bioremediation include *Coprinellus radians*, *Marasmiellus troyanus*, *Gloeophyllum trabeum*, *Pleurotus ostreatus*, *Fomitopsis pinicola*, *Penicillium simplicissimum* and *Phanerochaete chrysosporium* (Biswas et al. 2015).

Fungal cell walls contain polysaccharides and proteins that both possess many functional groups (carboxyl, amine, hydroxyl and phosphate) that are known to be involved in metal chelation and adsorption (Rudakiya et al. 2019). Table 6.1 lists the efficiency of some fungi for removing potentially toxic elements (metals and semi-metals). These organisms use biosorption and/or bioaccumulation processes for the remediation of metals or semi-metals, but biosorption is the main process that is used. Additionally, the ability to remove metals varies according to organism type and the type of element that must be removed. It should be noted that there are organisms capable of acting on various metals and other contaminants with unique remedial capacity.

Certain genera of filamentous fungi, including *Aspergillus*, *Curvularia*, *Acremonium* and *Pythium*, have been demonstrated to be effective in the bioremediation of metals due to their tolerance to these chemical species. These filamentous fungi exhibited tolerance to concentrations as high as 1,500 mg/dm^3 of PTM ions such as Cd^{2+}, Cu^{2+} and Ni^{2+}, thus highlighting their usefulness for the bioremediation of contaminated soils and waters (Akhtar et al. 2013). According to a study by Zapana-Huarache et al. (2020), 14 species of filamentous fungi, including *Penicillium citrinum* and *Trichoderma viride*, were completely tolerant to chromium concentrations of up to 100 mg/L and exhibited significant growth of up to 250 mg/L. This indicates that these microorganisms are capable of functioning in environments contaminated with chromium, where they can remove a significant portion of this metal from the environment.

TABLE 6.1

Efficiency in Removing Potentially Toxic Elements (Metals and Semi-Metals) by Different Fungi

Fungus	%Removal (Element)
Beauveria bassiana	74 (Cu), 61 (Cr), 63 (Cd), 67 (Zn) and 75 (Ni)
Phanerochaete chrysosporium	98 (As), 29 (Cd) and 56 (Cr)
Penicillium simplicissimum	64 (Cu), 31 (Zn), 32 (Cd), 71 (Pb) and 92 (Cr)
Aspergillus sp.	98 (Cr)
Aspergillus niger	95 (Au)
Geotrichum sp. dwc-1	96 (U)

Source: Based on Rudakiya, D.M., Tripathi, A., Gupte, S., and Gupte, A, *Advancing Frontiers in Mycology & Mycotechnology*, Springer, Singapore, 2019.

6.4.2 Bacteria

Microorganisms in general and bacteria in particular stand out in their use of organic substances that can be either natural or synthetic as sources of carbon and energy. In this way, bacteria can play a significant role in mitigating or removing both organic and inorganic contaminating molecules from the environment. Bacteria have a wide range of mechanisms to allow for the absorption and transformation of various molecules for their immobilization or mobilization, and these mechanisms can play an important role in the ability of the bacteria to detoxify environments (Abatenh et al. 2017).

Bacterial isolates or consortia are widely used in the degradation of pesticides and hydrocarbons, as they use these compounds as a source of carbon and energy. Specifically, they can use polluting agents as 'food', thus removing them from the environment (Biswas et al. 2015; Ahmad et al. 2019). Similarly, associations between microorganisms can be found in nature. For example, a relationship exists between bacteria and fungi to allow for the degradation of lignocellulose, suggesting cooperation and specialization among the decomposers of this biomass (Wilhelm et al. 2019). Bacteria are also capable of playing an important role in the bioremediation of environments contaminated with PTM ions such as chromium (Cr), manganese (Mn), copper (Cu), nickel (Ni) and zinc (Zn), as these are important micronutrients in their metabolic processes (Biswas et al. 2015). These microorganisms possess mechanisms of resistance to metals, and this makes them potentially suitable for use in biotechnological processes in which eliminating or decreasing the concentration of PTMs is necessary (Niño-Martínez et al. 2019).

PTMs alter bacterial physiology and can modify their growth, morphology and metabolism. Additionally, these metals can cause a loss of biomass and a decrease in microbiological diversity, ultimately causing bacteria to develop mechanisms of tolerance and resistance to metals. These adaptive characteristics highlight bacteria as an important and promising tool for the bioremediation of contaminated environments. The primary mechanisms of toxicity and detoxification of metals that are used by bacteria to survive in environments possessing the presence of PTMs are shown in Figure 6.3.

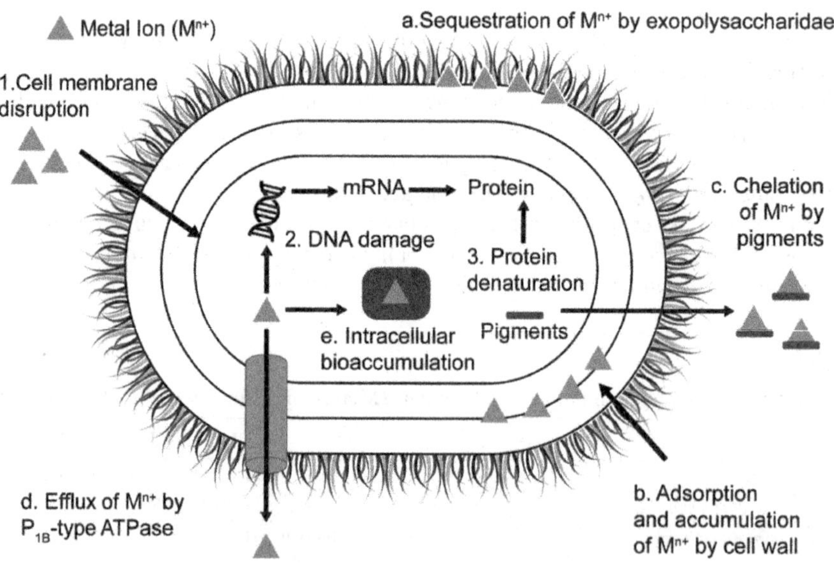

FIGURE 6.3 Toxicity caused by PTMs and their detoxification in bacteria. Toxicity in response to metals can include: (1) rupture of the cell membrane; (2) DNA damage; and (3) protein denaturation. Detoxification mechanisms may include (a) sequestration of metals by exopolysaccharides, (b) adsorption and accumulation of metal by the cell wall, (c) pigment chelation, (d) efflux by type P1B ATPase and (e) intracellular bioaccumulation. (Adapted from Kushwaha, A., Hans, N., Kumar, S., and Rani, R, *Ecotoxicol. Environ. Safety*, 147, 1035–1045, 2018.)

The primary genera of bacteria that are used both in the process of bioremediation of PTMs and for bioaccumulation and biotransformation of these elements are *Bacillus*, *Pseudomonas*, *Flavobacterium*, *Azotobacter*, *Xanthomonas*, *Nocardia*, *Streptomyces*, *Mycobacterium*, *Aerococcus*, *Citrobacter*, *Micrococcus*, *Enterobacter*, *Gemella* and others (Biswas et al. 2015; Igiri et al. 2018).

6.4.3 Algae

Algae, specifically microalgae, can offer an ecologically more economical and efficient method for removing metal ions from wastewater and aquatic environments. Due to their abundance in aquatic environments (freshwater and marine environments) and their ability to tolerate different conditions in wastewater, microalgae can be used in several situations for bioremediation. These organisms can be used primarily for the sorption of potentially toxic and radioactive metal ions (Leong and Chang 2020). The biosorption process is one of the technologies used to remove ions from contaminated waters. This process can be performed using biomasses consisting of live or dead algae that are composed of a set of polymers such as cellulose, pectins and glycoproteins. This biomass is capable of adsorbing metal ions present in water, thus enabling a low-cost treatment process.

The removal of ions by microalgae involves two phases that include an initial phase that constitutes the extracellular binding of the metal to the algal cell wall followed by a subsequent phase that involves sequestration of that metal into the cell. The first phase occurs quickly and is dependent on electrostatic interactions of the metal with functional groups on the cell surface; however, this phase can be reversible and can occur using living or non-living organisms. The second phase of the removal of metals by algae involves a metabolic process and requires specific carriers within the cell membranes. The metal ions internalized in the microalgae can accumulate in organelles or become attached to certain compounds, and this type of metal uptake is generally irreversible, is slower and requires living cells.

Resistance to metal toxicity can be highly specific for certain species of algae, where the remediation capacity can be determined to be more effective using a specific algal strain for the metal to be remedied. Some microalgae used in the bioremediation of PTMs are *Anabaena subcylindrica*, *Phormidium valderium*, *Chlorella vulgaris*, *Nostoc muscorum*, *Tetraselmis chuii*, *Tetraselmis suecica*, *Spirogyra hyalina*, *Chlorella pyrenoidosa*, *Scenedesmus abundans*, *Scenedesmus quadricauda*, *Spirulina* sp., *Spirulina maxima* and *Lyngbya spiralis* (Biswas et al. 2015). The major mechanism of the removal of various types of metals is the adsorption process, and the removal efficiency by these organisms varies greatly and is dependent upon conditions such as pH, temperature, salinity, atomic mass of the element, reduction potential and ion concentration.

Other species of macroalgae capable of adsorption of metals are also reported in the literature, and these include *Fucus spiralis*, *Ascophyllum nodosum*, *Chondrus crispus*, *Asparagopsis armata*, *Spirogyra insignis* and *Sargassum* sp. that are used to remove copper and other potentially toxic metals. The use of association between microalgae and macroalgae to remove metallic contaminants such as Ni(II), Zn(II), Cd(II) and Cu(II) from the environment and bioenergy production has recently been proposed by Piccini et al. (2019). The authors used two species of microalgae [*Chlorella vulgaris* and *Arthrospira platensis* (*Spirulina*)] and two of macroalgae (*Ulva lactuca* and *Sargassum muticum*) as bioremediation agents, and the resulting biomasses were used to produce oil with bioenergy potential. Such associations can contribute to remediation and sustainability processes simultaneously.

6.4.4 Plants

The use of plants and their associated microorganisms to decrease the toxicity of contaminated environments (usually soils) is called phytoremediation. This remediation process is suitable for low-permeability soils and is economically more efficient and more viable than other processes such as excavation, soil incineration, washing and solidification. The tolerance of the plant to a certain metallic ion will be greater and is related to a greater capacity for accumulation of this ion without it being able to influence the health of the organism. The tolerance potential depends on mechanisms of binding of the metal ion to the cell wall, active transport of the ion to the vacuoles and chelation of the ion by proteins (Sarwar et al. 2017).

Phytoremediation is an emerging green technology that is based on the use of plants to decontaminate soil and water. It is among the main methodologies currently available for the remediation of contaminated soils, and it incorporates the use of plants to remove, immobilize or render harmless to the ecosystem the organic and inorganic contaminants present in the soil and water. The success of phytoremediation aimed at removing PTMs depends on the potential of plants to produce high biomass and to withstand the stress caused by the metal. The bioavailability of the metal in rhizospheric soil is considered another critical factor that determines the efficiency of the translocation and phytostabilization process of the metal (Ma et al. 2011). The phytoremediation process can occur through phytoextraction, phytofiltration and phytostabilization (Ekta and Modi 2018).

A summary diagram of the relationships between different types of phytoremediation in the soil is presented in Figure 6.4.

Phytoextraction, also known as phytoaccumulation or phytoabsorption, is the process of absorbing contaminants from soil or water by plant roots and the subsequent accumulation of these contaminants in the above-ground biomass. Phytofiltration corresponds to the removal of pollutants from water, and this can be accomplished by roots, seedlings or stems. Phytostabilization makes use of certain plants to stabilize soil contaminants, and it is used to reduce the mobility and availability of pollutants in the environment.

Among the advantages of phytoremediation are the low cost, the possibility for application *in situ* in large areas and for different types of pollutants, the easy monitoring of plants, soil maintenance, the ability of this method to stimulate the life of organisms and the possibility of being used in combination with other decontamination methods. *Cannabis sativa* is an example of a plant that possesses the potential to remove toxic and radioactive compounds from contaminated soils and that can be successfully used to remove strontium-90 (Sr-90) and caesium-137 (Cs-137), which are commonly used radioactive elements on weapons and nuclear reactors (Hoseini et al. 2012).

Some plants are used to remove metal from soil, and these plants are called metallophytes. The three types of metallophytes are metal indicators, metal excluders and metal hyperaccumulators (Awa and

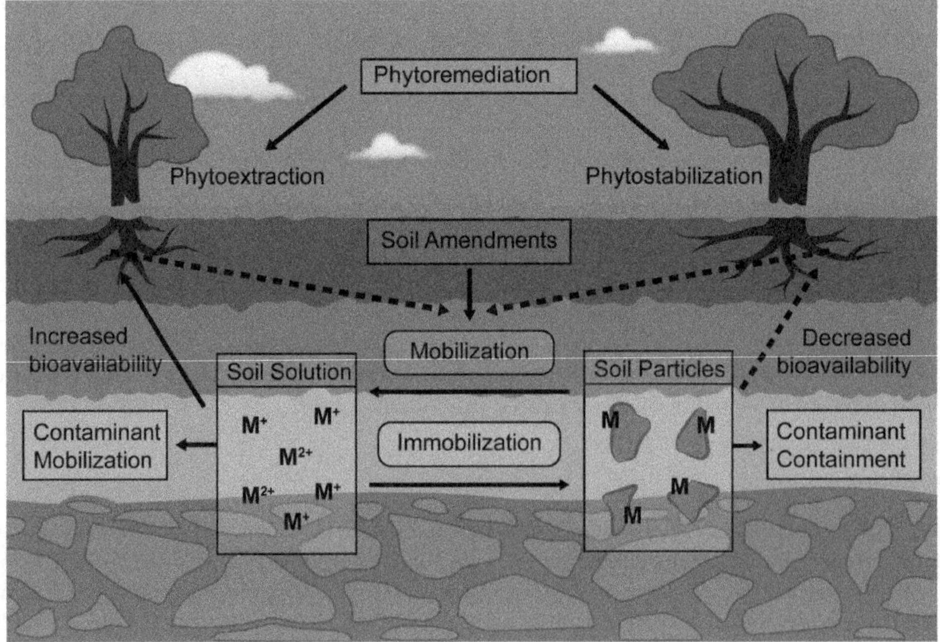

FIGURE 6.4 Illustrative diagram of the relationships between immobilization, bioavailability and phytoremediation of PTMs (M, M^+ and M^{2+}). (Adapted from Bolan, N., Kunhikrishnan, A., Thangarajan, R., Kumpiene, J., Park, J., Makinof, T., Kirkhamg, M.B., and Scheckel, K, *J. Hazard. Mater.*, 266, 141–166, 2014.)

Hadibarata 2020). Several plants can be used in these phytoremediation processes, and they include *Helianthus annuus, Brassica juncea, perennial ryegrass, Medicago sativa* L., *Nicotiana tabacum, Tagetes patula, Arundo donax, Typha domingensis, Solanum nigrum* and *Lemna minor* (Biswas et al. 2015). Certain bamboo species are resistant to soils contaminated by metals, and they also possess the capacity to absorb and accumulate these elements. Additionally, bamboo possesses high biomass productivity, short rotation and high economic value, and it can therefore be used in phytoremediation. However, the mechanisms underlying the capture of heavy metals, including the transport, sequestration and detoxification of these metals by these different species, still require further exploration (Bian et al. 2020).

Recent studies also indicate that native plants that can grow in contaminated soils can not only provide an important tool for the ecorestoration of contaminated areas (mining tailings, for example), but can also facilitate the removal of metals from these contaminated areas through phytoremediation effects that involve phytostabilization, rhizodegradation and phytodegradation (Gajić et al. 2018).

6.4.4.1 Microorganisms Associated with Plants

Plants that grow on soils contaminated by PTMs are home to a diverse group of microorganisms capable of tolerating high concentrations of metals and of providing a series of benefits to both the soil and plants. Among the microorganisms involved in phytoremediation of PTMs, rhizosphere bacteria deserve special attention (Figure 6.5). These microorganisms can directly improve the phytoremediation process by altering the bioavailability of a given metal by changing the pH of the soil, causing the release of

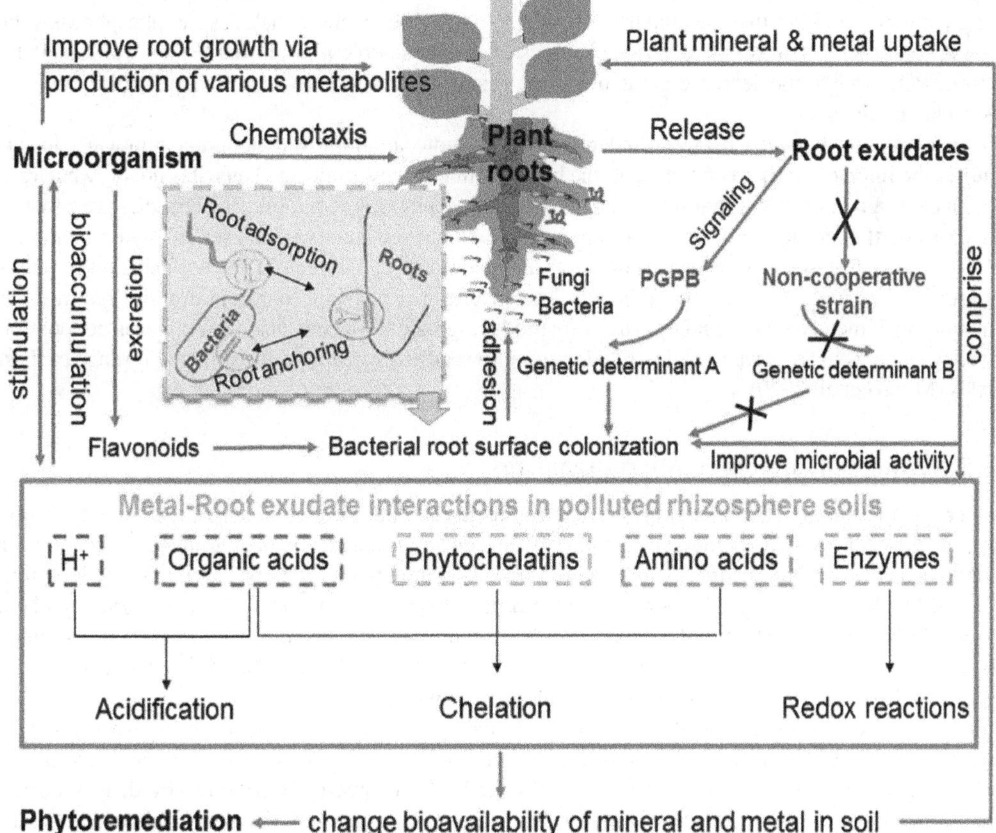

FIGURE 6.5 Schematic overview of mechanisms regulating plant–microbe–metal interactions. PGPB: plant growth-promoting bacteria. (Adapted from Ma, Y., Oliveira, R.S., Freitas, H., and Zhang, C, *Front. Plant Sci.*, 7, 918, 2016.)

chelators (organic acids and siderophores) and contributing to oxidation–reduction reactions (Ma et al. 2016). These microorganisms associated with plants shorten the phytoremediation process in soils contaminated by metals by increasing the mobilization or immobilization of metals.

Mycorrhizal fungi can also act as a filtering obstacle against the translocation of PTMs from plant roots to sprouts. Experiments examining pine seedlings showed that inoculation with the ectomycorrhizal fungi *Scleroderma citrinum*, *Amanita muscaria* and *Lactarius rufus* reduced the translocation of zinc (Zn), cadmium (Cd) or lead (Pb) from the roots to sprouts compared to that observed in controls. This effect was attributed to the increased metal biosorption by the external and internal components of the mycelium (Krupa and Kozdrój 2007). The microbial activities in the rhizosphere increase the effectiveness of phytoremediation processes in soil contaminated by metals in two complementary ways. These include (i) direct promotion of phytoremediation, where microorganisms associated with the plant increase the translocation of metals (facilitate phytoextraction) or decrease the mobility/availability of metallic contaminants in the rhizosphere (phytostabilization) and (ii) the indirect promotion of phytoremediation, where microorganisms confer tolerance to metals in plants and/or increase their biomass production to stop pollutants or even remove them (Rajkumar et al. 2012).

Some plant growth-promoting bacteria (PGPB) that are tolerant to PTMs are known to promote plant growth in soils contaminated with metals. These bacteria decrease the bioavailability of toxic metals and their absorption by the plant, ultimately resulting in healthy food production (Ma et al. 2016). PGPB include different species of bacteria capable of increasing plant growth by increasing the content of nutrients such as nitrogen and phosphorus (Martínez and Dussán 2018), and they can also induce the production of phytohormones (Khan et al. 2012).

The microorganisms associated with plants improve the acquisition of nutrients by plants by mobilizing them and making them available to the plant roots. One example includes the phosphorus-solubilizing bacteria (*P*) that dissolve various sources of poorly soluble phosphorus (P) such as $Ca_3(PO_4)_2$ and $Zn_3(PO_4)_2$ due to the decrease in the pH of the rhizospheric soil to increase the availability of *P* for absorption by the plants.

The interaction between bacteria and plants to promote an improved pollutant removal capacity requires the interaction between the plant, the bacteria and the environment. Thus, it is not only bacterial resistance to a certain agent that makes bacteria useful in phytoremediation. Bacteria that are resistant to PTMs and that produce phytohormones possess an excellent capacity to be used in phytoremediation processes; however, the characteristics of plants are also important. Recently, studies using an isolate of *Cupriavidus* sp. have revealed that although this bacterium is resistant to chromium and cadmium by removing such metals *in vitro* and by efficiently producing indole-3-acetic acid (IAA), it was not possible to verify changes in the development of *Cajanus cajan* seeds using *in vitro* tests in the presence of these metals (Minari et al. 2020).

6.4.5 Genetically Modified Microorganisms

Numerous microorganisms present in the environment can degrade pollutants naturally at their production sites; however, their metabolic pathways are not as efficient for degrading these compounds on a large scale. Several factors, such as the presence of complex mixtures of pollutants, the long time required for degradation and the non-biocompatibility of microorganisms, restrict the capacity of the bioremediation process. Despite the presence of highly diversified and specialized microbial communities in the environment that can transform many pollutants, the natural remediation process can be quite slow, thus allowing the accumulation of materials in the environment.

With the advent of the latest molecular and biotechnological techniques, rhizospheric and endophytic bacterial strains can be genetically engineered to produce specific enzymes that promote the degradation of toxic organic substances (Pandotra et al. 2018). The specificity acquired by these modified microorganisms allows for a focused action that is important for the removal of toxic elements that are difficult to degrade. Environmental contamination by PTMs is one of the problems that can be solved more efficiently using this technology. The application of genetic engineering to remove PTMs has been of great interest.

The use of genetically modified bacteria to eliminate the disposal of PTMs exhibits the characteristics of strong adaptability and high treatment efficiency, and this technology has been one of the main research foci in recent years. When they exceed a certain concentration limit, metals such as copper (Cu), cobalt (Co), chromium (Cr), nickel (Ni), molybdenum (Mo), iron (Fe), zinc (Zn) and manganese (Mn) can produce various biochemical, physiological and genotoxic effects in all types of microorganisms. Certain microorganisms can develop resistance to PTMs through detoxification mechanisms.

The recombinant *E. coli* JM109 strain with the expression of transport protein (merT-merP) and metallothionein possesses pronounced potential for bioaccumulation of mercury and cell propagation rate in culture medium with a higher concentration of Hg^{2+} (up to 7,402 mg/L). To remove Hg^{2+} from wastewater, a modified BL21–7 strain was developed that contains the synthetic operon P16S-g10-merT-merP-merB1-merB2-ppk-rpsT to allow it to remove approximately 43.7% of Hg^{2+} from wastewater (Liang et al. 2015). The lack of progress in research to commercialize and use GMMs for bioremediation *in situ* is attributed to several factors, including challenges in obtaining approval from regulatory agencies, public acceptability and safety risks associated with these modified bacteria (Liu et al. 2019).

Although the use of GMMs for bioremediation can provide a highly effective and ecological approach, certain issues associated with the use of these organisms must be addressed. The most important issue is the unpredictable effects on various forms of life within the environment that are not direct targets of GMMs. GMMs contain viable genetic material, and there is a chance of genetic contamination caused by transferring this genetic material horizontally or vertically to similar or compatible species, and this could negatively affect the entire ecosystem.

There is the possibility that they can compete with natural species, thus allowing them to spread to new habitats and cause ecological and economic damage. Once GMMs are released into the environment, they may be impossible to dispose of and may possess the potential to cause harm to other species, including plants and humans. In this way, the ethical problems associated with the horizontal transfer of genes from genetically modified organisms to other organisms were created based on the possible threats to the integrity and intrinsic value of species and ecosystems.

Transgenic plants and associated bacteria constitute a new generation of genetically modified organisms for bioremediation. The engineering of a phytoremediator plant requires the optimization of several processes, including trace element mobilization in the soil, uptake into the root, detoxification and allocation within the plant. A small number of transgenic plants have been generated by introducing resistance genes to modify the tolerance, uptake or homeostasis of trace elements to improve the capacity of bioremediation (Ibañez et al. 2016). Transgenic approaches that have successfully been employed to promote the phytoextraction of metals (mainly Cd, Pb and Cu) and metalloids (As and Se) from the soil through their accumulation in the above-ground biomass have primarily involved implementation of metal transporters, improved production of enzymes of sulphur metabolism and the production of metal-detoxifying chelators (Kaur et al. 2019).

6.5 Bioremediation Techniques

As mentioned previously, bioremediation can be performed *in situ* or *ex situ*, but using the characteristics and properties of different organisms in nature and how the microorganism works. Some authors have used other classifications to describe the different types of remediation: bioaugmentation, biostimulation, biosorption, bioaccumulation, bioreduction and biomineralization.

6.5.1 Bioaugmentation

Bioaugmentation is one of the bioremediation techniques that are based on the inoculation of microorganisms in environments that are to be remedied, and these environments include water, soils, sediments and silt. These microorganisms have previously been identified and characterized, and they must possess the desired catalytic functions for the inactivation of the target contaminant (Cycoń et al. 2017). The bioaugmentation process is briefly illustrated in Figure 6.6, and this process is performed primarily in environments contaminated with oil (Goswami et al. 2018).

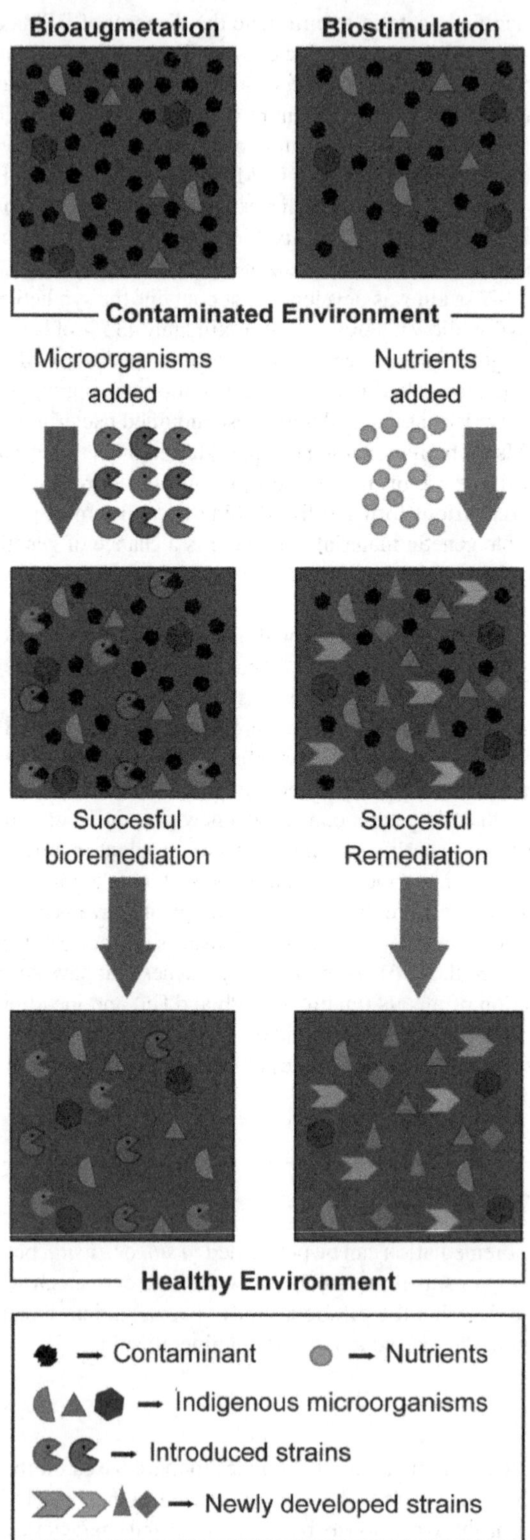

FIGURE 6.6 Diagram representing the bioaugmentation and biostimulation processes. (Adapted from Goswami, M., Chakraborty, P., Mukherjee, K., Mitra, G., Bhattacharyya, P., Dey, S., and Tribedi, P., *J. Microb. Exp.*, 6, 223–231, 2018.)

The use of bioaugmentation is primarily recommended for places where the number of native microorganisms that perform the degradation within their habitat is insufficient or the microorganisms already present do not possess the necessary metabolic pathways and enzymes to degrade a given contaminating agent. Thus, to improve efficiency, more microorganisms must be inoculated at the site. Al-Gheethi et al. (2016) demonstrated the potential of the use of bioaugmentation in the treatment of effluents from pharmaceutical industries. In this study, strains of *Bacillus subtilis* were added to contaminated water samples. These strains demonstrated a high degrading potential for faecal compounds, PTMs and antibiotics. Emenike et al. (2017) concluded in their research that bacteria present in contaminated soils possess the potential to remove metals; however, the addition of bacteria that were already characterized as capable of removing specific metals increases the efficiency of bioremediation. In this study, they demonstrated that *Bacillus* sp., *Lysinibacillus* sp. and *Rhodococcus* sp. are good candidates for the bioaugmentation of soils contaminated with metals such as lead, copper and aluminium.

6.5.2 Biostimulation

Biostimulation is one of the bioremediation techniques that can be performed *in situ* through the addition of one or more nutrients to accelerate the rate of biodegradation of contaminants (Figure 6.6). Not only adding nutrients to the environment is necessary for biostimulation, but it also requires optimizing conditions such as aeration, pH and temperature that favour bioremediation. The major advantage of this technique is the use of microorganisms already present at the target site of the process, as these organisms are already adapted to the environment and are well distributed on that surface.

One of the primary challenges encountered is that the addition of nutrients can promote not only the growth of desired microorganisms, but also the growth of others that are not hydrocarbon degraders, thus creating competition in the resident microbiota. A study developed by Abioye et al. (2012) demonstrated that several sources of organic nutrients can be used. These researchers performed biostimulation using organic sources such as oil, food scraps and banana peels to achieve greater growth of microorganisms, and they obtained a 40% greater degradation result than that achieved using only oil as an organic source.

Inorganic sources are also used in the biostimulation process. This is demonstrated in a study by Chorom et al. (2010), who observed the effectiveness of inorganic fertilizers (NPK – nitrogen, phosphorus and potassium) in increasing the microbial degradation of various petroleum derivatives within the soil. The concentrations of paraffinic and isoprenoid compounds decreased 40%–60%, demonstrating the effectiveness of inorganic fertilizers in the processes of biostimulation.

6.5.3 Bioaccumulation, Biomineralization, Bioreduction and Biosorption

We can find efficient solutions to the decontamination of environments by observing the processes that occur naturally within the environment (Figure 6.7). Various soluble chemicals are found in aqueous environments or even within the biomass. These compounds can interact with the microorganisms present by binding on the cell wall surface in a process called biosorption, or they accumulate inside living cells through a process called bioaccumulation. Furthermore, microorganisms can transform the compounds (primarily metals) through the processes of bioreduction and biotransformation.

Bioaccumulation (Figure 6.7a) is an active process in which molecules are absorbed by the cell, and only living biomass can perform this process. Biomineralization is a general term for the processes by which living organisms form minerals, and this can result in metal removal from a given solution, thus providing a means of detoxification and biorecovery (Figure 6.7b). The most common biominerals precipitated by microbes include oxides, phosphates, sulphides and oxalates, and these can exhibit special chemical properties such as high metal sorption capacities and redox catalysis (Gadd and Pan 2016).

The ability of microbes to reductively transform a variety of metals has wide-reaching implications for controlling the mobility of contaminants within the subsurface, as this can result in the degradation of toxic organics or the reductive immobilization of metals. These processes are termed bioreduction (Figure 6.7c) and have been used for the bioremediation of soluble toxic metal contaminants, including Cr^6, Hg^{2+}, V^{5+}, Co^{3+}, U^{6+}, Tc^{7+} and Np^{5+}. These elements are used as terminal electron acceptors during anoxic respiration and can be transformed by various enzymes (Watts and Lloyd 2013). Biosorption is a passive process in which metal ions are adsorbed on the surface of the cell wall (Figure 6.7d). Live or dead biomass can act as a biosorbent (Inoue et al. 2017).

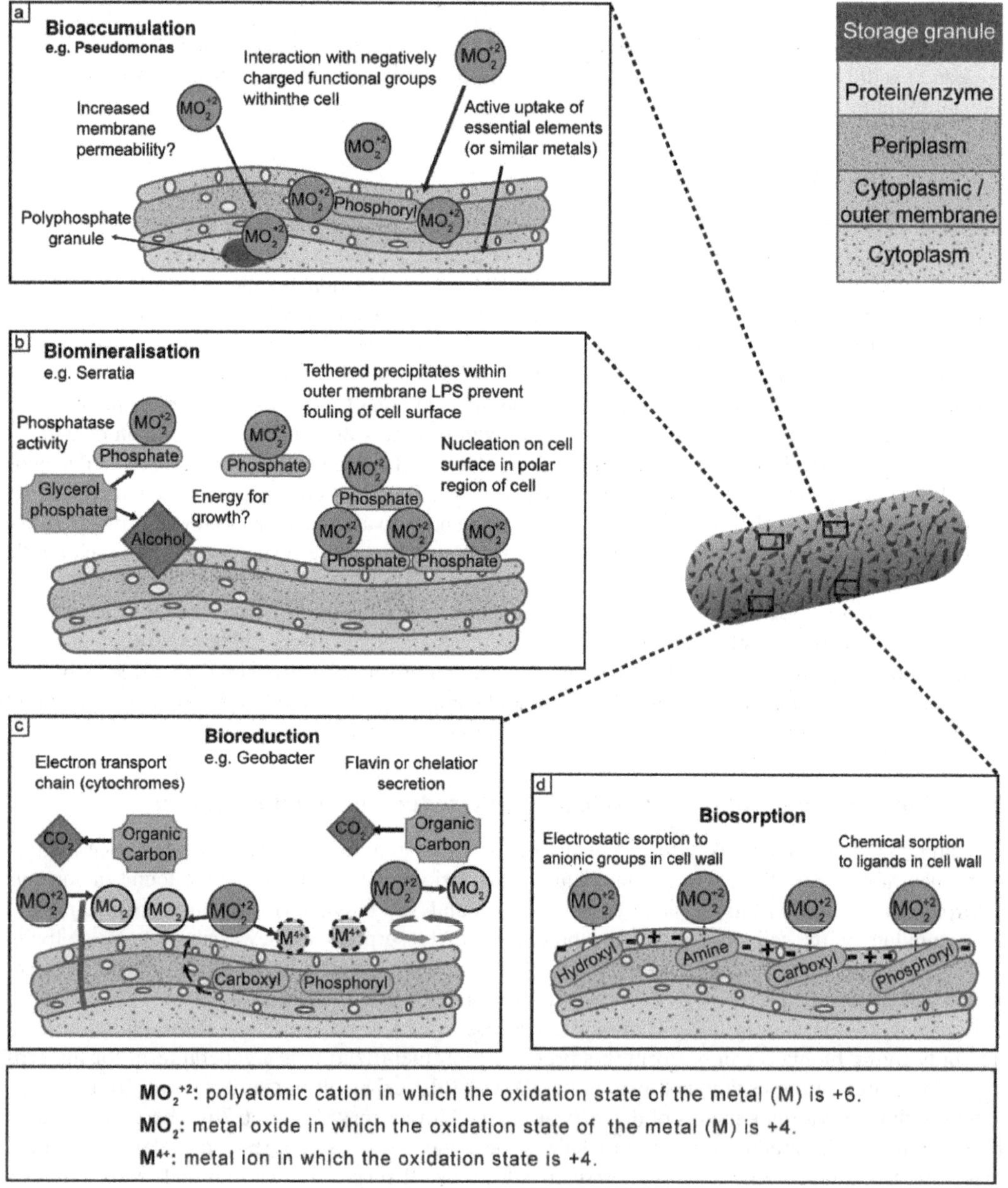

FIGURE 6.7 Illustration of the environmental decontamination processes that occur in nature: (a) bioaccumulation; (b) biomineralization; (c) bioreduction; and (d) biosorption. (Adapted from Newsome, L., Morris, K., and Lloyd, J.R., *Chem. Geol.*, 363, 164–184, 2014.)

6.6 Important Biomolecules in Bioremediation

6.6.1 Enzymes

Enzymes are proteins that act as specific biological catalysts for chemical reactions such as pollutant degradation reactions that are promoted by living beings (Sharma et al. 2018). Ghosh et al. (2017) pointed out that although the use of microorganisms in the degradation of contaminants is efficient, it is a slow process and thus limits the practical feasibility of bioremediation. Based on this, enzymes have been used to overcome the practical limitations related to the use of microbial cells in bioremediation processes (Thatoi et al. 2014). These enzymes can be obtained from certain organisms and are naturally found in nature; however, they can also originate from genetically modified organisms.

Enzymatic engineering has used molecular techniques to explore the possibility of improving the function of enzymes to allow for their high-capacity application in removing PTMs and pollutants. Enzyme engineering consists of altering, through the use of recombinant DNA techniques, the primary structure of amino acids by changing their positions or replacing one amino acid with another, and this can improve the properties of a given enzyme to provide it with greater activity yield and tolerance to wide pH and temperature range (Rayu et al. 2012). Therefore, when genomes and transcriptomes of microorganisms that can perform bioremediation activities are known, it is possible to modify the enzymes and alter them according to the need for degradation or environmental conditions, thus making the process much more efficient.

Most of the microflora on our planet remains unknown due to the limited growing conditions of these microorganisms. However, with the advent of metagenomic, metatranscriptomic, and metaproteomic methodologies, the knowledge limitations of these organisms can be overcome. Using '-omic' analyses and bioinformatics tools, researchers can identify and characterize new microorganisms found in nature and their enzymes without the need for laboratory cultivation. This set of techniques and methodologies increases the prospects for more effective use of enzymes by altering their capacity in bioremediation processes. Bioremediation using purified enzymes does not depend upon the growth of microorganisms at the site of contamination; instead, it depends upon the activity of enzymes secreted by them that possess greater substrate specificity and greater mobility due to their smaller size (Sharma et al. 2018). Table 6.2 describes the groups of enzymes that exhibit potential for the degradation of pollutants and their most used specimens in bioremediation according to the action of living organisms or by direct application to the contaminated material.

6.6.2 Biosurfactants

Surfactants are amphipathic compounds that possess hydrophilic and hydrophobic portions and are found between liquid interfaces such as the oil/water interface that possess different polarities. The apolar portion of surfactants is typically formed by hydrocarbon chains, while the polar region is ionic, non-ionic or amphoteric. These compounds increase the solubility of hydrophobic molecules and decrease the surface tension between the interfaces. Most of the produced surfactants are derived from oil, making them synthetic agents that are toxic and difficult to break down within the environment. For this reason, as a sustainable alternative, biosurfactants produced by microorganisms have been sought (Vijayakumar and Saravanan 2015).

Studies examining biosurfactants began in the 1960s, and their use has been expanding in recent decades due to their advantages that include structural diversity; low toxicity; greater biodegradability; performance over a wide pH, temperature and salinity range; greater selectivity; lower critical micellar concentration (CMC); and ability to be produced from renewable sources (Da Rosa et al. 2015). The CMC is the surfactant concentration at which molecular arrangements, called micelles, are formed, and this value corresponds to the point at which the surfactant reaches the lowest stable surface tension (Campos et al. 2013).

The hydrophobic portion of biosurfactants is composed of fatty acids, and the hydrophilic portion can be formed by carbohydrates, amino acids, carboxylic acids or alcohols. These substances possess

TABLE 6.2

Main Classes of Enzymes Used for Bioremediation

Class (EC)	Main Reaction	Uses
Oxidoreductases (EC 1)	Oxidoreduction reactions	Biological importance in metabolic pathways of organisms and protection against substances toxic to cells. Catalyse the reduction of PTM ions and the degradation of industrial dyes and phenolic compounds.
Oxygenases (EC 1.13) – monooxygenases and dioxygenases	Catalyse the aerobic degradation of aromatic compounds	Degrade the aromatic compounds and are involved in the processes of dehalogenation, desulphurization, denitrification, cleavage and hydroxylation of aromatic compounds. Important in the degradation of some pesticides, fatty acids and various types of aromatic compounds.
Laccases (EC 1.10.3.2) Part of the 'multicopper oxidases' family	Oxidize a range of phenolic and non-phenolic substrates, with the reduction of oxygen in water	Degrade ortho- and para-diphenols, phenols, lignin and diamines. One of the main applications of laccases is their use in the decolourization of industrial effluents that contain dyes.
Peroxidases (EC 1.11.1.7)	Catalyse the degradation reactions only in the presence of hydrogen peroxide and some mediators	Excellent degraders of aromatic contaminants, toxic compounds and lignin. Applications in wastewater treatment.
Hydrolases – lipases (E.C.3.1.1.3)	Participate in important reactions such as hydrolysis, interesterification, esterification, alcoholysis and aminolysis	Used in the cosmetics, food and pharmaceutical industries, and as bioremediators. Application in places contaminated by oil spills is highly effective.
Hydrolases – cellulases (EC 3.2.1.4)	Hydrolysis of glycosidic linkage as an enzymatic cocktail (endoglucanases, exoglucanases and β-glucosidases)	Degradation of cellulose. Cellulose is a biopolymer that can help us reduce the use of petroleum products, cellulosic fibers in the manufacture of 'green' compounds and paper.
Hydrolases – carboxylesterases (EC 3.1.1.1)	Catalyse the degradation of synthetic compounds (organophosphates, carbamates and chlorine-containing organic compounds)	They have widely been used in the bioremediation of pesticides, insecticides and fungicides.

Source: Based on Sharma, B., Dangi, A.K., and Shukla, P, *J. Environ. Manag.*, 210, 10–22, 2018.

molecular mass between 500 and 1,500 Da, and most are neutral or anionic. Unlike artificial surfactants that are classified according to the functional group polarity, biosurfactants are categorized according to their microorganism of origin and their chemical composition. The main surfactants produced by microorganisms are glycolipids, lipopeptides and phospholipids that belong to the group of low-mass biosurfactants. Additionally, particulates and polymeric compounds are produced and are characterized as high molecular mass surfactants.

Biosurfactants produced by bacteria are presented in the biofilm form, and this aids in their interaction with the interface where the microorganism is found by altering the surface properties, including the mobility of molecules. A study performed using the bacterium *Pseudomonas aeruginosa* that was isolated from seawater polluted with oil revealed that after incubating the bacteria for 28 days, the bacteria could degrade hexadecane, octadecane and heptadecane in addition to other types of hydrocarbons such as 2-methylnaphthalene due to the production of biosurfactant (Karlapudi et al. 2018).

As different technologies can be used either separately or in combination for the treatment of organic pollutants or PTMs, biosurfactants can also be applied, as they can influence the solubility of the contaminants to lead to an increase in the bioavailability contact area between hydrophobic contaminants and the microorganisms (Lászlová et al. 2018). Biosurfactants are surface-active molecules that are produced by a variety of microorganisms that are capable of reducing surface and interfacial tension, thus facilitating mobilization and removal of metal ions by forming biosurfactant–metal complexes (Lal et al. 2018). These molecules can be important for soil remediation by plants, and the integration of several molecules and organisms can be observed in this process (Figure 6.8).

The removal of PTM ions can be performed by ionic biosurfactants that can be either cationic or anionic and are produced by various microorganisms with relatively low production costs and high removal efficiency (Da Rocha et al. 2019). In general, the removal process occurs according to a series of steps that include (i) sorption of the biosurfactant on the soil surface and complexation with the metal ion, (ii) release of the metal ion into the solution and (iii) association with micelles. In this way, PTM ions are 'stuck' to the micelles by electrostatic interactions, and they can be easily recovered by precipitation or by membrane separation.

The biosurfactant industry has undergone remarkable growth in recent decades; however, the large-scale production of these biomolecules remains challenging from an economic point of view due to small financial investment and a lack of viable industrial production. However, the use of biosurfactants in the context of bioremediation is a viable solution, as these substances are not required to be pure and can be synthesized from mixtures of cheap carbon sources, thus allowing for the creation of an economically and environmentally viable technology for bioremediation.

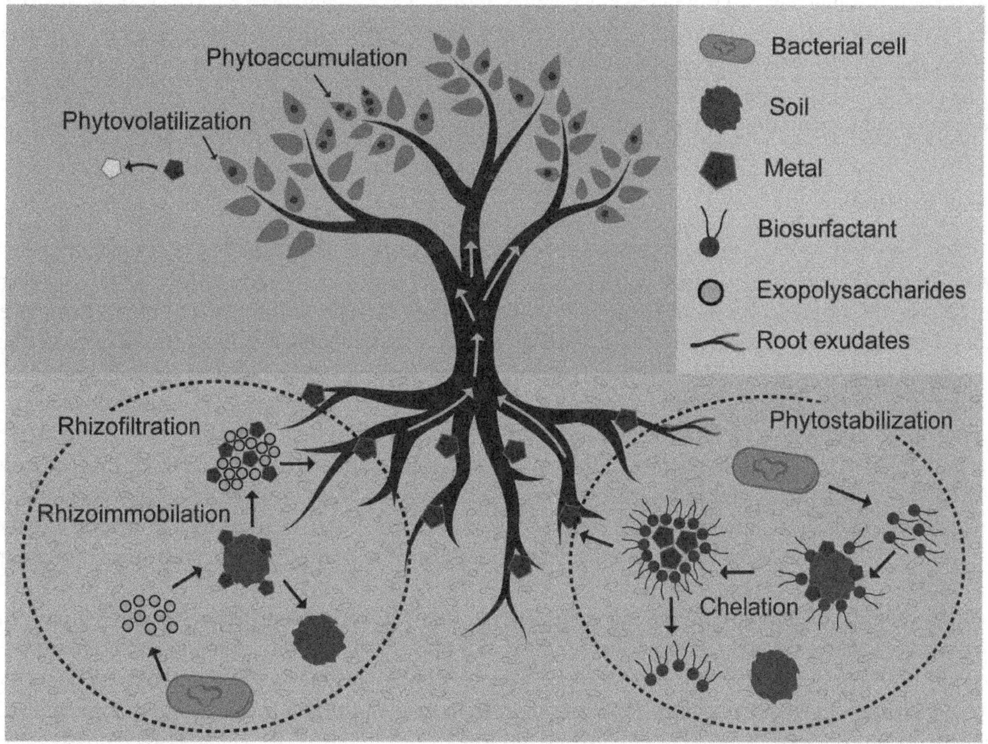

FIGURE 6.8 Microbial surfactants and exopolysaccharide-assisted mechanism of 'heavy metal' remediation by plants. (Adapted from Lal, S., Ratna, S., Said, O.B., and Kumar, R, *Environ. Technol. Innov.*, 10, 243–263, 2018.)

6.6.3 Extracellular Polysaccharides (EPS)

Extracellular polysaccharides (EPS) are organic compounds that possess high molecular mass and are produced from an environment under strong stress and in response to different nutritional conditions. They are the structural and functional components of microbial biofilms that play a vital role in the formation of bio-aggregates and the survival of bacterial cells. EPS are the primary macromolecular components in microbial collections and are comprised of polysaccharides, proteins, nucleic acids, lipids, uronic acids and organic and inorganic compounds. These glycopolymeric biomolecules are soluble in water and possess exemplary physiological, rheological and physicochemical properties that make them suitable for various clinical, industrial and environmental applications.

EPS are extensively studied as bacterial by-products, and they can be used for several applications. Despite their applications in the food, pharmaceutical, nutraceutical, herbicide and cosmeceutical industries, they are also well known for their efficiency in the bioremediation of water and soil contaminated with PTMs (Mohite et al. 2017; Sardar et al. 2018). Additionally, EPS play important roles in biological systems, such as in receptor–ligand interaction, cell signalling and biofilm formation. For decades, extensive efforts have been continuously dedicated to the isolation, identification, characterization and functionalization of new and valuable EPS from different microorganisms for different purposes and applications.

The effects of EPS on soil contamination caused by mining activities were tested using *Azotobacter* bd39, which is a microorganism that is well known to produce phytohormone and non-symbiotic nitrogen fixation. *Azotobacter* not only is responsible for the growth of the plant, but is also associated with mercury removal. The microorganism can survive in an environment contaminated with mercury, and it can remove the mercury due to the formation of EPS (Hindersah et al. 2017). The effect of EPS on soil contamination due to agricultural activities is highlighted by the ability of *Achyranthes aspera* strains to form biofilms that can be used in the purification of soil contaminated with Ni^{2+} (Karthik et al. 2016). The *Pseudomonas* Psd strain is not just a plant growth-promoting strain. The EPS formed by this strain is also capable of zinc biosorption. The main component for the formation of EPS by this strain was the *algE8* gene, a subunit of polymerase. A knockout strategy revealed that in the absence of the *algE8* gene, there was a subsequent drop in zinc adsorption (Upadhyay et al. 2017).

6.6.4 Biofilms

Biofilms are clusters of microbial cells that are attached to a surface. They typically occur in humid environments where there is a sufficient nutrient flow. They can be formed by one or more species of organisms consisting of algae, bacteria, fungi and protozoa. Biofilms represent stable and highly structured ecosystems, and they are generally surrounded by a matrix that adheres to a living or inactive surface (Farasin et al. 2017).

The metal ion removal mediated by biofilms is crucial, as a high metal ion concentration is harmful to planktonic organisms. Planktonic communities often experience problems when they come in contact with PTM ions. Biofilms increase the tolerance of microorganisms to environments contaminated by PTM ions and by acids due to improved adaptation and survival, as they are better protected within a matrix. In addition to improving the survival of organisms, biofilms improve the rate and extent of the transformation of contaminants when compared to pure and planktonic cultures (Mangwani et al. 2014).

The proximity of the biofilm and the physiological interactions among the organisms increase the tolerance and the metal removal. The surrounding EPS protects the biofilm microbial cells from metal stress by preventing the diffusion of metals within the biofilm. In this manner, biofilms play an important role in the fixation and colonization of microorganisms in response to different minerals. The negatively charged carboxyl, phosphoryl and sulphhydryl groups in the biofilm EPS, such as that produced by *Escherichia coli* PHL628, have been suggested to be responsible for copper sorption (Hu et al. 2007). This study revealed that the absorption of Cu by biofilm is sixfold greater than that of planktonic cells of *E. coli* PHL628.

The development of biofilm-based remediation technology remains a challenge. The fixing material of a biofilm must be biodegradable, insoluble, non-toxic to cells and the environment, easy to access, inexpensive, able to be obtained in large quantities, stable and suitable for regeneration (Catania et al. 2020). In this sense, the use of biodegradable polymers derived from natural sources (such as lignin) has caused great interest in this area (Huang et al. 2019; Xu et al. 2019).

6.7 Final Considerations

As noted in this review, there are many mechanisms by which various living organisms can be used in the remediation of polluted areas, particularly soil and water. However, bioremediation has not yet been fully explored. It can be efficient and be performed at low costs; however, in many situations, several parameters require further study. Specifically, regarding pollution by PTMs, the best approach is to reduce the use and to ensure non-contamination of the environment by treating and correctly disposing of contaminating waste. When waste is obtained in large volumes such as that produced during mining, by reducing the waste or treating it, the possibility of major accidents such as those occurred in Brumadinho and other places throughout the world can be reduced.

The increased frequency of contamination of the soil and water by PTMs is a matter of great concern, particularly regarding human health and environmental preservation. Conventional methods of treatment are time-consuming and inefficient. In contrast, bioremediation is a sustainable alternative of great efficiency that can be performed by various microorganisms and biopolymers. Given the range of advantages provided by the use of bioremediation, studies should be performed to identify new microorganisms that possess high potential for decontamination or to identify molecules of microbial origin that can aid in the process of transforming pollutants by making this process more rapid and more sustainable. It must be noted that sustainable development requires that we avoid negative environmental impacts instead of simply using more ecological processes of remediation, as is the case for bioremediation processes used for the recovery of areas that are already impacted. In this sense, we must reduce the rampant purchase of consumer goods to minimize the generation of waste, as many countries, due to political, geographical, economic and social issues, do not have adequate treatment and final disposal methods.

In addition to minimizing waste generation, we must also be concerned with the life cycle of the products we consume. Many products such as certain electronic devices (cell phones, for example) and/or their components are sources of potentially toxic elements and are 'quickly discarded', thus increasing the levels of pollutants that are released into the environment. Therefore, in parallel with the development of remediation processes that are economically and environmentally sustainable, it is urgent to change the consumption paradigms that are commonly present in several 'modern societies'.

Acknowledgements

The authors thank the Coordination of Superior Level Staff Improvement (CAPES) and National Council for Scientific and Technological Development (CNPq proc 302085/2017-3) for scholarship funding and the Agricultural and Livestock Graduation Program in School of Agricultural and Veterinarian Sciences (UNESP)

REFERENCES

Abatenh, E., Gizaw, B., Tsegaye, Z., and Wassie, M. 2017. The role of microorganisms in bioremediation: a review. *Open Journal of Environmental Biology* 2(1): 038–046. https://doi.org/10.17352/ojeb.000007.

Abioye, O.P., Agamuthu, P., and Abdul Aziz, A.R. 2012. Biodegradation of used motor oil in soil using organic waste amendments. *Biotechnology Research International* 2012: 587041. https://doi.org/10.1155/2012/587041.

Ahmad, F., Ashraf, N., Da-Chuan, Y., Jabeen, H., Anwar, S., Wahla, A.Q., and Iqbal, S. 2019. Application of a novel bacterial consortium BDAM for bioremediation of bispyribac sodium in wheat vegetated soil. *Journal of Hazardous Material* 374: 58–65. https://doi.org/10.1016/j.jhazmat.2019.03.130.

Akhtar, S., Mahmood-ul-Hassan, M., Ahmad, R., Suthor, V., and Yasin, M. 2013. Metal tolerance potential of filamentous fungi isolated from soils irrigated with untreated municipal effluent. *Soil Environment* 32(1): 55–62. https://www.researchgate.net/publication/284625238_Metal_tolerance_potential_of_filamentous_fungi_isolated_from_soils_irrigated_with_untreated_municipal_effluent.

Al-Gheethi, A.A., Mohamed, R.M.S.R., Efaq, A.N., Norli, I., Abd Halid, A., Amir, H.K., and Ab Kadir, M.O. 2016. Bioaugmentation process of secondary effluents for reduction of pathogens, heavy metals and antibiotics. *Journal of Water Health* 14(5): 780–795. https://doi.org/10.2166/wh.2016.046.

Angelucci, D.M., and Tomei, M.C. 2016. Ex situ bioremediation of chlorophenol contaminated soil: comparison of slurry and solid-phase bioreactors with the two-step polymer extraction-bioregeneration process. *Journal of Chemical Technology Biotechnology* 91: 1577–1584. https://doi.org/10.1002/jctb.4882.

Awa, S.H., and Hadibarata, T. 2020. Removal of heavy metals in contaminated soil by phytoremediation mechanism: a review. *Water Air Soil Pollution* 231: 47. https://doi.org/10.1007/s11270-020-4426-0.

Azubuike, C.C., Chikere, C.B., and Okpokwasili, G.C. 2016. Bioremediation techniques–classification based on site of application: principles, advantages, limitations and prospects. *World Journal of Microbiology and Biotechnology* 32: 180. https://doi.org/10.1007/s11274-016-2137-x.

Bian, F., Zhong, Z., Zhang, X., Yang, C., and Gai, X. 2020. Bamboo—an untapped plant resource for the phytoremediation of heavy metal contaminated soils. *Chemosphere* 246: 125750. https://doi.org/10.1016/j.Chemosphere.2019.125750.

Biswas, K., Paul, D., and Sinha, S.N. 2015. Biological agents of bioremediation: a concise review. *FEM* 1(3): 39–43. https://doi.org/10.11648/j.fem.20150103.11.

Bolan, N., Kunhikrishnan, A., Thangarajan, R., Kumpiene, J., Park, J., Makinof, T., Kirkhamg, M.B., and Scheckel, K. 2014. Remediation of heavy metal (loid)s contaminated soils–to mobilize or to immobilize? *Journal of Hazardous Materials* 266: 141–166. https://doi.org/10.1016/j.jhazmat.2013.12.018.

Campos, J.M., Stamford, T.L., Sarubbo, L.A., de Luna, J.M., Rufino, R.D., and Banat, I.M. 2013. Microbial biosurfactants as additives for food industries. *Biotechnology Progress* 29(5): 1097–1108. https://doi.org/10.1002/btpr.1796.

Catania, V., Lopresti, F., Cappello, S., Scaffaro, R., and Quatrini, P. 2020. Innovative, ecofriendly biosorbent-biodegrading biofilms for bioremediation of oil-contaminated water. *New Biotechnology* 58: 25–31. https://doi.org/10.1016/j.nbt.2020.04.001.

Chorom, M., Sharifi, H., and Motamedi, H. 2010. Bioremediation of a crude oil-polluted soil by application of fertilizers. *Iranian Journal of Environmental Health Science and Engineering* 7(4): 319–326. ISSN: 1735-1979. http://www.bioline.org.br/request?se10037.

Cycoń, M., Mrozik, A., and Piotrowska-Seget, Z. 2017. Bioaugmentation as a strategy for the remediation of pesticide-polluted soil: a review. *Chemosphere* 172: 52–71. https://doi.org/10.1016/j.Chemosphere.2016.12.129.

Da Rocha Junior, R.B., Meira, H.M., Almeida, D.G., Rufino, R.D., Luna, J.M., Santos, V.A., and Sarubbo, L.A. 2019. Application of a low-cost biosurfactant in heavy metal remediation processes. *Biodegradation* 30: 215–233. https://doi.org/10.1007/s10532-018-9833-1.

Da Rosa, C.F.C., Freire, D.M.G., and Ferraz, H.C. 2015. Biosurfactant microfoam: application in the removal of pollutants from soil. *Journal of Environment and Chemical Engineering* 3(1): 89–94. https://doi.org/10.1016/j.jece.2014.12.008.

Dangi, A.K., Sharma, B., Hill, R.T., and Shukla, P. 2019. Bioremediation through microbes: systems biology and metabolic engineering approach. *Critical Reviews of Biotechnology* 39: 79–98. https://doi.org/10.1080/07388551.2018.1500997.

DNPM. 2017. Anuário mineral brasileiro 2017. Relatório. Departamento Nacional de Produção Mineral, Ministério de Minas e Energia, Brasília. https://www.gov.br/anm/pt-br/centrais-de-conteudo/publicacoes/serie-estatisticas-e-economia-mineral/anuario-mineral/anuario-mineral-brasileiro.

Ekta, P., and Modi, N.R. 2018. A review of phytoremediation. *Journal of Pharmacology and Phytochemistry* 7: 1485–1489. http://www.phytojournal.com/archives/2018/vol7issue4/PartY/7-3-558-344.pdf.

Emenike, C.U., Liew, W., Fahmi, M.G., Jalil, K.N., Pariathamby, A., and Hamid, F.S. 2017. Optimal removal of heavy metals from leachate contaminated soil using bioaugmentation process. *Clean Soil Air Water* 45(2): 1500802. https://doi.org/10.1002/clen.201500802.

Farasin, J., Koechler, S., Varet, H., Deschamps, J., Dillies, M.A., Proux, C., Erhardt, M., Huber, A., Jagla, B., Briandet, R., Coppee, J.-Y., and Arsène-Ploetze, F. 2017. Comparison of biofilm formation and motility processes in arsenic-resistant *Thiomonas* spp. strains revealed divergent response to arsenite. *Microbial Biotechnology* 10(4): 789–803. https://doi.org/10.1111/1751-7915.12556.

Gadd, G.M., and Pan, X. 2016. Biomineralization, bioremediation and biorecovery of toxic metals and radionuclides. *Geomicrobiology Journal* 33: 175–178. https://doi.org/10.1080/01490451.2015.1087603.

Gajić, G., Djurdjević, L., Kostić, O., Jarić, S., Mitrović, M., and Pavlović, P. 2018. Ecological potential of plants for phytoremediation and ecorestoration of fly ash deposits and mine wastes. *Frontiers in Environmental Science* 6: 124. https://doi.org/10.3389/fenvs.2018.00124.

Ghosh, A., Dastidar, M.G., and Sreekrishnan, T.R. 2017. Bioremediation of chromium complex dyes and treatment of sludge generated during the process. *International Journal of Biodeterioration & Biodegradation* 119: 448–460. https://doi.org/10.1016/j.ibiod.2016.08.013.

Goswami, M., Chakraborty, P., Mukherjee, K., Mitra, G., Bhattacharyya, P., Dey, S., and Tribedi, P. 2018. Bioaugmentation and biostimulation: a potential strategy for environmental remediation. *Journal of Microbiol Experiments* 6(5): 223–231. https://doi.org/10.15406/jmen.2018.06.00219.

Gupta, S., Wali, A., Gupta, M., and Annepu, S.K. 2017. Fungi: an effective tool for bioremediation, In: *Plant-Microbe Interactions in Agro-ecological Perspectives*. Springer, Singapore, pp. 593–606. https://doi.org/10.1007/978-981-10-6593-4_24.

Hindersah, R., Mulyani, O., and Osok, R. 2017. Proliferation and exopolysaccharide production of *Azotobacter* in the presence of mercury. *Biodiversity Journal* 8(1): 21–26.

Hoseini, P.S., Poursafa, P., Moattar, F., Amin, M.M., and Rezae, A.H. 2012. Ability of phytoremediation for absorption of strontium and cesium from soils using *Cannabis sativa*. *International Journal of Environmental Health and Engineering* 1(2): 17. https://doi.org/10.4103/2277-9183.96004.

Hu, Z., Jin, J., Abruña, H.D., Houston, P.L., Hay, A.G., and Ghiorse, W.C. 2007. Spatial distributions of copper in microbial biofilms by scanning electrochemical microscopy. *Environmental Science and Technology* 41(3): 936–941. https://doi.org/10.1021/es061293k.

Huang, C., Shi, X., Wang, C., Guo, L., Dong, M., and Hu, G. 2019. Boosted selectivity and enhanced capacity of As(V) removal from polluted water by triethylenetetramine activated lignin-based adsorbents. *International Journal of Biology and Macromolecules* 140: 1167–1174. https://doi.org/10.1016/j.ijbiomac.2019.08.230.

Ibañez, S., Talano, M., Ontañon, O., Suman, J., Medina, M.I., Macek, T., and Agostini, E. 2016. Transgenic plants and hairy roots: exploiting the potential of plant species to remediate contaminants. *New Biotechnology* 33(5): 625–635. https://doi.org/10.1016/j.nbt.2015.11.008.

Igiri, B.E., Okoduwa, S.I.R., Idoko, G.O., Akabuogu, E.P., Adeyi, A.O., and Ejiogu, I.K. 2018. Toxicity and bioremediation of heavy metals contaminated ecosystem from tannery wastewater: a review. *Journal of Toxicology* 2018: 2568038. https://doi.org/10.1155/2018/2568038.

Inoue, K., Parajuli, D., Ghimire, K.N., Biswas, B.K., Kawakita, H., Oshima, T., and Ohto, K. 2017. Biosorbents for removing hazardous metals and metalloids. *Materials* 10(8): 1–33. https://doi.org/10.3390/ma10080857.

Karlapudi, A.P., Venkateswarulu, T.C., Tammineedi, J., Kanumuri, L., Ravuru, B.K., Dirisala, V.R., and Kodali, V.P. 2018. Role of biosurfactants in bioremediation of oil pollution: a review. *Petroleum* 4(3): 241–249. https://doi.org/10.1016/j.petlm.2018.03.007.

Karthik, C., Oves, M., Thangabalu, R., Sharma, R., Santhosh, S.B., and Indra Arulselvi, P.I. 2016. *Cellulosimicrobium funkei*-like enhances the growth of Phaseolus vulgaris by modulating oxidative damage under chromium(VI) toxicity. *Journal of Advanced Research* 7(6): 839–850. https://doi.org/10.1016/j.jare.2016.08.007

Kaur, R., Yadav, P., Kohli, S.K., Kumar, V., Bakshi, P., Mir, B.A., Thukral, A.K, and Bhardwaj, R. 2019. Chapter 4. Emerging trends and tools, In: *Transgenic Plant Technology for Remediation of Toxic Metals and Metalloids, Transgenic Plant Technology for Phytoremediation of Toxic Metals and Metalloids*. Academic Press, Cambridge, ISBN: 9780128143896, pp. 63–88. https://doi.org/10.1016/B978-0-12-814389-6.00004-3.

Kaza, S., Yao, L.C., Bhada-Tata, P., and Van Woerden, F. 2018. What a waste 2.0: a global snapshot of solid waste management to 2050. *Urban Development*. World Bank, Washington, DC. https://openknowledge.worldbank.org/handle/10986/30317.

Khalid, S., Shahid, M., Niazi, N.K., Murtaza, B., Bibi, I., and Dumat, C. 2017. A comparison of technologies for remediation of heavy metal contaminated soils. *Journal of Geochemical Explor (Part B)* 187: 182247–182268. https://doi.org/10.1016/j.gexplo.2016.11.021.

Khan, N., Mishra, A., and Nautiyal, C.S. 2012. *Paenibacillus lentimorbus* B-30488r controls early blight disease in tomato by inducing host resistance associated gene expression and inhibiting. *Alternaria solani. Biological Control* 62(2): 65–74. https://doi.org/10.1016/j.biocontrol.2012.03.010.

Krupa, P., and Kozdrój, J. 2007. Ectomycorrhizal fungi and associated bacteria provide protection against heavy metals in inoculated pine (*Pinus silvestris* L.) seedlings. *Water Air Soil Pollution* 182: 83–90. https://doi.org/10.1007/s11270-006-9323-7.

Kushwaha, A., Hans, N., Kumar, S., and Rani, R. 2018. A critical review on speciation, mobilization and toxicity of lead in soil-microbe-plant system and bioremediation strategies. *Ecotoxicology and Environmental Safety* 147: 1035–1045. https://doi.org/10.1016/j.ecoenv.2017.09.049

Lal, S., Ratna, S., Said, O.B., and Kumar, R. 2018. Biosurfactant and exopolysaccharide-assisted rhizobacterial technique for the remediation of heavy metal contaminated soil: an advancement in metal phytoremediation technology. *Environmental Technology and Innovations* 10: 243–263. https://doi.org/10.1016/j.eti.2018.02.011.

Lászlová, K., Dudášová, H., Olejníková, P., Horváthová, G., Velická, Z., Horváthová, H., and Dercová, K. 2018. The application of biosurfactants in bioremediation of the aged sediment contaminated with polychlorinated biphenyls. *Water Air Soil Pollution* 229: 219. https://doi.org/10.1007/s11270-018-3872-4.

Leong, Y.K., and Chang, J.S. 2020. Bioremediation of heavy metals using microalgae: recent advances and mechanisms. *Bioresource Technology* 303: 122886. https://doi.org/10.1016/j.biortech.2020.122886.

Liang, X.L., Zhao, F., Shi, R.J., Ban, Y.H., Zhou, J.D., Han, S.Q., and Zhang, Y. 2015. Construction and evaluation of an engineered bacterial strain for producing lipopeptide under anoxic conditions. *Ying Yong Sheng Tai Xue Bao* 26(8): 2553–2560. https://www.ncbi.nlm.nih.gov/pubmed/26685621.

Liu, L., Bilal, M., Duan, X., and Iqbal, H.M.N. 2019. Mitigation of environmental pollution by genetically engineered bacteria — current challenges and future perspectives. *Science of the Total Environment* 667: 444–454. https://doi.org/10.1016/j.scitotenv.2019.02.390.

Ma, Y., Oliveira, R.S., Freitas, H., and Zhang, C. 2016. Biochemical and molecular mechanisms of plant-microbe-metal interactions: relevance for phytoremediation. *Frontiers in Plant Science* 7: 918. https://doi.org/10.3389/fpls.2016.00918.

Ma, Y., Prasad, M.N.V., Rajkumar, M., and Freitas, H. 2011. Plant growth promoting rhizobacteria and endophytes accelerate phytoremediation of metalliferous soils. *Biotechnology Advances* 29(2): 248–258. https://doi.org/10.1016/j.biotechadv.2010.12.001.

Mangwani, N., Shukla, S.K., Kumari, S., Rao, T.S., and Das, S. 2014. Characterization of *Stenotrophomonas acidaminiphila* NCW-702 biofilm for implication in the degradation of polycyclic aromatic hydrocarbons. *Journal of Applied Microbiology* 117(4): 1012–1024. https://doi.org/10.1111/jam.12602.

Martínez, S.A., and Dussán, J. 2018. *Lysinibacillus sphaericus* plant growth promoter bacteria and lead phytoremediation enhancer with *Canavalia ensiformis*. *Environmental Progress and Sustainainable Energy* 37(1): 276–282. https://doi.org/10.1002/ep.12668.

Milhome, M.A.L., Holanda, J.W.B., De Araújo Neto, J.R., and Nascimento, R.F. 2018. Diagnóstico da Contaminação do Solo por Metais Tóxicos Provenientes de Resíduos Sólidos Urbanos e a Influência da Matéria Orgânica. *Rev Virtual Quím* 10(1): 59–72. https://doi.org/10.21577/1984-6835.201.80007.

Minari, G.D., Saran, L.M., Lima Constancio, M.T., Correia da Silva, R., Rosalen, D.L., Melo, W.J., and Carareto Alves, L.M. 2020. Bioremediation potential of new cadmium, chromium, and nickel-resistant bacteria isolated from tropical agricultural soil. *Ecotoxicology and Environmental Safety* 204: 111038. https://doi.org/10.1016/j.ecoenv.2020.111038.

Mohite, B.V., Koli, S.H., Narkhede, C.P., Patil, S.N., and Patil, S.V. 2017. Prospective of microbial exopolysaccharide for heavy metal exclusion. *Applied Biochemistry and Biotechnology* 183: 582–600. https://doi.org/10.1007/s12010-017-2591-4.

NASA 2019. Earth observatory images by Lauren Dauphin, using Landsat data from the U.S. Geological Survey. Caption by Adam Voiland. https://earthobservatory.nasa.gov/images/144501/another-deadly-dam-collapse-in-brazil.

Newsome, L., Morris, K., and Lloyd, J.R. 2014. The biogeochemistry and bioremediation of uranium and other priority radionuclides. *Chemical Geology* 363: 164–184. https://doi.org/10.1016/j.chemgeo.2013.10.034.

Niño-Martínez, N., Salas Orozco, M.F., Martínez-Castañón, G.A., Torres Méndez, F., and Ruiz, F. 2019. Molecular mechanisms of bacterial resistance to metal and metal oxide nanoparticles. *International Journal of Molecular Science* 20(11): 1–15. https://doi.org/10.3390/ijms20112808.

Pandotra, P., Raina, M., Salgotra, R.K., Ali, S., Mir, Z.A., Bhat, J.A., Tyagi, A., and Upadhahy, D. 2018. Plant-bacterial partnership: a major pollutants remediation approach, In *Modern Age Environmental Problems and their Remediation*. Springer, Cham, Germany, pp. 169–200. https://doi.org/10.1007/978-3-319-64501-8_10.

Piccini, M., Raikova, S., Allen, M.J., and Chuck, C.J. 2019. A synergistic use of microalgae and macroalgae for heavy metal bioremediation and bioenergy production through hydrothermal liquefaction. *Sustainable Energy Fuels* 3: 292–301 https://doi.org/10.1039/C8SE00408K.

Queiroz, H.M., Nóbrega, G.N., Ferreira, T.O., Almeida, L.S., Romero, T.B., Santaella, S.T., Bernardino, A.F., and Otero, X.L. 2018. The Samarco mine tailing disaster: a possible time-bomb for heavy metals contamination? *Science of the Total Environment* 637–638: 498–506. https://doi.org/10.1016/j.scitotenv.2018.04.370.

Quintella, C.M., Mata, A.M.T., and Lima, L.C.P. 2019. Overview of bioremediation with technology assessment and emphasis on fungal bioremediation of oil contaminated soils. *Journal of Environmental Management* 241: 156–166. https://doi.org/10.1016/j.jenvman.2019.04.019.

Rajkumar, M., Sandhya, S., Prasad, M.N.V., and Freitas, H. 2012. Perspectives of plant-associated microbes in heavy metal phytoremediation. *Biotechnology Advances* 30(6): 1562–1574. https://doi.org/10.1016/j.biotechadv.2012.04.011.

Rayu, S., Karpouzas, D.G., and Singh, B.K. 2012. Emerging technologies in bioremediation: constraints and opportunities. *Biodegradation* 23: 917–926. https://doi.org/10.1007/s10532-012-9576-3.

Rudakiya, D.M., Tripathi, A., Gupte, S., and Gupte, A. 2019. Fungal bioremediation: a step towards cleaner environment, In *Advancing Frontiers in Mycology & Mycotechnology*, Springer, Singapore, pp. 229–249. https://link.springer.com/chapter/10.1007%2F978-981-13-9349-5_9.

Sardar, U.R., Bhargavi, E., Devi, I., Bhunia, B., and Tiwari, O.N. 2018. Advances in exopolysaccharides-based bioremediation of heavy metals in soil and water: a critical review. *Carbohydrate Polymer* 199: 353–364. https://doi.org/10.1016/j.carbpol.2018.07.037.

Sarwar, N., Imran, M., Shaheen, M.R., Ishaque, W., Kamran, M.A., and Matloob, A. 2017. Phytoremediation strategies for soils contaminated with heavy metals: modifications and future perspectives. *Chemosphere* 171: 710–721 https://doi.org/10.1016/j.Chemosphere.2016.12.116.

Sharma, B., Dangi, A.K., and Shukla, P. 2018. Contemporary enzyme-based technologies for bioremediation: a review. *Journal of Environmental Management* 210: 10–22. https://doi.org/10.1016/j.jenvman.2017.12.075.

Thatoi, H., Das, S., Mishra, J., Rath, B.P., and Das, N. 2014. Bacterial chromate reductase, a potential enzyme for bioremediation of hexavalent chromium: a review. *Journal of Environmental Management* 146: 383–399. https://doi.org/10.1016/j.jenvman.2014.07.014.

Upadhyay, A., Kochar, M., Rajam, M.V., and Srivastava, S. 2017. Players over the surface: unraveling the role of exopolysaccharides in zinc biosorption by fluorescent *Pseudomonas strain* Psd. *Frontiers in Microbiology* 8: 284. https://doi.org/10.3389/fmicb.2017.00284.

Vijayakumar, S., and Saravanan, V. 2015. Biosurfactants-types, sources and applications. *Research Journal of Microbiology* 10(5): 181–192. https://doi.org/10.3923/jm.2015.181.192.

Watts, M.P., and Lloyd, J.R. 2013. Bioremediation via microbial metal reduction, In *Microbial Metal Respiration*. Springer, Berlin, Heidelberg, pp. 161–201. https://doi.org/10.1007/978-3-642-32867-1_7.

Wilhelm, R.C., Singh, R., Eltis, L.D., and Mohn, W.W. 2019. Bacterial contributions to delignification and lignocellulose degradation in forest soils with metagenomic and quantitative stable isotope probing. *The ISME Journal* 13: 413–429. https://doi.org/10.1038/s41396-018-0279-6.

Xu, G., Shi, Z., Zhao, Y., Deng, J., Dong, M., Liu, C., Murugadoss, V., Mai, X., and Guo, Z. 2019. Structural characterization of lignin and its carbohydrate complexes isolated from bamboo (*Dendrocalamus sinicus*). *International Journal of Biology and Macromolecules* 126: 376–384. https://doi.org/10.1016/j.ijbiomac.2018.12.234.

Zapana-Huarache, S.V., Romero-Sánchez, C.K., Dueñas Gonza, A.P., Torres-Huaco, F.D., and Lazarte Rivera, A.M. 2020. Chromium(VI) bioremediation potential of filamentous fungi isolated from Peruvian tannery industry effluents. *Brazillian Journal of Microbiology* 51: 271–278. https://doi.org/10.1007/s42770-019-00209-9.

7

Bacterial Biofilm Formation for the Remediation of Environmental Pollutants

Muhsin Jamal
Abdul Wali Khan University

Sayed Muhammad Ata Ullah Shah Bukhari
Quaid-i-Azam University

Liloma Shah, Sana Raza, and Redaina
Abdul Wali Khan University

Muhammad Asif Nawaz
Shaheed Benazir Bhutto University

Sidra Pervez
Shaheed Benazir Bhutto Women University

CONTENTS

DOI: 10.1201/9781003181224-7

7.1 Introduction

Biofilm is called a community of microbes that adhere to solid exteriors and are enclosed in an extracellular polymeric substance (EPS) matrix. The EPS matrix consists of nucleic acids, proteins and carbohydrates within a liquid-containing environment (De Souza et al. 2014). Biofilms could exist on abiotic and biotic exteriors. Antoni van Leeuwenhoek, a Dutch researcher, for the first time examined biofilms on tooth surface with the use of a simple microscope. Later on, in 1973, Characklis indicated that biofilms show greater resistance to antiseptics such chlorine. Costerton in 1978 used the term biofilm (Cortés et al. 2011). Humans are influenced by biofilms in several ways as they are formed within industrial, medical and natural settings. For example, biofilms form on medical instruments such as implants or catheters and are very hard to be treated (Donlan 2008). The architecture of biofilm comprises of two chief constituents: a water channel, which plays an important role in the transportation of nutrients, and an area of tightly crowded cells, which lacks pores.

Bioremediation is an eco-friendly procedure for the detoxification of water, air and soil by the usage of microbes (Prasad and Prasad 2012). Bioremediation offers several advantages over other methods, including no or minimum disruption of land or wildlife surrounding the treated area, reduction of dust during treatment as well as removal of harsh chemicals. Bioremediation is economical when applied to large areas compared to conventional decontamination methods. The process of bioremediation employs diverse microbes for the degradation or treatment of xenobiotic compounds, aromatic hydrocarbons, volatile organic compounds, pesticides, herbicides, heavy metals, radionuclides, crude oil, jet fuels, petroleum products and explosives (Gaur et al. 2014). Microbes, particularly bacteria, can be easily grown and genetically manipulated, which makes them suitable for bioremediation. Microbes are extremely efficient in degrading natural organic compounds or waste pollutants via diverse catabolic pathways as microbes adapt themselves to living in diverse environments (Balaji et al. 2014). Some microbes used in bioremediation are extremophiles that can withstand acidic or heavy metal-contaminated or radioactive environment. The success of bioremediation relies on the maintenance of conditions or factors that accelerate biodegradation of pollutants.

The diverse species of microbes are found within biofilms consortia, and each species has different metabolic degradation pathways and capabilities for the degradation of numerous contaminants either separately or cooperatively (Gieg et al. 2014). Bacteria that form biofilms are suitable for bioremediation. Biofilm-facilitated remediation is an eco-friendly and cost-effective choice for cleaning up pollutants from the environment. The usage of biofilms is efficient for bioremediation as biofilms absorb, immobilize and degrade various pollutants in the environment. Responses of biofilms such as twitching motility, chemotaxis and quorum sensing support the movement of microorganisms towards the contaminant to enhance biodegradation (Lacal et al. 2013).

Most of the bacterial species persist in the EPS of biofilms under normal ecological circumstances, which also delivers a valuable structure to biofilm-forming microorganisms in the process of bioremediation. As compared to planktonic cells, the matrix of biofilm provides greater resistance to microorganisms against acid stress, shear stress, ecological stress, UV damage, solvents, predation, desiccation, biocides, highly toxic chemical concentration, pollutants and antimicrobial substances (More et al. 2014). In contrast, free-floating planktonic cells decontaminate environmental contaminants through metabolism, but these cells are not stationary and are unable to survive the ecological and mechanical stresses.

Microbes which form biofilms, on the other hand, are skilled in bioremediation because such microbes are immobilized within the EPS and can cause immobilization of contaminants during the degradation process. Cyanobacteria EPS acts as a biosorbent, and it helps in the removal of heavy metals from aqueous phases (De Philippis et al. 2011). Certain extracellular enzymes are also found in the EPS of biofilms, which help in the decontamination of contaminants such as organic compounds and heavy metals. The EPS acts as a trapping agent for metalloids and metals because of the existence of various negatively charged functional groups that allow formation of complexes with organic pollutants and heavy metals and hence help in their successive removal (Li and Yu 2014).

7.2 Role of Biofilms in Bioremediation

Biofilms could be applied in the bioremediation of waste materials that are found in the environment. There are successful examples of the positive use of biofilms that are called beneficial biofilms, and these biofilms have numerous advantages; for example, they protect the environment from the dangerous effects of toxic contaminants (Lazarova and Manem 2000). For treating large volumes of municipal and industrial wastewaters, biofilm-based reactors are frequently employed. Biofilms could exist in natural and engineered systems. Biofilms can be found in natural and engineered systems. Common place natural environmental biofilms exist in soil, aquatic plants, sediments, covering rocks in streams and plants, lakes and rivers and in wetlands. Natural biofilms in the environment can degrade and remove pollutants from soil and rivers. Biofilms formed on the surface of water consist of bacteria, protozoa, fungi and algae (Kartal et al. 2010). Naturally developed biofilms by algae such as *Nitzschia* in organic sediments can aid in the recovery of sediments by producing oxygen, which in turn facilitates the biodegradation of aerobic bacteria present in the biofilm. Slow sand filters utilize biofilms formed on the sand surface to remove metals and organic substances from lakes, rivers and water reservoirs. Planctomycetes present in biofilms of seaweed in marine waters have the ability to remove nitrogen from wastewater because of their capacity to perform anammox reactions in which ammonia is anaerobically oxidized to dinitrogen (Rafida et al. 2011). Biofilms are also used for the removal of heavy metals. Recently, the bioremoval of Cr(III) by means of bacterial biofilms within a continuous-flow reactor has been described by Sundar et al. (2011). Biofilms established by *Bacillus cereus* and *Bacillus subtilis* consortium on rough sand were capable of removing 98% of Cr(III). The capability of *Escherichia coli* biofilms for the elimination of Fe(III), Ni(II), Cd(II) and Cr(VI) from wastewater was significant. With the increase in population and the development of technology, there are extensive applications of plastics in each aspect of life, which becomes a major source of environmental pollution (Mathur et al. 2011). Algal biofilms are useful to degrade polythene. Fifteen algal taxa, namely *Oedogonium, Phormidium, Chroococcus, Oocystis, Coleochaete scutata, Aphanochaete, Chaetophora, Aphanothece, Coleochaete soluta, Navicula, Cymbella, Fragilaria, Oscillatoria, Cocconis* and *Gloeotaenium*, were recognized to degrade polyethylene. The proliferation of *C. soluta* and *C. scutata* on the polythene surface was seen beneath a scanning electron microscope. The strain of *Rhodococcus* rubber (C208) was isolated from polyethylene, and the degree of degradation was 0.86% per week. The extensive and enormous use of organic substances such as polymers, explosives, solvents, flame retardants, biocides, chlorinated organic compounds and crude oil has polluted the environment, due to which the life of most of organisms living on the earth is threatened. Microorganisms have shown capabilities to degrade most organic compounds such as polycyclic aromatic hydrocarbons (PAHs) and polybrominated biphenyl ethers in the atmosphere (Fennell et al. 2019).

7.3 Microorganisms Involved in Bioremediation

Microbes that are responsible for biodegradation in numerous surroundings are recognized as active members of microbial consortiums. These microbes comprise *Arthrobacter, Flavobacterium, Methylosinus, Acinetobacter, Pseudomonas, Actinobacteria, Mycococcus, Alcaligenes, Nitrosomonas, Nocardia, Bacillins, Serratia, Phanerochaete, Rhizoctomia, Beijerinckia, Xanthobacter, Mycobacterium, Trametes* and *Penicillium*. Microbes cannot mineralize harmful complexes independently, so they act in association with others to completely degrade harmful substances. Natural microbial communities within numerous environments have an astonishing physical adaptability to metabolize and frequently mineralize a huge number of organic substances (Colberg and Young 1995).

7.4 Advantages of Biofilms in Bioremediation

The biofilms formed by microbes possess numerous benefits as compared to planktonic bacterial cells, for instance defence from harsh environments, exchange of genetic material, perseverance within diverse metabolic states, nutrient availability and communication with the environment and each other. Furthermore, biofilms made with multiple species comprise of microbes that evolve from one or more kingdoms, for instance bacteria, algae and archaea, with variable metabolic rates and necessities such as electron donors/acceptors (Baker et al. 2009). Due to such multiple species, there exists cooperativity amongst microorganisms for existence under unfavourable ecological conditions. Within biofilms, the bacteria are enclosed in EPS, which is produced through microorganisms. After maturation of the biofilm, the water channels are made, which help out in transportation of oxygen, nutrients and other substances (Chen et al. 2013). Common components of biofilms having multiple species or polymicrobial biofilms are polysaccharides, protein, lipids and e-DNA; however, several precise EPS components help in solubilizing recalcitrant or hydrophobic subtracts that were unreachable to microorganisms. In the process of biotransformation or bioremediation, biofilms are advantageous as compared to planktonic cells. Microorganisms in planktonic state can be killed by chemicals because of their higher concentrations at polluted locations. However, biofilms show tolerance to different hazardous and toxic chemicals. The floating biofilms or sessile biofilms possess the capability to tolerate alterations within the environmental settings, such as exposure to salt concertation, antibiotics, high pollutants, water contents, pH, nutrients and temperature. It has been proposed that within heavily polluted locations and with the existence of antibiotics, sub-MIC concentration bacterial species mostly grow in the form of biofilms (Gross et al. 2007). About 100–1,000× higher concentrations of drugs are needed for killing the bacteria found in the form of biofilms. Furthermore, membrane vesicles (MVs) are released under unfavourable conditions by the bacterial species found inside the biofilms. It is stated that MVs released by *E. coli* have the ability to neutralize environmental substances and provide protection from lysis of cells (Baumgarten et al. 2012).

7.5 Strategies for the Use of Biofilms in Remediation

Biofilm-mediated remediation may be performed in various ways. The process of natural degradation relies on natural processes without the use of engineered steps or interventions, including the addition of specific strains for bioremediation. For example, microbial biofilm community present in the soil can biotransform certain pollutants into less harmful components. Natural degradation is based on the evidence that under favourable conditions, certain contaminants can be degraded, transformed, immobilized and detoxified without any human intervention (Sayler et al. 1995). This passive remediation process requires resident profile of microbes that may be present in the biofilm and are capable of degrading pollutants. Extra nutrients such as carbon and phosphorus compounds, air to improve oxygen availability and additives may be added to increase the growth and degradation rate of pollutants within a procedure called bio-stimulation (Vogt and Richnow 2013). Natural degradation strategy is typically used when the amount of pollutant is comparatively little. Bioaugmentation or bio-enhancement depends on the precise inoculation of competent microbes in polluted spots for performing degradation. In bio-stimulation, stimuli such as nutrients, growth substrates, electron donors and electron acceptors are provided for enhancing the activity of microbes that are present near the site to better biodegrade the pollutants. A study on comparative bioremediation strategies shows that the use of bioaugmentation and bio-stimulation was more efficient in remediation for petroleum hydrocarbons. Bioreactor-based biofilms have commercially been used for cleaning up industrial wastewaters for decades (Vogt and Richnow 2013).

7.5.1 Strategy for the Biodegradation of Phenolic Compounds from Industrial Wastewaters

Industrial effluents contain different organic and/or inorganic contaminants; therefore, their bioremediation represents an important challenge. As it was previously mentioned, both anaerobic and aerobic biological treatment systems have been recognized as effective methods for degrading extremely polluted wastewater containing different phenolic compounds. In addition, phenolic compound removal is also evaluated in some effluents those derived from olive oil extraction mills, paper factories, tanneries, crude oil refineries, palm oil industries, phenolic resin-producing industries and coke and coal gasification processes (Maszenan et al. 2011). Certain authors have paid attention to olive mill wastewater bio-treatment.

This is important considering that the extraction of olive oil generates great quantities of highly toxic phenolic wastes. The phenol content (1,200 mg/L) of olive mill wastewater gradually decreased to 71.9% and 71.4% after treatment for about 25 days with the use of a mixture of bacteria such as *P. putida, P. fluorescens* and *Azotobacter vinelandii.* Furthermore, Gonçalves et al. also noticed extraordinary removal efficacies of phenols (up to 60%) in anaerobic reactors inoculated with active sludge, to which raw olive mill wastewater comprising 4,300 mg/L of phenols was added. Khoufi et al. (2009) defined a new process for the treatment of wastewater of olive mills, which combines ultrafiltration, anaerobic digestion and electro-Fenton process. By applying such method, phenolic substances were removed efficiently.

The ability of bacteria for the removal of phenolic substances from wastes of paper factories has expansively been considered. Chandra and Singh confirmed that a mixed bacterial culture formed by *Providencia rettgeri IITRP2, Pantoea sp. RCT2* and *Pseudochrobactrum glaciale IITRP1* degraded 91% and 61% of entire phenols and chlorophenols from the effluent of paper factories, respectively (Chandra and Singh 2012). Singh et al. stated that a mixed culture of bacterial strains, i.e. *Serratia marcescens* and *Bacillus* spp., was capable of degrading 94% of the pentachlorophenol (PCP) present in the effluent of paper factories. Furthermore, Karn et al. confirmed that *P. stutzeri* strain (CL7) was capable of removing 66.8% of PCP after 2 weeks from the effluents of paper factories, which contained 100 mg/L of PCP. Nair et al. applied certain immobilized and free cells of *Alcaligenes* spp. within a packed bed reactor for treating effluent from a paper factory, which contained phenol in concentration of 9.41 mg/L. About 99% removal of phenol was done by both the cells after 20 hours of treatment by means of a batch process (Karn et al. 2010). A combined chemical/biological method named activated sludge ozonation method was used for treating the effluents from paper factories. By this activated sludge ozonation method, about 70% of phenols were removed. Tannery wastewater usually comprises high amounts of organic matter, including other chemicals and phenols. These wastes cannot be released into the surroundings without pretreatment due to their high toxicity (Bajza and Vrcek 2001). Consequently, the efficacy of numerous strains of bacteria for treating the tannery contaminated water has been considered. For instance, Paisio et al. noticed that strain of *Rhodococcus* spp. was capable of growing in and degrading phenols from tannery wastes, which contained phenols in concentrations of about 17.5 mg/L, after 9 hours of incubation (Paisio et al. 2012).

7.5.1.1 Bacterial Immobilization – a Useful Strategy for Efficient Phenol Biodegradation

For phenol biodegradation, the immobilization of bacterial biomass is an operative method to defend bacteria from higher amounts of this compound. For this purpose, cells are immobilized in different support matrices such as polyacrylamide, polyvinyl alcohol (PVA), agar, agarose, polysulphone, polyacrylonitrile and calcium alginate (Ca-alginate). There are many evidences of the reduction in toxicity of high phenol concentrations and hence a higher degradative ability when bacteria are immobilized on Ca-alginate (Du et al. 2012). This increase in the efficiency was explained not only by a higher reaction ability due to the high density of immobilized cells, but also by a protective effect of the polymer, giving a more favourable microenvironment for the catalytic reaction. Two *Bacillus cereus* strains immobilized on Ca-alginate showed a higher phenol degradation efficiency than free cells at high concentrations of phenol (1,500–2,000 mg/L), and it was shown that immobilized cells were much more tolerant to the toxicity of phenol. Moreover, these immobilized cells degraded more than 50% of 2,000 mg/L phenol in 26 and 36 days (Banerjee and Ghoshal 2011) (Table 7.1).

7.5.2 Strategy for the Biodegradation of Polycyclic Aromatic Hydrocarbons

Polycyclic aromatic hydrocarbons (PAHs) get absorbed into the soil and are undissolvable, which hampers the process of bioremediation. Also, these compounds enter food chain and cause cancer and mutation. In order to successfully bioremediate PAHs, biofilms were employed. Johnson and Karlson stated that biofilms enhanced the solubility of PAHs and then their mass transfer from recalcitrant crystals to cells for biotransformation (Johnsen and Karlson 2004). Approximately 100 diverse kinds of PAHs are detected in the environment. However, the mixed or multiple-species biofilms can be applied for the

TABLE 7.1

Bioremediation of Phenolic Compounds

S. No.	Contaminant	Type of Reactor	Microorganism	Concentration (mg/L)	Temperature (°C)	Removal Efficiency (%)	pH	Degradation Time (hours)
1.	Synthetic phenol	Batch (bubble column bioreactor)	Immobilized *P. putida*	5–150	30	100	7.0	0.3–5
2.	Synthetic phenol	Continuous (spouted bed bioreactor)	Immobilized *P. putida*	10–150	30	88–97	7.0	0.67–2.55
3.	Synthetic phenol	Batch (shake flasks)	Free *P. putida* DSM 548	1–100	26±0.5	100 (for Co=23.4 mg/L)	6.8	14
4.	Synthetic phenol	Batch (shake flasks)	Free *Pseudomonas* sp. SA01	300–1,000	30	100	7.0	20–85
5.	Synthetic phenol	Batch (shake flasks)	Free *Ochrobactrum* sp.	50–400	30	45.2 (for Co=62 mg/L, upon addition of molasses)	8.0 (optimum)	96
6.	Synthetic phenol	Batch bioreactor	Freely suspended mixture of *P. aeruginosa* and *P. fluorescence*	100–500	30	100		24–96
7.	Synthetic phenol	Batch (shake flasks) and continuous packed bed	Immobilized *Ralstonia eutropha*	25–500	30	Batch: 68 (for Co=100 mg/L)	7.0	
8.	Synthetic phenol	Batch (shake flasks)	Free mixed bacterial consortium	23.5–658	25±2	90–100	7.2	1–10

Source: Adapted from Al-Khalid and El-Naas (2012), El-Naas, M.H., Al-Muhtaseb, S.A., and Makhlouf, S., *J. Hazard. Mater.*, 164, 720–725, 2009,), Bajaj, M., Gallert, C., and Winter, J, *Biochem. Eng. J.*, 46, 205–209, 2009, Tepe, O., and Dursun, A.Y., *J. Hazard. Mater.* 151, 9–16, 2008, Shourian, M., Noghabi, K.A., Zahiri, H.S., Bagheri, T., Karbalaei, R., Mollaei, M., Rad, I., Ahadi, S., Raheb, J., and Abbasi, H., *Desalination* 246, 577–594, 2009,

TABLE 7.2

Polycyclic Aromatic Hydrocarbon-Degrading Bacteria

S. No.	PAH	Bacteria
1.	Phenanthrene	*Alcaligenes denitrificans, Mycobacterium* sp., *Aeromonas* sp., *P. putida, Rhodococcus* sp., *Nocardia* sp., *P. paucimobilis, Alcaligenes faecalis, Arthrobacter polychromogenes, Vibrio* sp., *Flavobacterium* sp., *Acinetobacter sp., Streptomyces griseus* and *Streptomyces* sp.
2.	Naphthalene	*Alcaligenes denitrificans, P. fluorescens, Pseudomonas* sp., *Acinetobacter calcoaceticus, P. cepacia, Rhodococcus* sp., *Moraxella* sp., *P. paucimobilis, Mycobacterium* sp., *P. fluorescens, Corynebacterium renale, Bacillus cereus, P. testosteroni* and *Streptomyces* sp.
3.	Chrysene	*Rhodococcus* sp.
4.	Benzo[a]pyrene	*Mycobacterium* sp. and *Beijerinckia* sp.
5.	Pyrene	*Mycobacterium* sp., *Rhodococcus* sp. and *Alcaligenes denitrificans*
6.	Acenaphthylene	*Pseudomonas* sp., *P. fluorescens, Beijerinckia* sp., *P. cepacia*
7.	Benz[a]anthracene	*Beijerinckia* sp., *P. putida* and *Alcaligenes denitrificans*
8.	Anthracene	*Mycobacterium* sp., *P. paucimobilis, Rhodococcus* sp., *Beijerinckia* sp., *Flavobacterium* sp., *Arthrobacter* sp. and *P. cepacia.*
9.	Fluoranthene	*P. paucimobilis, Rhodococcus* sp., *P. putida, Alcaligenes denitrificans, Pseudomonas* sp., *Mycobacterium* sp. and *P. cepacian*

Source: Adapted from Bamforth, S.M., and Singleton, I. *J. Chem. Technol. Biotechnol.*, 80, 723–736, 2005, Ashok, B.T., and Saxena, S., *J. Sci. Ind. Res.*, 54, 443–451, 1995, Cerniglia, C.E., *Curr. Opin. Biotechnol.*, 4, 351–368, 1992..

process of bioremediation. When such mixed or multiple-species biofilms are used, each species has diverse metabolism, which plays a role in the transformation and solubilization of PAHs. For the bioremediation of chlorinated ethenes, the *Dehalococcoides* bacteria are used. Dechlorinating Chloroflexi is applied in order to dechlorinate polychlorinated biphenyls (PCBs). Pentachlorobiphenyls are degraded effectively when biofilms are used. The biofilms formed by multiple species of *Burkholderia sp. NK8* in combination with *P. aeruginosa PA01* are applied for chlorinated benzoates degradation. When the *Comamonas sp. strain KD7* biofilms was used, it effectively degraded the dioxins found within the soil (Wang and Oyaizu 2011) (Table 7.2).

7.5.3 Biofilm-Based Strategy for Wastewater Treatment

In order to treat municipal wastewater, the biological systems need (i) the active microbe's accumulation within a bioreactor and (ii) the microbe's isolation from the waste being treated. In activated sludge processes as in the case of the suspended growth reactors, microbes multiply and undergo bioflocculation; then, in the bulk phase the resulting flocs are freely suspended. By means of membranes or through process of sedimentation, the flocculated bacterial species are formerly separated from the bulk liquid. 'Clarifier-coupled suspended growth reactors' depend on underflow or return activated sludge from the coupled clarifier to offer the desired active biomass concentration within the bioreactor. Bacteria are retained by the biofilm reactors within a biofilm that is attached to the free or immovable moving carriers. The matrix of biofilm comprises water and different types of particulate (X) and soluble (C) components, which involve soluble products of microbes, EPS and inert materials. Categorization of biofilm reactors is basically on the basis of biofilm, which is attached to the moving or fixed carrier in the reactor, and it is also on the basis of the number of phases such as solid, liquid and gas. Also, they are categorized into seven basic kinds on the basis of how electron acceptors or donors are applied (Harremoës and Wilderer 1993). (i) Three-phase scheme – semifixed or fixed biofilm-laden carrier, air and bulk water. Water moves across the biofilm reactor and forms gas bubbles (for example, aerobic biologically active filter (BAF)). Polystyrene beads are semifixed, and gravel is a fixed medium. (ii) Three-phase scheme – fixed biofilm-laden carrier, air and bulk water. Water dribbles over the surface of biofilm, and air moves downwards or upwards within the third phase (for example, trickling filter (TF)).

(iii) Three-phase scheme – movable biofilm-laden carrier, air and bulk water. Water moves across the biofilm reactor (for example, aerobic moving bed biofilm reactor (MBBR)). (iv) Two-phase scheme, bulk water and fixed biofilm-laden carrier material. Electron acceptor and donor are present, and the water moves across the biofilm reactor (for example, denitrification filter). (v) Two-phase scheme, bulk water and moving biofilm-laden carrier. Water moves across the biofilm reactor. Electron acceptor and electron donor are present (for example, denitrification fluidized bed biofilm reactor (FBBR)). (vi) Two-phase membrane system, a proton exchange membrane which separates a compartmentalized cathode from the compartmentalized biofilm-laden anode. It has water on its two sides, but it has electron acceptor on one side and electron donor on the other (e.g. biofilm-based microbial fuel cell (MFC)). (vii) Three-phase membrane scheme, a hollow fibre-type membrane which is microporous and has biofilm and gas on one side and water on the other side, diffusing through the membrane to the biofilm (e.g. membrane biofilm reactor (MBfR)) (Harremoës and Wilderer 1993).

Biofilms are universal within engineered systems and in nature and could be applied positively in wastewater and municipal water treatment. Suspended growth reactors and biofilm could meet the treatment purposes such as desulphurization, nitrification, denitrification and carbon oxidation. Numerous oxidized pollutants such as bromate and perchlorate are also treated by biofilm reactors (Metcalf and Eddy 2003). Biofilm reactors used to treat wastewaters from industries are also used to encounter treatment purposes similar to those in industrial pretreatment and in the treatment of municipal wastewater. The pretreatment process aims to treat industrial waste streams till their features are similar to raw sewage. Consequently, the industry could formerly release the treated wastewater into municipal sewers, and after this process, the wastewater is further processed in municipal wastewater treatment plants. For various industrial applications, the biofilm reactors are used because the methods are easy to operate, robust, reliable and tough to shock or toxic loading (Metcalf and Eddy 2003).

7.5.3.1 Moving Bed Biofilm Reactors

The MBBR is an anoxic or aerobic stage scheme having a freely moving plastic biofilm carrier, which needs aeration or automated mingling for the distribution of carriers through the tank. The procedure comprises a biofilm reactor and a separation unit for the separation of solids and liquids (Ødegaard 2006). MBBRs are capable of treating a variety of wastewaters. The MBBR method meets the same treatment purposes as the activated sludge method, such as denitrification, nitrification and carbon oxidation, but the MBBR uses a much smaller tank as compared to the clarifier-coupled activated sludge system. Multiple reactors are arranged in sequence without having the requirement of return activated sludge pumping or intermediate pumping. Solids and liquids are separated by means of different processes such as dissolved air flotation, sedimentation basins, membrane filters and cloth discs (Ødegaard 2006). The MBBR is the best for treating wastewater in municipal wastewater treatment plants. It has a plastic biofilm carrier. Screens are characteristically connected with one wall of the MBBR and permit treated waste to move to the subsequent treatment stage while freely moving plastic biofilm carriers are retained. MBBRs which are aerobic use different aeration systems for distributing the plastic biofilm carriers and for meeting oxygen necessities. In contrast, MBBRs which are anoxic use mechanical mixers for distributing the plastic biofilm carriers due to the reason that there are no oxygen necessities. Anoxic regions consist of flat wall screens, and aerobic regions consist of cylindrical screens. Coarse bubble diffusers are particularly applied within moving bed reactors (Ødegaard 2006).

7.5.4 Environmental Toxicity of Selenium and Tellurium Compounds

The entire volume of both tellurium (Te) and selenium (Se) formed on yearly basis globally is 220 tons; correspondingly, with Japan, Germany, the USA, Canada, Sweden, Russia and Belgium, the build-up of Te or Se compounds within the environment mostly depends on their anthropogenic usage in numerous areas, which cause their emission into the environment (Winkel et al. 2015). Glass manufactures, electronics and agriculture applications, pigments production and metallurgic industries are all the main factors responsible for the accumulation of Se, whereas compounds that contain Te are used

TABLE 7.3

Bioremediation of Different Pollutants in Wastewater

S. No.	Pollutant	Biofilm-Forming Microorganism	Material Support	Type of Reactor	Treatment Efficiency
1.	Cr(III) and Cr(VI)	*Arthrobacter viscosus*	NaY zeolite	Suspended biofilm	14 mg/g zeolite for Cr(III); 3 mg/g zeolite for Cr(VI)
2.	Chlorophenol	*Pseudomonas* sp., *Rhodococcus* sp.	Silica-based spherical Celite R-633 (Celite Co.) microcarriers	Laboratory and pilot scale (FBR)	99.9%
3.	Cr(VI)	*Bacillus coagulans*	Granular activated carbon (GAC)	Column laboratory scale	5.34 mg g biosorbent for initial concentrations of 100 mg L
4.	Sulphate	*Desulfovibrio desulfuricans*	Activated carbon cloth	MFC	99%
5.	MCPP; 2,4-D (herbicides)	Mixed culture of herbicide-degrading bacteria	GAC	Biofilm column reactor	100% (2,4-D); partially MCPP
6.	COD	*Geobacter metallireducens* + bacteria present in the wastewater	Graphite	MFC	80%
7.	COD	Proteobacteria, Bacteroidetes, Nitrospirae, Cyanobacteria and Actinobacteria	Sponge, zeolite and ceramsite	Natural ventilation trickling filters (NVTFs)	37.33%, 53.83% and 47.87%
8.	Ammonia	Proteobacteria, Bacteroidetes, Nitrospirae, Cyanobacteria and Actinobacteria	Sponge, zeolite and ceramsite	Natural ventilation trickling filters (NVTFs)	84.90%, 65.28% and 63.77%

Source: Adapted from Asri et al. (2018), Quintelas, C., Fernandes, B., Castro, J., Figueiredo, H., and Tavares, T., *Chem. Eng. J.*, 136, 195–203, 2008, Oh, K.H., and Tuovinen, O.H., *Int. Biodeteriorat. Biodegrad.*, 33, 93–99, 1994, Zhang, X., Li, J., Yu, Y., Xu, R., and Wu, Z., *Biochem. Eng. J.*, 106, 87–96, 2016, Zhao, F., Rahunen, N., Varcoe, J.R., Chandra, A., Avignone-Rossa, C., Thumser, A.E., and Slade, R.C., *Environ. Sci. Technol.*, 42, 4971–4976, 2008., Puhakka et al. (1995).

in vulcanization of rubber, copper refining, solar panels and photovoltaic cells and tarnishing metals production of colour glass or ceramics (Belzile and Chen 2015). Se found within the environment in the form of SeO_3^{2-}, SeO_4^{2-} or Se^{2-} are usually present within ground and surface waters as contaminants, and these are the inorganic forms of Se, whereas the volatile and organic ones (selenoamino acids, trimethylselenonium ions and methyl selenides) are found in soils and air (Vesper et al. 2008). Likewise, Te compounds are present in waters or soils mostly in TeO_3^{2-} and TeO_4^{2-} forms and these are toxic and highly soluble. The occurrence of Te and Se compounds within water reservoirs is a major issue for both ecological wildlife and human health. Therefore, numerous strategies are needed to defend human and aquatic life, as poisoning events related to Se have happened in the last few decades globally, for instance in the Lake Sutton (the USA), the uranium mine in Saskatchewan (Canada) and the Kesterson Wildlife Reservoir (California), which caused mutations and physical deformities (Najafi et al. 2010).

7.5.4.1 Bioremediation of Chalcogen-Polluted Environments Based on Bacterial Biofilms

Most studies about the bioremediation of Te- and Se-polluted environments concentrated on the use of bacteria grown up as free planktonic cells (Turner et al. 2012). However, within natural surroundings microbes are mostly present in adjacent association with interfaces and surfaces in the form of communities, which are specified as biofilms. Inside the biofilms made by bacteria, cells are sheltered from the surrounding environment via the existence of a matrix called EPS (comprising a higher volume of water, proteins, polysaccharides, lipids and e-DNA (extracellular-DNA) (Flemming and Wingender 2010)). When bacterial species are in the form of biofilm, then it offers them numerous benefits, and because of the biofilm, these bacteria have the ability to populate a wide range of habitats, including those polluted by chalcogen oxyanions. Hence, peculiar topographies of biofilms (i.e. different cellular physiologies, quorum sensing signalling process, existence of the colony morphology variants and EPS) make them tolerant and/or resistant towards either Te or Se oxyanions. By the way, sulphate-reducing bacteria (SRB) inside a biofilm yield S_2 (sulphide) that could abiotically bio-convert SeO_3^{2-} and/or SeO_4^{2-}, which precipitate SeO within the EPS. Unlike sulphate-reducing bacteria, the biofilms of *S. oneidensis* made underneath anaerobic circumstances could cause reduction of SeO_3^{2-} and TeO_3^{2-}, accumulating SeO and TeO within both the EPS and cells. In detoxification approach, numerous biofilm-based reactors are applied for supporting the bio-conversion and biosorption of Te and Se oxyanions. Certainly, biofilms of *Burkholderia cepacia* grown up on the surface of alumina, in addition to a biofilm made by diverse species composed of *Thauera* sp. and *Dechloromonas* sp., are used for the bioremediation of Se oxyanions, resulting in the taking and bio-conversion of 'SeO_4^2 to SeO' via the cells of bacteria (Turner et al. 2012). Correspondingly, biofilms comprising sulphate-reducing and denitrifying microbes grown up on a 'hallow membrane biofilm reactor' are positively applied for removing SeO_4^{2-} from wastewater, whereas the already established biofilm of the sulphate-reducing bacteria, i.e. *Desulfomicrobium norvegicum*, was capable of abiotically reducing SeO_3^{2-} extracellularly. Additionally, biofilms made via isolates of non-sulphur marine photosynthetic bacteria which were resistant to TeO_3^{2-} exhibited their ability to bio-convert such Te oxyanion by intracellularly reducing it (Hockin and Gadd 2003).

7.5.4.2 The BSeR Systems

In order to treat wastewaters contaminated with Se, reactors comprising multispecies biofilms (BSeR) signify a hopeful strategy. In the current investigation, a multispecies biofilm comprised of bacterial strains (i.e. *Arthrobacter* sp., *Pseudomonas* sp., *Rhodococcus* sp. and *Bacillus* sp.) adapted to higher SeO_3^{2-} concentrations was examined to check its ability to transform such oxyanions to SeO, which is actually their elemental form (Yang et al. 2016). In the BSeR process, biofilms of bacteria are grown up on granular activated carbon within 'anaerobic fixed-film reactors' and this method has higher bio-process ability towards both SeO_3^{2-} and SeO_4^{2-}. This method results in about 97% recovery of SeO from wastewater of agriculture processes (Garfield Wetlands-Kessler Springs, Utah, the USA) (Yang et al. 2016).

7.5.4.3 The ABMet Reactor System

An ABMet® reactor is both a filtration and biological system. In this system, the consortia of microbes are grown up on porous granular activated carbon (GAC) beds, which creates anoxic environments for ideal reduction of SeO_3^{2-} and SeO_4^{2-} (Tan et al. 2018). This system comprises biofilter tanks in which oxyanions of Se are biotransformed to the elemental form (i.e. SeO). SeO is then removed from the biofilter by means of backwash cycle. A nutrient dose tank is also used in this system, which usually contains a syrup-based solution and acts like an electron donor for the consortia of microbes, which allows the biotransformation of oxyanions of Se (Sonstegard et al. 2007). Consequently, within such a reactor, the communities of microbes require just a smaller quantity of added nutrients, which reduces the costs of the whole system. Furthermore, the beds of GAC are employed as a substratum for sustaining the growth of bacteria, permitting the biofilm formation, which is structurally much strong relative to cells in planktonic state, and it resists the steps of washing used in the reactor (Tan et al. 2018). In another study, the communities of anaerobic microbes were inserted in an ABMet® biofilter system and the bio-conversion of oxyanions of Se was noticed in 16 hours. The removal efficacy was found to be 99.3% at the 'Duke Energy and Progress Energy' within North Carolina. Furthermore, co-pollutants found within such wastewaters, such as heavy metals and NO_3, together with oxyanions of Se were removed with a greater efficacy by the consortium of microbes inside the ABMet® biofilter system (Tan et al. 2018).

7.5.5 Bioremediation of Polychlorinated Biphenyls and Dioxins by Bacterial Biofilms

Dioxins and polychlorinated biphenyls (PCBs) are other persistent organic pollutants, which are amongst the extremely poisonous compounds. In spite of their prohibition over 30 years before, PCBs could yet be detected in sediments, soil and also air all around the world (Roots et al. 2010). Lately, attention has been paid to the microbial group known as *dechlorinating Chloroflexi*, due to its capability to reductively dechlorinate extremely chlorinated PCB congenators under anaerobic circumstances and hence leave lesser chlorinated structures accessible for aerobic degradation. Payne et al. (2011) stated that bioaugmentation through anaerobic dechlorinating bacterial species *D. chlorocoercia DF1* to sediment polluted with PCBs influenced doubly edged chlorines and also it activated natural dechlorinating groups of microbes to cause dichlorination of further PCBs. Linking the bioaugmentation idea with activated carbon sediment adjustment, it was revealed that matured dehalogenating bacterial biofilms could grow on surfaces of GAC (Mercier et al. 2013). Additionally, numerous investigations revealed that matured biofilm formation on the surface of GAC doesn't hinder the GAC adsorption characteristics and that PCBs adsorption does not disturb the formation of biofilm. Meanwhile the practices for transporting GAC to sediments by means of water columns have been ready for virtually a decade, the PCB bioremediation field may significantly benefit from combining the GAC adjustment with the engineering of the biofilm, which is assessed on a mesocosm scale for polluted sediments (Kjellerup and Edwards 2013). Presently, dioxins such as 2-chlorodibenzofuran or dibenzofuran are thought out to be the most lethal contaminants. Inadvertently, they are made as by-products when numerous chemicals are mixed in industrial processes such as herbicides production, incineration of municipal and medical wastes and wood pulp bleaching for paper. The favoured approaches of remediation comprise higher-energy chemical and thermal incineration, but the process of biodegradation seems to be much fascinating as investigations enhanced the understanding of microbes accessible to metabolize such substances (Wang and Oyaizu 2011). By means of quinone profiling for studying communities of microbes, we are able to get 22% polychlorinated dibenzo-p-dioxins (PCDDs) and difurans degradation in 3 months in extremely contaminated soil microcosms. *Burkholderia* sp. *NK8* in conjunction with *P. aeruginosa PA01* was used in investigation, and the biofilms formed by such two species displayed a boosted capability of degrading chlorinated benzoates. *Comamonas* sp. *strain KD7* biofilm was used, and a substantial reduction of dioxins was observed in samples of soil (Wang and Oyaizu 2011).

7.5.6 Biofilms and Petroleum Hydrocarbon Bioremediation

Soil pollution is caused when petrochemicals are released in soil. The remediation of soil by means of civil engineering approaches is costly, and also the fertility of the soil cannot be retained. Consequently, the bioremediation by means of hydrocarbon-degrading microorganisms together with rhizo- and phytoremediation can be applied. The following are the communities of bacterial species responsible for the degradation of hydrocarbon: *Brevibacterium, Burkholderia, Pseudomonas, Bacillus, Dietzia, Acinetobacter, Aeromicrobium, Mycobacterium and Gordonia* (Zhang et al. 2014). Genes that encode for enzymes that are needed for petroleum hydrocarbons degradation are present in these bacterial species. For instance, *Rhodococcus, Burkholderia* and *Pseudomonas* have alkane hydroxylases (for example, AlkB correlated); monooxygenases are present in *Methylocella, Methylococcus or Methylobacter*; and *Mycobacterium, Caulobacter* and *Acinetobacter* have 'bacterial P450 oxygenase', which can metabolize the hydrocarbons and change these hydrocarbons into non-toxic or least toxic forms (Ibrahim et al. 2013). Bioremediation by means of bacterial biofilm can be used for the degradation of hydrocarbons in non-aqueous phase. Biofilms can enhance the entry of microorganisms into the hydrocarbon surface, defend microorganisms from stress, increase gene transfer rate and improve the health of the entire community of microbes. Several bacteria with hydrocarbon-degrading ability have been isolated from sites contaminated by petroleum hydrocarbons (Balseiro-Romero et al. 2017).

7.5.7 Pollution of Toxic Dyes

Textile industries and pigment-based industries that comprise rubber, paints, plastics, photography, tannery, printing, paper, cosmetics and pharmaceuticals liberate different types of dyes or colour wastes as a main contaminant into the surroundings. Approximately 10,000 diverse pigments and dyes are commercially accessible globally, and yearly, more than 7×10^5 tons are produced. Approximately 20% of the entire pigments or dyes are lost throughout the printing and dyeing procedure, and nearly 50% of these dyes are released into the surroundings (Sinha et al. 2018). Numerous inorganic and organic dyes, particularly synthetic and azo dyes, are found in textile effluents. In the current dying procedures, the increased usage of azo dyes (30% of entire dye market) is the main issue for water contamination due to their higher solubility in water (Shukla et al. 2014).

7.5.7.1 Toxic Effects of Dye Molecules

Residues of dye molecules, particularly aromatic amines, and several molecules of dyes are carcinogenic, mutagenic, extremely toxic, teratogenic and allergic to living organisms, including human beings (Sinha et al. 2018). Those dyes which cause damage to DNA or genetic material of living organisms are considered serious chemicals. Numerous dyes are formed from several toxic chemicals such as anthraquinone and benzidine and compounds containing nitro and azo groups, and they create health problems. Cancers are caused by several azo dyes in humans, such as hepatocarcinomas, splenic sarcomas, chromosomal aberrations and nuclear anomalies (Padhi 2012). Breathing difficulties, headache, diarrhoea, muscle and joint pain, dizziness, skin itchiness, fatigue and nausea are caused by extended exposure to dyes. Within children, the exposure could lead to red ears and cheeks, dark circles under the eyes, learning problems and hyperactivity. The existence of dyes within the aquatic setting for long time is the reason for bio-magnification and bio-accumulation in the food chain (Sinha et al. 2018). Dye molecules have toxic nature, and because of it, they threaten the life of humans, animals and aquatic fauna and flora. Long-chain synthetic molecules of dyes have a complex organic structure, which makes them non-degradable, much stable and therefore lethal to the natural setting (Padhi 2012). Therefore, substantial release of untreated dye-containing industrial effluents into the environment, including water bodies, is a serious threat to the biological ecosystem.

7.5.7.2 Application of Biofilms for Bioremediation of Dyes

Numerous biomasses of bacteria were used for the bioremoval of lethal dyes from the dye-polluted waste, as stated by several investigators. Several bacterial species such as *Lysinibacillus* sp. AK2, *Enterococcus faecalis, Bacillus amyloliquefaciens, Vibrio fischeri, Bacillus* sp. *AK1, Aeromonas* spp., *Kersteasia* sp. *VKY1, Staphylococcus aureus, P. aeruginosa, Pseudomonas* sp. *strain DY1, E. coli, B. subtilis and Pseudomonas luteola* were able to efficiently remove several main dyes, for instance Acid Black 172, amaranth dye, crystal violet, malachite green, methyl orange, Reactive Violet 2, Acid Orange 7, Reactive Blue 5, Reactive Yellow 2 and Reactive Red 22. The biomass of *P. aeruginosa* was shown to be an effective adsorbent for removing malachite green (MG) from effluents of industrial wastes (Lim et al. 2014). From wastewater, the Acid Black 172 was removed effectively by biomass of *Pseudomonas sp.* DY1 and it was shown to be an operative adsorbent for it. Reactive Yellow 2, Reactive Violet 2, Reactive Blue 5 and Reactive Red 22 were removed by the biomass of *S. aureus, B. subtilis, P. luteola, E. coli* and *Aeromonas spp.* The biofilm formed by *E. faecalis* efficiently removed AO7 (Acid Orange 7), and it was an efficient adsorbent for it and more than 80% of COD was removed (Lim et al. 2014).

7.5.8 Bioremediation of Palm Oil Mill Effluent by Bacterial Biofilms

For producing 1 ton of crude palm oil (CPO), about 5–7.5 tons of water is needed by palm oil industries and along with it about 2.5–3.5 tons of palm oil mill effluent (POME) is generated by these industries. POME is actually a viscous brownish liquid, which has higher amounts of total suspended solids (TSS), biochemical oxygen demand (BOD), chemical oxygen demand (COD) and turbidity. When POME is discharged into the environment without proper treatment, serious aquatic problems are caused as it has extensive variety of contaminants (Wu et al. 2010). Hence, there is an extreme need to design a system for treatment and to control the environmental pollution.

In the treatment of POME, microbes use the nutrients present in the POME as an energy source for their development and growth. And along with them, the detrimental elements such as heavy metals are also decomposed by these microbes. Microbes continue to multiply and undergo maturation of complex matrix until they attain extreme thickness. Formerly, it could be thought out to be active and stable to provide several benefits to process palm oil mill effluents. It is suggested to optimize the growth speed and start-up time of the treatment using biofilms (Escudié et al. 2011). Consequently, bioremediation technique by means of biofilms is a suitable approach to processing POME via the removal of BOD, COD, turbidity and suspended solids. Biofilm reactors are categorized into unfixed biofilm bed and fixed biofilm bed. As compared to the fixed biofilm bed reactor, the unfixed biofilm reactor is commonly used. It involves numerous technologies such as expanded granular sludge blanket (EGSB), up-flow sludge blanket, internal circulation reactors, biofilm air-lift suspension (BAS) and fluidized biofilm bed. The performance of bioremediation and effluent quality are decided by the concentration of active biomass and rate of influent. Moreover, a positive active control on thickness of biofilm makes the treatment process stable and effectual and it also prevents washout of biofilms (Escudié et al. 2011).

For wastewater that have higher contamination or higher concentration of biomass, the fixed biofilm reactor is a much appropriate process. Inside the reactor, an immobilized inert medium is present, which increases the likelihood of attachment of biofilm-forming bacterial species on the surface of the medium. An investigator used a two-stage anaerobic biofilm treatment, which involved one acidogenic and one methanogenic reactor for 'food waste-recycling' wastewater. About 73%–85.9% of COD was removed by this process. Apart from the above-discussed reactors, other reactors are also used, such as an anaerobic–aerobic granular biofilm bioreactor. In this, the biofilms mineralize numerous contaminants inside an up-flow anaerobic sludge bed (UASB) reactor. Chlorinated contaminants are removed by means of this reactor. The anaerobic–aerobic fixed film bioreactor is also used. This reactor is capable of treating higher concentrations of grease and oils (Lim et al. 2014).

7.5.9 Application of Biofilms for Bioremediation of Heavy Metals

Heavy metals contamination within aquatic habitats, soil and sediments has increased because of the unintended management of waste and also due to changes in industrial sector (Hossain and Rao 2014). Although different approaches are established for heavy metals removal depending upon the availability and need, new bioremediation methods using biofilms formed by bacteria have gained attention because of their more benefits. Numerous investigators have reported and investigated the biosorption of heavy metals such as Mn(II), Zn(II), Cr(VI), Hg(II), Cu(II), Cd(II), Ni(II) and Pb(II) by means of different kinds of bacterial biomasses. use of indigenous biomass of bacteria such as *Pseudomonas putida CZ1, E. coli, Bacillus subtilis, Pseudomonas sp. Lk9, Staphylococcus epidermidis, Arthrobacter* sp. *SUK 1205, Shewanella oneidensis, Penicillium simplicissimum, Leptothrix cholodnii SP-6SL, Bacillus cereus* and *Stenotrophomonas* spp. isolated from different metal-polluted locations in the bioremediation approaches seems to be operative, and it efficiently removes heavy metals from the polluted waste (Hossain and Rao 2014) (Tables 7.4 and 7.5).

TABLE 7.4
Bioremediation of Heavy Metals by Different Biofilm-Forming Bacteria

S. No.	Heavy Metals	Biofilm-Forming Organisms/ Culture
1.	Cr(VI)	*Escherichia coli*
2.	Cu(II), Zn(II) and Cd(II)	Activated sludge from a sewage
3.	Cd(II), Cu(II), Zn(II) and Ni(II)	*Pseudomonas* sp. NCIMB(11592)
4.	Cr(III)	*Bacillus subtilis* and *Bacillus cereus*
5.	Cr(VI), Cd(II), Fe(III) and Ni(II)	*Escherichia coli*
6.	Zn(II)	*Pseudomonas putida*
7.	Cr(VI), Cd(II), Fe(III) and Ni(II)	*Escherichia coli*

Source: Adapted from Chellaiah, 2018, Kumari and Das 2019, Costley, S.C., and Wallis, F.M., Water Res., 35, 3715–3723, 2001, Rabei, M.G., Sanna, M.F., Gad, N.N., Abskharon, E.R., *World J. Microbiol. Biotechnol.*, 25, 1695–1703, 2009.

TABLE 7.5
Bioremediation of Heavy Metals Using Biofilms in Bioreactors

S. No.	Heavy Metals Remediated	Reactor or Experimental Conditions	Methods of Remediation
1.	Cd, Zn, Cu, Pb, Y, Co, Ni, Pd and Ge	Bacteria-immobilized composite membrane reactor	Bio-precipitation
2.	Cu, Zn, Ni and Co	Biofilm formed on moving bed sand filter	Biosorption and bio-precipitation
3.	Cd, Cu, Zn and Ni	Biofilm developed over granular activated carbon	Adsorption
4.	Zn, Cd and Ni	Anaerobic–anoxic–oxic (A_2O) biofilm process	Biosorption

Source: Adapted from Singh, R., Paul, D., and Jain, R.K., *Trend. Microbiol.*, 14, 389–397, 2006, Scott, J.A., and Karanjkar, A.M., Water Sci. Technol., 38, 197–204, 1998. Babu et al. 2020.

7.6 Genetically Engineered Biofilms: A Future Prospective for Bioremediation

Genetically modified organisms (GMOs) could be made when we manipulate the organism genes according to the requirements by using the technology of genetic engineering. Bacterial species that exist within the polluted places are adapted to such environment via evolving an interior resistance. When can create a recombinant strain from such a bacterium which is present in natural habitats, it could be a potential candidate for the process of bioremediation. The recombinant strain is made by the introduction of a cluster of genes or a single gene. The common recombinant methods could be attained through genetic engineering of an operon or single gene, pathway switching and modifications of gene sequences of existing genes (Shukla et al. 2017). However, further research is needed to form effective genetically engineered biofilms and to analyse their potential for bioremediation of toxic contaminants.

7.7 Conclusions

a. Most bacteria within natural habitats are present in complex assemblies, which comprise of one or more bacterial species. Biofilms represent the perfect example of such assemblies. Studying the communities of biofilms and gene transfer in biofilms would enable the advancement of better methods for the bioremediation of polluted wastewaters and other sites.

b. Biofilm-based processes are unceasingly drawing consideration for research. On the basis of several reports, it might be concluded that such biofilms possess the capability of removing xenobiotic compounds and heavy metals from different wastewaters and other naturally occurring waters that have contaminants.
c. Bioremediation is the eco-friendly procedure by means of which harmful pollutants are removed from water, air and soil by the usage of microbes. The process of bioremediation could be only operative where ecological conditions allow the growth and activity of microbes. In diverse sites, the process of bioremediation is used and it has different success rates.
d. The development of novel techniques and use of microbes for treating different soils and wastewater contaminated by heavy metals and other pollutants is of environmental and economic interest.
e. As microbes have different catabolic abilities for several synthetic and natural compounds, they are considered unfailing. Bioremediation separately or in combination with different microbial activities is a cost-effective and eco-friendly procedure alternative to numerous chemical and physical techniques of remediating a contaminated site.
f. The presence of pollutants in the environment pose a threat to the life of living organisms, so there is a need for effective procedures for the remediation of different pollutants. Furthermore, there is a need for further research to address the role of biofilms in the remediation of different environmental pollutants to ensure clean environment.

Acknowledgement

We are grateful to Dr Saadia Andleeb, Associate Professor, Atta-ur-Rahman School of Applied Biosciences, National University of Sciences and Technology (NUST), Islamabad, Pakistan, for her guidance and support in writing this chapter.

REFERENCES

Al-Khalid, T., and El-Naas, M. H. 2012. Aerobic biodegradation of phenols: a comprehensive review. *Critical Reviews in Environmental Science and Technology*, 42(16):1631–1690.

Asri, M., Elabed, S., Koraichi, S.I. and El Ghachtouli, N. 2018. Biofilm-based systems for industrial wastewater treatment. *Handbook of Environmental Materials Management*, Springer Cham, Hillerod, Denmark, pp. 1–21.

Ashok, B.T. and Saxena, S. 1995. Biodegradation of polycyclic aromatic-hydrocarbons-A review. *Journal of Scientific & Industrial Research* 54:443–451.

Bajaj, M., Gallert, C. and Winter, J. 2009. Phenol degradation kinetics of an aerobic mixed culture. *Biochemical Engineering Journal* 46:205–209.

Bajza, Z. and Vrcek, I.V. 2001. Water quality analysis of mixtures obtained from tannery waste effluents. *Ecotoxicology and Environmental Safety* 50:15–18.

Baker, B.J., Tyson, G.W., Goosherst, L. and Banfield, J.F. 2009. Insights into the diversity of eukaryotes in acid mine drainage biofilm communities. *Applied and Environmental Microbiology* 75:2192–2199.

Balaji, V., Arulazhagan, P. and Ebenezer, P. 2014. Enzymatic bioremediation of polyaromatic hydrocarbons by fungal consortia enriched from petroleum contaminated soil and oil seeds. *Journal of Environmental Biology* 35:521–529.

Balseiro-Romero, M., Gkorezis, P., Kidd, P.S., Van Hamme, J., Weyens, N., Monterroso, C. and Vangronsveld, J. 2017. Characterization and degradation potential of diesel-degrading bacterial strains for application in bioremediation. *International Journal of Phytoremediation* 19:955–963.

Bamforth, S.M. and Singleton, I. 2005. Bioremediation of polycyclic aromatic hydrocarbons: current knowledge and future directions. *Journal of Chemical Technology & Biotechnology: International Research in Process, Environmental & Clean Technology* 80:723–736.

Banerjee, A. and Ghoshal, A.K. 2011. Phenol degradation performance by isolated *Bacillus cereus* immobilized in alginate. *International Biodeterioration & Biodegradation* 65:1052–1060.

Baumgarten, T., Vazquez, J., Bastisch, C., Veron, W., Feuilloley, M.G., Nietzsche, S., Wick, L.Y. and Heipieper, H.J. 2012. Alkanols and chlorophenols cause different physiological adaptive responses on the level of cell surface properties and membrane vesicle formation in *Pseudomonas putida* DOT-T1E. *Applied Microbiology and Biotechnology* 93:837–845.

Belzile, N. and Chen, Y.W. 2015. Tellurium in the environment: A critical review focused on natural waters, soils, sediments and airborne particles. *Applied Geochemistry* 63:83–92.

Babu, R.P., Pandit, S., Khanna, N., Chowdhary, P., Mathuriya, A.S. and Fosso-Kankeu, E., 2020. Importance of Bacterial Biofilm in Bioremediation. In *Contaminants and Clean Technologies* (pp. 149–164). CRC Press, Florida, United States.

Cerniglia, C.E. 1992. Biodegradation of polycyclic aromatic hydrocarbons. *Current Opinion in Biotechnology* 4:351–368.

Chandra, R. and Singh, R. 2012. Decolourisation and detoxification of rayon grade pulp paper mill effluent by mixed bacterial culture isolated from pulp paper mill effluent polluted site. *Biochemical Engineering Journal* 61:49–58.

Chellaiah, E.R. 2018. Cadmium (heavy metals) bioremediation by Pseudomonas aeruginosa: a minireview. *Applied Water Science* 8(6):1–10.

Chen, X., Suwarno, S.R., Chong, T.H., McDougald, D., Kjelleberg, S., Cohen, Y., Fane, A.G. and Rice, S.A. 2013. Dynamics of biofilm formation under different nutrient levels and the effect on biofouling of a reverse osmosis membrane system. *Biofouling* 29:319–330.

Colberg, P.J.S. and Young, L.Y. 1995. Anaerobic degradation of nonhalogenated homocyclic aromatic compounds coupled with nitrate, iron, or sulfate reduction. *Microbial Transformation and Degradation of Toxic Organic Chemicals* 307330.

Cortés, M.E., Bonilla, J.C. and Sinisterra, R.D. 2011. Biofilm formation, control and novel strategies for eradication. *Science against Microbial Pathogens: Communicating Current Research and Technological Advances* 2:896–905.

Costley, S.C. and Wallis, F.M. 2001. Bioremediation of heavy metals in a synthetic wastewater using a rotating biological contactor. *Water Research* 35(15):3715–3723.

De Philippis, R., Colica, G. and Micheletti, E. 2011. Exopolysaccharide-producing cyanobacteria in heavy metal removal from water: molecular basis and practical applicability of the biosorption process. *Applied Microbiology and Biotechnology* 92:697–708.

De Souza, P.R., De Andrade, D., Cabral, D.B. and Watanabe, E. 2014. Endotracheal tube biofilm and ventilator-associated pneumonia with mechanical ventilation. *Microscopy Research and Technique* 77:305–312.

Donlan, R.M. 2008. Biofilms on central venous catheters: is eradication possible? In Rafi Ahmed, Shizuo Akira, Arturo Casadevall, Jorge E. Galan, Adolfo Garcia-Sastre, Bernard Malissen, and Rino Rappuoli (Eds.), *Bacterial Biofilms* (pp. 133–161). New York: Springer Link

Du, L.N., Wang, B., Li, G., Wang, S., Crowley, D.E. and Zhao, Y.H. 2012. Biosorption of the metal-complex dye Acid Black 172 by live and heat-treated biomass of *Pseudomonas* sp. strain DY1: kinetics and sorption mechanisms. *Journal of Hazardous Materials* 205:47–54.

El-Naas, M.H., Al-Muhtaseb, S.A. and Makhlouf, S. 2009. Biodegradation of phenol by *Pseudomonas putida* immobilized in polyvinyl alcohol (PVA) gel. *Journal of Hazardous Materials* 164:720–725.

Escudié, R., Cresson, R., Delgenès, J.P. and Bernet, N. 2011. Control of start-up and operation of anaerobic biofilm reactors: an overview of 15 years of research. *Water Research* 45:1–10.

Fennell, D.E., Du, S., Liu, H., Häggblom, M.M. and Liu, F. 2019. Dehalogenation of polychlorinated dibenzo-p-dioxins and dibenzofurans, polychlorinated biphenyls, and brominated flame retardants, and potential as a bioremediation strategy. In *Comprehensive Biotechnology* (pp. 143–157). Elsevier. https://www.researchwithrutgers.com/en/publications/dehalogenation-of-polychlorinated-dibenzo-p-dioxins-and-dibenzofu-2

Flemming, H.C. and Wingender, J. 2010. The biofilm matrix. *Nature Reviews Microbiology* 8:623–633.

Gaur, N., Flora, G., Yadav, M. and Tiwari, A. 2014. A review with recent advancements on bioremediation-based abolition of heavy metals. *Environmental Science: Processes & Impacts* 16:180–193.

Gieg, L.M., Fowler, S.J. and Berdugo-Clavijo, C. 2014. Syntrophic biodegradation of hydrocarbon contaminants. *Current Opinion in Biotechnology* 27:21–29.

Gross, R., Hauer, B., Otto, K. and Schmid, A. 2007. Microbial biofilms: new catalysts for maximizing productivity of long-term biotransformations. *Biotechnology and Bioengineering* 98:1123–1134.

Harremoës, P. and Wilderer, P.A. 1993. Fundamentals of nutrient removal in biofilters. In *9th EWPCA-ISWA Symposium*. https://orbit.dtu.dk/en/publications/fundamentals-of-nutrient-removal-in-biofilters

Hockin, S.L. and Gadd, G.M. 2003. Linked redox precipitation of sulfur and selenium under anaerobic conditions by sulfate-reducing bacterial biofilms. *Applied and Environmental Microbiology* 69:7063–7072.

Hossain, K. and Rao, A.R. 2014. Environmental change and it's affect. *European Journal of Sustainable Development* 3:89–89.

Ibrahim, M.L., Ijah, U.J.J., Manga, S.B., Bilbis, L.S. and Umar, S. 2013. Production and partial characterization of biosurfactant produced by crude oil degrading bacteria. *International Biodeterioration & Biodegradation* 81:28–34.

Johnsen, A.R. and Karlson, U. 2004. Evaluation of bacterial strategies to promote the bioavailability of polycyclic aromatic hydrocarbons. *Applied Microbiology and Biotechnology* 63:452–459.

Karn, S.K., Chakrabarty, S.K. and Reddy, M.S. 2010. Pentachlorophenol degradation by *Pseudomonas stutzeri* CL7 in the secondary sludge of pulp and paper mill. *Journal of Environmental Sciences* 22:1608–1612.

Kartal, B., Kuenen, J.V. and Van Loosdrecht, M.C.M. 2010. Sewage treatment with anammox. *Science* 328:702–703.

Khoufi, S., Aloui, F. and Sayadi, S. 2009. Pilot scale hybrid process for olive mill wastewater treatment and reuse. *Chemical Engineering and Processing: Process Intensification* 48:643–650.

Kjellerup, B. and Edwards, S. 2013. *Application of Biofilm Covered Activated Carbon Particles as a Microbial Inoculum Delivery System for Enhanced Bioaugmentation of PCBs in Contaminated Sediment*. Goucher College, Baltimore, MD.

Kumari, S. and Das, S. 2019. Expression of metallothionein encoding gene bmtA in biofilm-forming marine bacterium Pseudomonas aeruginosa N6P6 and understanding its involvement in Pb (II) resistance and bioremediation. *Environmental Science and Pollution Research* 26(28):28763–28774.

Lacal, J., Reyes-Darias, J.A., García-Fontana, C., Ramos, J.L. and Krell, T. 2013. Tactic responses to pollutants and their potential to increase biodegradation efficiency. *Journal of Applied Microbiology* 114:923–933.

Lazarova, V. and Manem, J. 2000. Innovative biofilm treatment technologies for water and wastewater treatment. *ChemInform* 31.

Li, W.W. and Yu, H.Q. 2014. Insight into the roles of microbial extracellular polymer substances in metal biosorption. *Bioresource Technology* 160:15–23.

Lim, C.K., Aris, A., Neoh, C.H., Lam, C.Y., Majid, Z.A. and Ibrahim, Z. 2014. Evaluation of macrocomposite based sequencing batch biofilm reactor (MC-SBBR) for decolorization and biodegradation of azo dye acid orange 7. *International Biodeterioration & Biodegradation* 87:9–17.

Maszenan, A.M., Liu, Y. and Ng, W.J. 2011. Bioremediation of wastewaters with recalcitrant organic compounds and metals by aerobic granules. *Biotechnology Advances* 29:111–123.

Mathur, G., Mathur, A. and Prasad, R. 2011. Colonization and degradation of thermally oxidized high-density polyethylene by *Aspergillus niger* (ITCC No. 6052) isolated from plastic waste dumpsite. *Bioremediation Journal* 15:69–76.

Mercier, A., Wille, G., Michel, C., Harris-Hellal, J., Amalric, L., Morlay, C. and Battaglia-Brunet, F. 2013. Biofilm formation vs. PCB adsorption on granular activated carbon in PCB-contaminated aquatic sediment. *Journal of Soils and Sediments* 13:793–800.

Metcalf, Eddy 2003. *Wastewater Engineering: Treatment and Reuse*, 4th ed. McGraw-Hill, New York.

More, T.T., Yadav, J.S.S., Yan, S., Tyagi, R.D. and Surampalli, R.Y. 2014. Extracellular polymeric substances of bacteria and their potential environmental applications. *Journal of Environmental Management* 144:1–25.

Najafi, N.M., Tavakoli, H., Alizadeh, R. and Seidi, S. 2010. Speciation and determination of ultra trace amounts of inorganic tellurium in environmental water samples by dispersive liquid–liquid microextraction and electrothermal atomic absorption spectrometry. *Analytica Chimica Acta* 670:18–23.

Ødegaard, H. 2006. Innovations in wastewater treatment:–the moving bed biofilm process. *Water Science and Technology* 53:17–33.

Oh, K.H. and Tuovinen, O.H. 1994. Biodegradation of the phenoxy herbicides MCPP and 2, 4-D in fixed-film column reactors. *International Biodeterioration & Biodegradation* 33:93–99.

Puhakka, J.A., Melin, E.S., Järvinen, K.T., Koro, P.M., Rintala, J.A., Hartikainen, P., Shieh, W.K., and Ferguson, J.F. 1995. Fluidized-bed biofilms for chlorophenol mineralization. *Water Science and Technology*, 31(1):227–235.

Padhi, B.S. 2012. Pollution due to synthetic dyes toxicity & carcinogenicity studies and remediation. *International Journal of Environmental Sciences* 3:940.

Paisio, C.E., Talano, M.A., González, P.S., Busto, V.D., Talou, J.R. and Agostini, E. 2012. Isolation and characterization of a Rhodococcus strain with phenol-degrading ability and its potential use for tannery effluent biotreatment. *Environmental Science and Pollution Research* 19:3430–3439.

Payne, R.B., May, H.D. and Sowers, K.R. 2011. Enhanced reductive dechlorination of polychlorinated biphenyl impacted sediment by bioaugmentation with a dehalorespiring bacterium. *Environmental Science & Technology* 45:8772–8779.

Prasad, M.N.V. and Prasad, R. 2012. Nature's cure for cleanup of contaminated environment–a review of bioremediation strategies. *Reviews on Environmental Health* 27:181–189.

Quintelas, C., Fernandes, B., Castro, J., Figueiredo, H. and Tavares, T. 2008. Biosorption of Cr (VI) by a *Bacillus coagulans* biofilm supported on granular activated carbon (GAC). *Chemical Engineering Journal* 136:195–203.

Rabei, M.G., Sanna, M.F., Gad, N.N., Abskharon, E.R. 2009. *World Journal of Microbiology and Biotechnology* 25:1695–1703.

Rafida, A.I., Elyousfi, M.A. and Al-Mabrok, H. 2011. Removal of hydrocarbon compounds by using a reactor of biofilm in an anaerobic medium. *World Academy of Science Engineering and Technology* 73:153–156.

Roots, O., Roose, A., Kull, A., Holoubek, I., Cupr, P. and Klanova, J. 2010. Distribution pattern of PCBs, HCB and PeCB using passive air and soil sampling in Estonia. *Environmental Science and Pollution Research* 17:740–749.

Sayler, G.S., Layton, A., Lajoie, C., Bowman, J., Tschantz, M. and Fleming, J.T. 1995. Molecular site assessment and process monitoring in bioremediation and natural attenuation. *Applied Biochemistry and Biotechnology* 54:277–290.

Scott, J.A. and Karanjkar, A.M. 1998. Immobilized biofilms on granular activated carbon for removal and accumulation of heavy metals from contaminated streams. *Water Science and Technology* 38:197–204.

Shourian, M., Noghabi, K.A., Zahiri, H.S., Bagheri, T., Karbalaei, R., Mollaei, M., Rad, I., Ahadi, S., Raheb, J. and Abbasi, H. 2009. Efficient phenol degradation by a newly characterized *Pseudomonas* sp. SA01 isolated from pharmaceutical wastewaters. *Desalination* 246:577–594.

Shukla, S.K., Mangwani, N., Karley, D. and Rao, T.S. 2017. 19 Bacterial biofilms and genetic. In Surajit Das and Hirak Ranjan Dash (Eds.), *Handbook of Metal-Microbe Interactions and Bioremediation* (p. 317). Taylor and Francis Group.

Shukla, S.K., Mangwani, N., Rao, T.S. and Das, S. 2014. Biofilm-mediated bioremediation of polycyclic aromatic hydrocarbons. In Surajit Das (Ed.), *Microbial Biodegradation and Bioremediation* (pp. 203–232). Amsterdam: Elsevier.

Singh, R., Paul, D. and Jain, R.K. 2006. Biofilms: implications in bioremediation. *Trends in Microbiology* 14(9):389–397.

Sinha, S., Behera, S.S., Das, S., Basu, A., Mohapatra, R.K., Murmu, B.M., Dhal, N.K., Tripathy, S.K. and Parhi, P.K. 2018. Removal of congo red dye from aqueous solution using Amberlite IRA-400 in batch and fixed bed reactors. *Chemical Engineering Communications* 205:432–444.

Sonstegard, J., Harwood, J. and Pickett, T. 2007. Full scale implementation of GE ABMet biological technology for the removal of selenium from FGD wastewaters. In *Proceedings of 68th International Water Conference* (Vol. 2, p. 580). Engineer's Society of Western Pennsylvania, Pittsburgh, PA.

Sundar, K., Sadiq, I.M., Mukherjee, A. and Chandrasekaran, N. 2011. Bioremoval of trivalent chromium using *Bacillus* biofilms through continuous flow reactor. *Journal of Hazardous Materials* 196:44–51.

Tan, L.C. 2018. *Anaerobic treatment of mine wastewater for the removal of selenate and its co-contaminants.* CRC Press, Florida, United States.

Tepe, O. and Dursun, A.Y. 2008. Combined effects of external mass transfer and biodegradation rates on removal of phenol by immobilized *Ralstonia eutropha* in a packed bed reactor. *Journal of Hazardous Materials* 151:9–16.

Turner, R.J., Borghese, R. and Zannoni, D. 2012. Microbial processing of tellurium as a tool in biotechnology. *Biotechnology Advances* 30:954–963.

Vesper, D.J., Roy, M. and Rhoads, C.J. 2008. Selenium distribution and mode of occurrence in the Kanawha Formation, southern West Virginia, USA. *International Journal of Coal Geology* 73:237–249.

Vogt, C. and Richnow, H.H. 2013. Bioremediation via in situ microbial degradation of organic pollutants. In Axel Schippers, Franz Glombitza, and Wolfgang Sand (Eds.), *Geobiotechnology II* (pp. 123–146). New York: Springer Link.

Wang, Y. and Oyaizu, H. 2011. Enhanced remediation of dioxins-spiked soil by a plant–microbe system using a dibenzofuran-degrading *Comamonas* sp. and *Trifolium repens* L. *Chemosphere* 85:1109–1114.

Winkel, L.H., Vriens, B., Jones, G.D., Schneider, L.S., Pilon-Smits, E. and Bañuelos, G.S. 2015. Selenium cycling across soil-plant-atmosphere interfaces: a critical review. *Nutrients* 7:4199–4239.

Wu, T.Y., Mohammad, A.W., Jahim, J.M. and Anuar, N. 2010. Pollution control technologies for the treatment of palm oil mill effluent (POME) through end-of-pipe processes. *Journal of Environmental Management* 91:1467–1490.

Yang, S.I., George, G.N., Lawrence, J.R., Kaminskyj, S.G., Dynes, J.J., Lai, B. and Pickering, I.J. 2016. Multispecies biofilms transform selenium oxyanions into elemental selenium particles: studies using combined synchrotron X-ray fluorescence imaging and scanning transmission X-ray microscopy. *Environmental Science & Technology* 50:10343–10350.

Zhang, X., Li, J., Yu, Y., Xu, R. and Wu, Z. 2016. Biofilm characteristics in natural ventilation trickling filters (NVTFs) for municipal wastewater treatment: comparison of three kinds of biofilm carriers. *Biochemical Engineering Journal* 106:87–96.

Zhang, X., Liu, X., Wang, Q., Chen, X., Li, H., Wei, J. and Xu, G. 2014. Diesel degradation potential of endophytic bacteria isolated from *Scirpus triqueter. International Biodeterioration & Biodegradation* 87:99–105.

Zhao, F., Rahunen, N., Varcoe, J.R., Chandra, A., Avignone-Rossa, C., Thumser, A.E. and Slade, R.C. 2008. Activated carbon cloth as anode for sulfate removal in a microbial fuel cell. *Environmental Science & Technology* 42:4971–4976.

8

Sustainable Development in Agriculture by Revitalization of PGPR

Nandkishor More, Anjali Verma, and Ram Naresh Bharagava
Babasaheb Bhimrao Ambedkar University (A Central University)

Arun S Kharat
Jawaharlal Nehru University

Rajnish Gautam and Dimuth Navaratna
Victoria University

CONTENTS

DOI: 10.1201/9781003181224-8

8.1 Introduction

Since the dawn of civilization, agriculture has been the largest financial source. About 7.41 billion people inhabit the earth, occupying 6.38 billion hectares of earth surface, and about 1.3 billion people are directly dependent on agriculture. For sustainable agriculture maintenance, the dynamic nature of soil is one of the most important factors (Paustian et al. 2016). Approximately 99.7% of the food for the earth's population comes from the terrestrial environment alone as shown by the Food Balance Sheet 2004 of Food and Agriculture Organization of the United Nations (FAO). In India, 60.6% of land is used for agricultural purposes to grow several forms of vegetables, cereals and pulses by half of its population. A number of biotic and abiotic factors along with organic carbon content, moisture, nitrogen, phosphorous and potassium, regulate the soil processes.

Sustainable management practices with some of the current techniques involve agricultural intensification (Shrestha 2016), genetically engineered crops to form nitrogen-fixing symbioses, fixing nitrogen without microbial symbionts (Mus et al. 2016), use of microbes or genetically engineered microbes to promote plant growth (Perez et al. 2016) and uses of biofertilizers (Suhag 2016).

Many other socio-economic and scientific techniques in addition to contribution towards sustainable development of agriculture include drought tolerance, disease resistance, heavy metals resistance, stress tolerance, salt tolerance and better nutritional value, and they are attained by the use of efficient soil microorganisms such as bacteria, fungi and algae (Vejan et al. 2016).

Plant growth-promoting rhizobacteria (PGPR) play a vital role in converting barren, poor-quality land into cultivable land, and they greatly affect the soil characteristics. PGPR show antagonistic and synergistic interactions resulting in the enrichment of plant growth and can explain the association of plants with microbes (Rout and Callaway 2012). PGPR as a biofertilizer increase the accessibility and uptake of nutrients and, by doing so, improve the plant growth; they also neutralize the stress effects of plant, including both biotic and abiotic stresses. PGPR also prevent the deleterious effects of one or more phytopathogenic organisms and indirectly involve in promoting plant growth (Rout and Callaway 2012).

PGPR have potential in crop production, usually by developing sustainable systems for hormones production as they are phytopathogenic organisms and biocontrol agents; they are able to protect the infections of plants. These rhizosphere microbes are beneficial for the metabolites secreted by the plant roots that can be utilized as nutrients. The rhizosphere is a centre of intense biological activities due to the food supply provided by the root exudates. Due to the constant and diverse secretion of antimicrobial root exudates, most soil microorganisms do not interact with plant roots; possibly, rhizosphere microorganisms produce plant hormones, vitamins, antibiotics and communication molecules that encourage plant growth.

Siderophores are the green fluorescent pigment of pseudomonas bacteria, which include salicylic acid, pyoverdine yellow and pyochelin, which contribute to disease suppression and act as biocontrol agents for the limited supply of essential trace minerals in the natural habitat by conferring a competitive advantage. Some siderophores such as pseudobactin and pyoverdine have affinity for ions of trivalent iron and contain high antimicrobial activity. Siderophores increase the availability of these minerals to the bacteria, may indirectly stimulate the biosynthesis of other antimicrobial compounds by antibiotics and additionally function as stress factors or signal inducing local and systemic host resistance.

8.2 PGPR Diversity and Sustainability – Its Nature and Scope

The term 'rhizobacteria' was first introduced by Kloepper and Smith in 1978 as soil bacterial community that competitively colonized plant roots and stimulated growth, thereby reducing the occurrence of plant diseases.

8.2.1 Plant Growth-Promoting Rhizobacteria (PGPR)

Plants have always been in symbiotic relationship with soil microbes during their growth and development. PGPR are symbiotic free-living soil microorganisms inhabiting the rhizosphere of many plant species and have diverse beneficial effects through different mechanisms such as nitrogen fixation and nodulation (Raza et al. 2016a, b). They tend to defend the health of plants in an eco-friendly manner (Akhtar et al 2012; Bhardwaj et al. 2014). PGPR and their interactions with plants are exploited commercially and have scientific applications for sustainable agriculture (Gonzalez et al. 2015). PGPR are also involved in various biotic activities of the soil ecosystem to make it dynamic for turnover and sustainable for crop production (Gupta et al. 2015). By different mechanisms, they competitively colonize plant roots system and enhance plant growth, including phosphate solubilization (Ahemad and Khan 2012); nitrogen fixation (Glick et al. 1998); production of indole-3-acetic acid (IAA), siderophores (Jahanian et al. 2012) and hydrogen cyanate (Liu et al. 2016); degradation of environmental pollutants; and production of hormones and antibiotics or lytic enzymes (Liu et al. 2016). In addition, some PGPR may also infer more specific plant growth-promoting traits, such as heavy metal-detoxifying activities, salinity tolerance and biological control of phytopathogens and insects (Egamberdieva and Lugtenberg 2014).

8.2.2 Rhizosphere

Rhizosphere also known as the storehouse of microbes is the soil zone surrounding the plant roots where the chemical and biological features of the soil are influenced by the roots. In the rhizosphere, bacteria may be symbiotic or non-symbiotic, which is determined by whether their mode of action is directly beneficial to the plant or not (Kundan et al. 2015). The main components of the rhizosphere are the location of the PGPR in plant roots, rhizospheric soil, rhizoplane and the root itself. Among them, the rhizosphere is the soil zone regulated by roots through the release of substrates that affect microbial activity. In the rhizosphere, the concentration of bacteria is approximately 10–1,000 times higher than in bulk soil, but less than that in a laboratory medium.

To maintain their beneficial effects in the root environment, bacteria must compete well with other rhizospheric microbes for nutrients secreted by the root. The interactions between the plant and the rhizosphere are essential for procuring water and nutrients from soil as well as for the interaction between the plants and the soil-borne microorganisms. Thus, the rhizosphere is a narrow zone of soil surrounding the root that is directly affected by the root system.

8.2.3 Different Forms of PGPR

On the basis of their origin, PGPR can be classified into two types: (i) extracellular plant growth-promoting rhizobacteria (ePGPR) or rhizospheric; (ii) intracellular plant growth-promoting rhizobacteria (iPGPR) or endophytic (Beneduzi et al. 2012). The ePGPR inhabit in the spaces between the cells of the root cortex in the rhizosphere, whereas iPGPR are mainly found inside the specialized nodular structures of the root cells. The bacterial genera included in the ePGPR category are *Azotobacter, Azospirillum, Serratia, Bacillus, Caulobacter, Chromobacterium, Agrobacterium, Erwinia, Flavobacterium, Arthrobacter, Micrococcus* and *Pseudomonas*. They have great agricultural importance (Prithiviraj et al. 2003). The endophytic microbes belonging to iPGPR include *Allorhizobium, Bradyrhizobium, Mesorhizobium, Rhizobium* and *Frankia* species, which can fix atmospheric nitrogen specifically for higher plants. This group of rhizobacteria is mostly Gram-negative and rod-shaped with a lower proportion being Gram-positive, rod-shaped, cocci and pleomorphic.

On the basis of their functions and taxonomical status, PGPR have been categorized into many groups such as Cyanobacteria, Firmicutes, Actinobacteria, Bacteroidetes and Proteobacteria. In the rhizospheric region, PGPR exudates secreted by plant roots are the most important factors responsible for high microbial diversity.

8.3 Roles of PGPR in Rhizosphere

PGPR play a vital role in soil processes that determine plant productivity. They can reduce the use of agrochemicals in crop production and support eco-friendly sustainable food production as they are naturally occurring soil bacteria that aggressively colonize plant roots and benefit plant health by increasing seedling emergence; phytohormones; root hair proliferation; leaf surface area; biomass, nutrient, air and water uptake; early nodulation; nodule functioning; accumulation of carbohydrates; and yield. Soil moisture content affects the colonization of the plant rhizosphere by the PGPR after inoculation (Shrivastava 2014).

In the rhizosphere, rhizobacteria not only benefit from the nutrients secreted by the plant roots, but also beneficially affect the plant in a direct or indirect way, resulting in a stimulation of its growth. PGPR enhance the resistance to different phytopathogens and plant growth through direct and indirect mechanisms, which involve enhancing plant physiology (Beneduzi et al. 2012) and also neutralizing biotic and abiotic stresses, nutrient fixation and producing volatile organic compounds (VOCs) and enzymes to prevent diseases.

8.3.1 Direct Mechanisms

PGPR facilitate plant's growth and development directly through mechanisms such as nitrogen fixation, increased nutrient uptake or nutrient availability, mineralization of organic compounds, solubilization of mineral nutrients and production of phytohormones. Such mechanisms directly affect the plant growth activity. In the presence of PGPR, the direct enrichment of mineral uptakes occurs due to increases in individual ion fluxes at the root surface.

8.3.1.1 Nutrient Fixation

PGPR have the tendency to increase the accessibility and concentration of nutrients by fixing or locking their supply for plant growth and productivity and act as direct growth enhancers to plants. Plants absorb nitrogen from the soil, which is an essential nutrient for growth, in the form of nitrate (NO_3^-) and ammonium (NH_4^+). Some PGPR can be easily taken up by plants and have the capability for phosphate solubilization, resulting in an increased number of phosphate ions available in the soil (Paredes and Lebeis 2016). Other microbes such as Klebsiella pneumoniae Fr, Klebsiella sp. Brl, Bacillus pumilus Slrl, Acinetobacter sp. and Bacillus subtilis UPMB10 have the capacity to fix atmospheric N_2, delay N remobilization and acts as potential source for nutrient fixation.

8.3.1.2 Nitrogen Fixation

The atmospheric nitrogen is converted by nitrogen-fixing microorganisms using a complex enzyme system known as nitrogenase into plant-utilizable forms by biological nitrogen fixation, which changes nitrogen to ammonia. Strains of symbiotic PGPR *Rhizobium sp., Azoarcus sp., Beijerinckia sp.* and *K. pneumonia* are the most frequently reported to fix atmospheric N_2 in soil (Ahemad and Kibret 2014), while in *cyanobacteria, Azotobacter* and non-symbiotic bacteria species, N_2 fixation is carried out a by a particular gene called *nif* gene.

8.3.1.3 Phosphate Solubilization

Phosphorus plays an important role in almost all major metabolic processes and is the second most essential nutrient required by plants in adequate amount for optimum growth, macromolecular biosynthesis, energy transfer, signal transduction, respiration and photosynthesis (Anand et al. 2016). Plants absorb phosphorus in only two soluble forms – the dibasic (HPO_4^{-2}) ions and the monobasic ions ($H_2PO_4^-$). Solubilization and mineralization of phosphorus is an important trait that can be achieved by phosphate-solubilizing bacteria PGPR. Bacterial genera such as *Erwinia, Rhizobium, Flavobacterium, Rhodococcus, Mesorhizobium, Azotobacter, Bacillus, Beijerinckia, Burkholderia, Enterobacter, Microbacterium, Pseudomonas* and *Serratia* are the most significant phosphate-solubilizing bacteria that have attracted the attention of agriculturists as soil inoculate improve plant growth and yield.

8.3.1.4 Potassium Solubilization

Potassium is the third major essential macronutrient that exists in the soil in the form of insoluble rocks and silicate minerals and is essential for plant growth. PGPR such as *Paenibacillus sp., Bacillus mucilaginosus, Pseudomonas sp., Acidothiobacillus sp., Bacillus edaphicus* and *Ferroxidans sp.* have been reported as potassium-solubilizing bacteria that release inaccessible potassium from potassium-bearing minerals in soils; more than 90% of potassium deficiency has become a major constraint in crop production (Liu et al. 2016). Thus, potassium-solubilizing PGPR can reduce the use of agrochemicals and increase the crop production as well as for agriculture improvement and support of eco-friendly and act as biofertilizers.

8.3.1.5 Phytohormone Production

Phytohormones or plant growth regulators are organic substances that promote, modify or inhibit growth and development of plants at low concentrations (<1 mM) (Damam et al. 2016). Phytohormones are produced by some PGPR such as abscisic acid, ethylene, gibberellins, cytokinins and auxins that can affect cell proliferation in the root architecture. Plant growth regulators are also called exogenous plant hormones, as they can be applied exogenously as extracted hormones and synthetic analogues to plants or plant tissues.

8.3.1.6 Siderophore Production

Siderophores, Greek word for iron carrier, enhances the iron uptake capacity; heterologous siderophores produced by other microorganisms such as *Pseudomonas putida* enhance the level of iron available in the natural habitat (Rathore 2015). A potent siderophore complex plays an important role in iron uptake by plants in the presence of other metals such as the ferric-siderophore such as Cd and Ni (Beneduzi et al. 2012). Plants are able to take up the labelled iron by a large number of PGPR including *Bacillus, Azadirachta* and *Azotobacter.*

8.3.2 Indirect Mechanism

8.3.2.1 Exopolysaccharides (EPSs) Production

EPSs are high molecular weight, biodegradable polymers that are biosynthesized by a wide range of bacteria, algae and plants and are formed of monosaccharide residues and their derivatives (Sanlibaba and Cakmak 2016). The production of an exopolysaccharide is generally important in biofilm formation. It plays a vital role in maintaining water potential, ensuring obligate contact between plant roots and rhizobacteria and aggregating soil particles. These are responsible for crop production and plant growth (Pawar et al. 2016). EPSs-producing PGPR such as *Bacillus drentensis, Enterobacter, Agrobacterium sp., Rhizobium leguminosarum, Azotobacter vinelandii, Xanthomonas sp. and Rhizobium sp.* have an important role in contributing to sustainable agriculture and increasing soil fertility (Mahmood et al. 2016).

8.3.2.2 *Induced Systemic Resistance*

It may be defined as an enhanced defensive capacity and physiological state improved or evoked in response to a specific environmental stimulus. A defence mechanism during pathogenic invasion is activated via the vascular system, which results in the activation of a huge number of defence enzymes such as polyphenol oxidase, chitinase, phenylalanine ammonia lyase, β-1,3-glucanase, peroxidase, SOD, lipoxygenase, CAT and APX along with some proteinase inhibitors. Although the vast majority of PGPR induce ISR in plants, modern tools and techniques that use them to revolutionize agriculture to support plants from laboratory to field have been lacking till date.

8.4 PGPR as a Natural Growth Enhancer

PGPR play an important role in enhancing plant growth via several modes of action, such as (i) promoting plant growth regulators, (ii) nutrient fixation for easy uptake by plant, (iii) abiotic stress tolerance in plants, (iv) production of siderophores, (v) production of VOCs and (vi) production of proteolytic enzymes such as ACC deaminase, chitinase and glucanase for the prevention of plant diseases. Nitrogen is the most limiting nutrient for plants, which is essential for the synthesis of amino acids and proteins. The processes by which atmospheric nitrogen is added into organic forms that can be assimilated by plants are exclusive to prokaryotes. *Azospirillum*, a notable example of a free-living nitrogen-fixing organism, is frequently associated with cereals in temperate zones and also useful to improve the rice crop yields. Plant growth regulators are synthetic substances and are also termed as plant exogenous hormones that are comparable to natural plant hormones. Auxin regulates most plant processes directly or indirectly. Auxins-producing *Bacillus sp.* inflicts a positive effect on Solanum tuberosum's growth. Gibberellin (GA) is responsible for the processes of seed germination, floral induction, emergence, fruit and flower development, stem and leaf growth, etc. Rhizobacteria secreted generally by IAA interfere with many plant developmental processes, and IAA plays an important role in rhizobacteria–plant interactions.

PGPR allow the control of various plant diseases by depriving the pathogen of iron nutrition, thus resulting in increased crop yield, and have been demonstrated to enhance the plant growth by producing very efficient extracellular siderophores. Soil fertility improvement by phosphate-solubilizing bacteria (PSB) is one of the most common strategies to increase agricultural production. The biological nitrogen fixation is most important in enhancing the soil fertility. Phosphorus (P) is the most essential macronutrient for biological growth and development. It is an important trait in a PGPR for increasing plant yields. Some microorganisms have the ability to convert insoluble phosphorus (P) to an accessible form, such as orthophosphate. Generally, the host plants can act as biological weed control agents and are not negatively affected by inoculation with cyanide-producing bacterial strains and host-specific rhizobacteria. Through enzymatic activities, growth enhancement by lytic enzymes is another mechanism used by PGPR. Plant growth-promoting rhizobacterial strains such as Serratia marcescens B2can produce certain enzymes such as chitinases against the soil-borne pathogens *Rhizoctonia solani* and *Fusarium oxysporum*. The mycelia of the fungal pathogens co-inoculated with this strain showed abnormalities such as partial bursting of the hyphal tip and swelling in the hyphae.

8.5 Nutrient Uptake Enhancers

There are 13 nutrients that are very important for plants to grow and thrive. These can be divided into micronutrients (required in small doses; less vital for plant growth) and macronutrients (required in large quantities). The macronutrients include nitrogen (N), phosphorus (P), potassium (K), calcium (Ca), magnesium (Mg) and sulphur (S), and the micronutrients include iron (Fe), chlorine (Cl), molybdenum (Mo), zinc (Zn), manganese (Mn), boron (B) and copper (Cu).

Nitrogen (N) is present in very large amount in the atmosphere, approximately 78%, and is the most essential nutrient for plant growth, but it remains unavailable for direct uptake by the plant. Nitrogen-fixing bacteria are of two types: symbiotic (*Rhizobia and Frankia*) (Ahemad and Khan 2012) and non-symbiotic (*Cyanobacteria, Azospirillum* and *Azotobacter*) (Roper and Gupta 2016). It is reported that only 5% or less of total phosphorus (P) present in soil is available for plant uptake, which is the second major nutrient required for plant growth and development (Khan et al. 2016).

Due to its high fixation rate in insoluble forms, most of the soils are deficient in phosphorus and a very less amount is present in soluble form and hence, in general, is not available for uptake by plants. Phosphate-solubilizing bacteria belong to various genera such as *Azotobacter, Pseudomonas, Rhizobium, Bacillus, Enterobacter, Erwinia* and *Serratia*, and they help in phosphate uptake by plants via one or more of the aforementioned mechanisms. Potassium (K) is essential for the growth and development of plants and is a very important nutrient; however, more than 90% of K in the soil exists in insoluble forms. K-solubilizing bacteria caused high productivity of tea plants by enhancing nutrient uptake efficiency as reported by Bagyalakshmi et al. (2012).

Various bacterial genera are now considered as Zn solubilizers, which include *Mycobacterium, Enterobacter, Rhizobium, Burkholderia, Bacillus, Pseudomonas* and *Xanthomonas* (Naz et al. 2016). PGPR solubilize Zn by several mechanisms such as production of chelating agents, excretion of organic acids (2-ketogluconic acid and gluconic acid) and protein extrusion (Nahas 1996; Seshadre et al. 2002). Zinc-solubilizing PGPR play an important role in agriculture and food security (Shaikh and Saraf 2017). Fe is the fourth most abundant and essential element present in the soil. Sulphur is a very important nutrient for plant growth and is taken up by plants mainly in the form of sulphate and sulphur dioxide (Marschner 1995). A major part of sulphur (95%) is unavailable for plant uptake, and only 5% is available (Kertesz and Mirleau 2004). PGPR have gained more attention in agriculture. Temperature, soil moisture, osmotic potential and soil texture are some factors affecting nutrient release. There are various ingredients, i.e. used for plant growth enhancers such as: Kelp is a type of marine algae are very important when fostering plant health that contains trace elements of magnesium, potassium, iron, zinc, nitrogen, etc. Kelp enhances the uptake of plant nutrients, plant cell division and seed germination and contains plant hormones, specifically cytokinins. Armament Technology unlocks nutrients in the soil and guards against nutrient chemical changes and is a biodegradable nutrient enhancer that stimulates the uptake of nutrients by the plant. Humates generally derived from leonardite – a type of coal, e.g. humic and fulvic acids – are carbon-rich soil amendments, and they enhance seed germination, support respiration of the microorganisms and promote root development. Plant Growth Protecting Bacteria (PGPB) is a special class of bacteria that have the ability to promote plant growth and plant defences and inhabit the root zone of the plant, including nitrogen-fixing bacteria and phosphate-solubilizing bacteria (Table 8.1).

8.6 Phytohormone and Metabolite Production

Plant hormones or phytohormones are signal molecules produced within the plants in extremely low concentrations and control all aspects of growth and development, stress tolerance through reproductive development, embryogenesis and the regulation of organ size and pathogen defence. Phytohormones are the most important chemicals and are produced by plants to facilitate various functions. The plant hormones are often diffused by utilizing four types of movements, not always localized hormones are transported within plant tissues are such as auxins, gibberellin, cytokinins, ethylene and abscisic acid.

Auxin, indole-3-acetic acid is produced primarily in the shoot tips and then in developing flowers, and seed was the first plant hormone identified. Auxins with other hormones are responsible for many factors of plant growth alone or in combination, and their transport from cell to cell occurs through the parenchyma. Cytokinins are present in the sites of active or functioning cell division in plants and are the class of plant growth substances that promote cell division, for example, in seeds, fruits, root, shoot tips and the leaves. They work in the presence of auxin hormones to promote cell division and are transported by

TABLE 8.1

Plant Nutrient Uptake and Signs of Deficiency

Nutrient Uptake	Mobility in Soil	Mobility in Plant	Role in Plant Growth	Sign of Deficiency
Nitrogen, NO_3^-	NO_3^-; mobile	NH_4^+; immobile, mobile amino acid	Chlorophyll, proteins	Leaf is yellowing in the middle, reduced and red-brown new growth
Potassium, K^+	Somewhat mobile	Very mobile	Plant metabolism, stress response, regulation of water loss	Yellowing of leaf margins and veins, rolling or crinkling leaves, poor growth
Calcium, Ca^{2+}	Somewhat mobile	Immobile	Cell wall formation	Yellowing new growth, localized tissue necrosis
Magnesium, Mg^{2+}	Immobile	Somewhat mobile	Photosynthesis, chlorophyll	Interveinal chlorosis (yellow leaves with green veins)
Copper, Cu^{2+}	Immobile	Immobile	Lignin production, photosynthesis, plant metabolism	Pale green, withered new growth, yellowing, wilting
Iron Fe^{2+}, Fe^{3+}	Immobile	Immobile	Chlorophyll and enzyme production	Yellowing in new growth
Manganese, Mn^{2+}	Mobile	Immobile	Photosynthesis, respiration, nitrogen assimilation	Interveinal chlorosis on new growth, sunken tan spots on leaves
Zinc, Zn^{2+}	Immobile	Immobile	Chlorophyll, enzymes, proteins, growth hormones	Interveinal chlorosis on new growth
Chlorine, Cl^-	Mobile	Mobile	Opening and closing stomata (respiration)	Yellowing of leaf margins on old growth

xylem cells. The gibberellins are vastly spread throughout the plant kingdom and are rich in seeds and young shoots, where stem elongation is controlled by stimulating both cell division and elongation, and about more than 75 per cent have been isolated. Amino acids produce ethylene, which is a simple gaseous hydrocarbon and appears in most plant tissues. It regulates shredding of leaves, opening of a flower and ripening of fruits and induces seed germination and root hair growth in large amounts when the plants face stressful conditions. Abscisic acid is synthesized in plastids and diffuses in all directions through parenchyma and vascular tissues, and despite the name being abscisic acid, it does not initiate abscission (shredding). Its main effect is inhibition of cell growth with developmental processes, including developing seeds and promoting dormancy in many plants. At low concentration, phytohormones or plant growth regulators (<1 mM) promote, inhibit or modify growth and development of plants (Damam et al. 2016). Plant growth regulators can be applied as exogenously extracted hormones or synthetic analogues to plants or plant tissues and are also called exogenous plant hormones.

The term metabolite is usually bounded to small molecules and the intermediate end product of metabolism. Metabolites have several functions such as catalytic activity of their own defence, including fuel, signalling, stimulatory structure, inhibitory effects on enzymes and interactions with other organisms, i.e. pigments, odourants and pheromones. A primary metabolite is directly involved in reproduction and normal growth development. Ethylene is produced in large scale by industrial microbiology and is an example of a primary metabolite. Secondary metabolites usually have an important ecological function and are directly involved in these processes. Profiling metabolites is an important or essential part of pharmaceutical compounds, and its production is by one of the vigorous and broad-spectrum mechanisms of biocontrol, such as drug metabolism and drug discovery, leading to an understanding of any undesirable side effects. Besides siderophores, PGPR also produce several antagonistic compounds against pathogens, such as volatile compounds, e.g. aldehydes, alcohols, hydrogen cyanide, ketones and sulphides, and non-volatile compounds, e.g. polyketides and heterocyclic nitrogenous compounds (Table 8.2).

TABLE 8.2

Secondary Metabolites

Class	Example
Alcohol	Ethanol
Amino acids	Glutamic acid, aspartic acid
Nucleotides	5′-Guanylic acid
Antioxidants	Isoascorbic acid
Organic acids	Acetic acid, lactic acid
Polyols	Glycerol
Vitamins	B_2

8.7 Competitive Biocontrol Process

Biological control (biocontrol) is the long-term, self-sustaining treatment method, which is the use of an invasive plant's natural enemies as agents (mainly insects, parasites and pathogens) to reduce its population below a desired level and for managing invasive plants. It can be divided into two categories: classical and inundative:

1. Classical biocontrol would fluctuate in a natural predator/prey relationship and uses agent populations.
2. Pathogens such as rusts and nematodes are normally used by inundative biocontrol and are applied to the target weed at high rates in a manner similar to herbicide application.

The long-term solution to managing invasive plants is biocontrol. It is applied to the agricultural environment and has potential to reduce the amount of herbicide in the future and to decrease the economic losses and costs for control. Biocontrol agents attack their host plant in a specific manner, and the visible symptoms of attack may be stems or roots damage, damaged leaves, dropping of leaves, flowers, wilting, discolouration, viability of seeds, etc.

8.7.1 Biocontrol Activities

Phytopathogens cause severe diseases in various crops and decrease the crop yield. Contamination of food grains is also considered responsible for the diminishing quality of food with phytopathogens and high economic losses. Biocontrol of phytopathogens is an effective method without causing harm to the environment so as to treat the plant diseases in an eco-friendly manner (Compant et al. 2016) by using natural organisms. By PGPR, phytopathogens is boon to modern as well as conventional agriculture by biological control of at present, several strains of PGPR in genera such as *Arthrobacter, Azotobacter, Agrobacterium, Erwinia, Flavobacterium, Azospirillum, Bacillus, Caulobacter, Chromobacterium, Enterobacter, Micrococcus, Rhizobium, Pseudomonas* and *Serratia* that are being used to control the diseases of agriculturally essential crops (Ahemad and Kibret 2014.)

8.7.2 Competition

PGPR control the growth of pathogen by reducing its nutrient uptake and by causing a reduction in nutrient availability around the host plant by creating competitive environments. In this competition, non-pathogens and pathogens are responsible for controlling the population of phytopathogens (Pal and Gardener 2006). Siderophore-producing bacteria suppress the growth of plant pathogens and are very essential for this type of interaction as they create iron-limiting conditions in soil (Verma et al. 2019).

8.8 Biotic and Abiotic Stress Management

Stress is a factor that has a negative effect on plant growth (Kamkar, 2016). It increases the formation of reactive oxygen species (ROS) such as H_2O_2, O_2^- and OH^- radicals. Excess ROS production causes damages to plants by oxidizing photosynthetic pigments, proteins, membrane lipids and nucleic acids and by creating oxidative stress. Some active PGPR participation and their role in stress management in plants are shown below.

8.8.1 Abiotic Stress

Abiotic stress (high wind, extreme temperature, salinity, drought, floods, etc.) has a highly negative impact on survival, biomass production and production of staple food crops by up to 70%, which threatens food security worldwide. Tolerance to this stress is multigenic and quantifiable in nature and includes collection of certain stress metabolites such as poly-sugars, glycine, proline and abscisic acid and upregulation in the synthesis of enzymatic and non-enzymatic antioxidants such as superoxide dismutase (SOD), ascorbate peroxidase (APX), catalase (CAT), glutathione reductase, ascorbic acid and glutathione (Agami et al. 2016) and improves crop tolerance by external application of PGPR in compatible osmolytes. Through bacterial strains, such as *Pseudomonas putida* and *Pseudomonas fluorescens* that neutralize the toxic effect of cadmium pollution on barley plants due to their ability to scavenge cadmium ions from soil (Baharlouei et al. 2011), the use of PGPR in plant abiotic stress management has been studied.

8.8.2 Biotic Stress

Stress encourages breeding of resistant crops due to the huge economic loss and big challenge of crop yield. Biotic stress has unfavourable impacts on plants, ecosystem nutrient cycling, natural habitat ecology, co-evolution, population dynamics and horticultural plant health (Gusain et al. 2015). Different pathogens result in a significant reduction in agricultural yield and cause biotic stress, such as bacteria, viruses, fungi, nematodes, protists, insects, and viroids (Haggag et al. 2015). Biotic stress are naturally happening, often intangible and inanimate factors abiotic stress factors, stressors, such as intense sunlight, temperature or wind and involve living disturbances such as fungi or harmful insects, that may cause harm to the plants and animals in the affected area. Abiotic stress includes the potentially adverse effects of salinity, drought, metal toxicity, flood, nutrient deficiency and high and low temperature. In some (rare) cases, such as the supply of water, too little (drought) or too much (flooding) can both impose stress on plants. In reality, abiotic and biotic stresses are often inseparably linked. Abiotic stress management is one of the most important challenges facing agriculture and can continuously limit the choice of crops and agricultural production over large areas, and extreme events can lead to total crop failures.

8.9 Soil Reclamation and Restoration

Land restoration is the process of ecological restoration of a site to a natural landscape and habitat, safe for humans, wildlife and plant communities. For reducing run-off and erosion in sodic soils, reclamation is a common practice. Reclamation of sodic soils involves replacement of exchangeable sodium by calcium. The invention relates to methods for land improvement (soil reclamation), first for soil remediation comprising the steps of (i) placing a material useful for improving soil and an explosive in the soil, and (ii) mixing the materials which are useful for reclaiming the soil and the polluted soil by explosion. The contaminants are separated from all soil particles by a soil reclamation system and generate reusable soil and contaminant products.

8.9.1 Applications of PGPR

PGPR can be found in the rhizosphere at root surfaces and are a heterogeneous group of bacteria that can improve the quality of plant growth directly or indirectly; mainly, PGPR are largely facilitated by the rhizospheric soil between the plant and the microbes. Ca, P, S and other micronutrients are also enriched by them in soil. They add organic matter to it and increase the physical and biological properties of soil. They have a long-term (residual) effect on the crops. Besides all these, PGPR also have effects on soil reclamation through a wide variety of mechanisms such as the production of siderophores, the production of VOCs, abiotic stress tolerance in plants, nutrient fixation for easy uptake by plants, production of plant growth regulators, production of protection enzymes such as chitinase, glucanase and ACC deaminase for the prevention of plant diseases.

8.10 Versatile Applications of PGPR

8.10.1 Biofertilizer

A biofertilizer is a substance that increases the supply or availability of primary nutrients to living microorganisms when applied to seeds, plant surfaces or soil and colonizes the rhizosphere or the interior of the plant, which promotes the growth of the host plant. The use of synthetic fertilizers and pesticides can be reduced by biofertilizers. Biofertilizers stimulate plant growth through the synthesis of growth-promoting substances and solubilize by adding nutrients through the natural processes of nitrogen fixation. Biofertilizers such as *Azotobacter, Azospirillum, Rhizobium* and blue-green algae (BGA) have been in use for a long period. *Azotobacter* can be used with crops such as mustard, potato, cotton, maize, wheat and other vegetable crops. *Azospirillum* inoculations are suggested mainly for sorghum, maize, millets, sugarcane and wheat. Chemical fertilizers are immovable in the soil, so plants absorbed less than 20% of the added fertilizer. Leguminous crops use *Rhizobium* inoculants. Some other types of bacteria such as *Pseudomonas putida* strain P13 and *Pantoea agglomerans* strain P5 are able to solubilize the insoluble phosphate from organic and inorganic phosphate sources. In fact, due to the immobilization of phosphate by organic acids and mineral ions such as Fe, Al and Ca in soil, the rate of available phosphate is well below the needs of plants. Benefits of biofertilizers are the means by which it can symbiotically connect with plant roots by *Rhizobium*. Symbiotic nitrogen fixation with legumes contributes substantially to total nitrogen fixation and fixing the nutrient availability in the soil. On the other hand, the microorganisms involved are easily taken up by the plants and safely convert complex organic materials into simple compounds.

8.10.2 Biopesticides

Biopesticides are certain types of pesticides that are usually less toxic than conventional pesticides. They are generally effective in very small quantities and often decompose quickly, resulting in lower exposures and largely avoiding the pollution problems caused by conventional pesticides. Biopesticides are obtained from such natural materials such as plants, animals, bacteria and certain minerals. For example, canola oil and baking soda have pesticidal applications and affect only the pest and closely related organisms, in contrast to conventional pesticides that may affect organisms as different as birds, mammals and insects. Biopesticides can mainly reduce the use of conventional pesticides, while crop yields remain high as a component of integrated pest management (IPM) programmes. Microbial pesticides have a tendency to be more targeted in their activity than conventional chemicals. For example, a certain fungus might control certain weeds and another fungus might control some insects. The most common microbial biopesticide is *Bacillus thuringiensis*. Natural biopesticides include some insect hormones that regulate mating, moulting and the food-finding behaviours as they have a tendency to control the pests without killing them. For example, they might repel pests or stunt their growth (Raza et al. 2019).

8.11 Future Perspectives of PGPR in Sustaining Agriculture

Through different mechanisms and processes, future prospects and perspective, PGPR has been enhancing the agriculture productivity. However, PGPR vary in the performance due to various environmental factors such as weather conditions or the composition or activity of the indigenous microbial flora of the soil, including climate and soil characteristics (Compant et al. 2005). Soil is the richest medium of natural nanoparticles, both as primary particles and as agglomerates or aggregates. Nanotechnology announces the surfacing of probable narrative applications in the field of agriculture and life sciences with the expansion of new nanodevices and nanomaterials (Dixshit et al. 2013). In the present scenario, biofertilizers offer a great opportunity in the context of nanotechnology to develop eco-friendly compounds that can be easily used in place of chemical pesticides (Caraglia et al. 2011). Bioremediation has emerged as a potential tool to clean up the metal-contaminated or polluted environment. In the rhizosphere, PGPR could significantly speed up the growth of plant and indeed its productivity by reducing the bioavailability of metal contaminants as well as improving plant establishment, health and growth (Ma et al. 2011).

Currently, some of the developed countries such as the USA, China, Japan, Germany, France, Switzerland and South Korea are involved in practices that use nanoscale-based products and technology for agricultural growth. The large-scale implementation of such products for golden rice, seedless bananas, Bt cotton, cucumber, Bt brinjal, etc., cultivation in India is still confined to selected biotechnological products and therefore requires major console to satisfy the needs of the growing population. Future research in rhizosphere biology will depend on the development of biotechnological and molecular approaches to achieving a combined management of soil microbial populations and to increasing our knowledge of rhizosphere biology. The application of multistrain bacterial consortium could be an effective way for reducing the harmful impact of stress on plant growth over single inoculation. Apart from that future research, PGPR products for use of agricultural farmer will be also helpful in optimizing growth condition and not phytotoxic to crop plants, tolerate adverse environmental condition, higher yield and cost-effective.

8.11.1 Applications of PGPR in Agriculture

As inoculants, PGPR are an essential and large component of biofertilizers technology to improve the productivity of agricultural systems in the long run. Great promise has been shown by many PGPR, which can play very critical roles in maintaining the sustainability of agro-ecosystems and can act as potential inoculants for agriculture uses and environmental protection. Presently, the use of PGPR is poor despite numerous reports on their good performance under laboratory conditions. Sustainable agriculture is a type of agriculture that focuses on producing livestock and long-term crops while having least effects on the environment. This type of agriculture tries to find a good balance within the environment between the need for food production and the preservation of the ecological system. There are many goals associated with the production of food – sustainable agriculture, reduction in the use of fertilizers, conservation of water and pesticides and promotion of biodiversity in crops grown in the ecosystem.

Sustainable agriculture also focuses on helping farmers improve their quality of life with maintaining economic stability of farms. The most common techniques include growing plants and many farming strategies or methods that can create their own nutrients to reduce the use of fertilizers and rotating crops in fields, which reduce pesticide use because the change of crops helps make agriculture more sustainable. Another technique is mixing crops, which decreases the need for pesticides and herbicides with the risk of a disease destroying a whole crop and many benefits of sustainable agriculture, which can be divided into two groups, i.e. human health benefits and environmental benefits. Sustainable agriculture uses 30% less energy per unit of crop yield in comparison with industrialized agriculture, one of the most important benefits to the environment. Sustainable agriculture increases area biodiversity and also benefits the environment by reducing soil degradation, balancing soil quality, preventing erosion and saving water. Additionally, sustainable agriculture provided these benefits in three dimensions – ecological, economic and social sustainability – with a variety of organisms with healthy and natural environments to live. The sustainable development of agriculture decreases the use of hazardous chemicals, controls pests and facilitates increased quality of top soil by storing and conserving the rain water. An improper use of pesticides and unacceptable storage may also lead to health problems.

8.11.2 Economic Sustainability

The sustainable development of agriculture tries to accomplish this objective. Governments aim to enhance the production of such products, which are export-oriented, and make agricultural sector sustainable over a long period. Its specialization will help in enhancing the efficiency and production. Developed countries have more demand as they are good in quality, cheap in production and environment-friendly also.

8.11.3 Social Sustainability

The sustainable development aims to increase the level of employment in the country as well as the productivity. Many modern technologies fail because of their complexities and limitations such as many in use and are not easily accessible to poor farmers. The development that meets the need of the present without compromising the ability of future generations to meet their needs is the goal of sustainable development. There are three main objectives in sustainable agriculture – a healthy environment, economic profitability and social and economic equity. In sustainable agriculture and sustainable food systems, there are many practices commonly used by working people. Growers may use methods to minimize water use, promote soil health and lower pollution levels on the farm. Thus, sustainable agriculture is farming in sustainable ways to study the relationships between organisms and their environment based on an understanding of ecosystem services.

8.12 Conclusions and Discussion

This chapter demonstrated the usefulness of PGPR in healthy growth and development of plant species for all PGPR traits. PGPR enhance plant growth due to the production of IAA and aid in phosphate solubilization and ammonia production. In addition to scientific techniques, the sustainable development of agriculture includes disease resistance, drought tolerance, heavy metal stress tolerance, salt tolerance and better nutritional value. Agriculture has been the largest financial source with the use of soil microorganisms such as plant growth-promoting rhizobacteria (PGPR) that aggressively explain the association of plants with microbes, show antagonistic and synergistic interactions, and play a vital role in converting barren, poor-quality land into cultivable land. PGPR improve the plant growth by biofertilization and are a widely available tool for sustainable development and an effective method to treat several plant diseases in an eco-friendly manner. PGPR show various biocontrol activities by production of various metabolites and induction of systemic resistance by controlling the growth of pathogen around the host plant through direct and indirect mechanisms, which involve enhancing plant physiology and resistance to different phytopathogens through various modes and actions.

The PGPR isolates may also be used as biofertilizers to enhance the growth and productivity of commercial plants, thereby improving the economic stability of a developing economy such as India. PGPR use in agriculture is an environment-friendly and cost-effective method. There are several tools commonly used for improving the yield of agricultural crops promoting biodiversity in crops grown in ecosystem and focuses on maintaining economic stability to improve their techniques and the quality of life. PGPR have been enhancing the agriculture productivity with the future prospects and perspectives through modern tools and techniques such as nano-fertilizers, nanomaterials and biosensors from the fields of nanotechnology and biotechnology.

REFERENCES

Agami, R.A., Medani, I.A., Abd El-Mola, R.S. 2016. TahaExogenous application with plant growth promoting rhizobacteria (PGPR) or proline induces stress tolerance in basil plants (*Ocimum basilicum* L.) exposed to water stress. *International Journal of Environment and Agriculture Research* 2(5): 78–92.

Ahemad, M., Khan, M. 2012. Evaluation of plant-growth promoting activities of rhizobacterium *Pseudomonas putida* under herbicide stress. *Annals of Microbiology* 62: 1531–1540.

Ahemad, M., Kibret, M. 2014. Mechanisms and applications of plant growth promoting rhizobacteria: current perspective. *Journal of King Saud University of Science* 26: 1–20.

Akhtar, N., Qureshi, M.A., Iqbal, A., Ahmad M.J., Khan K.H. 2012. Influence of Azotobacter and IAA on symbiotic performance of Rhizobium and yield parameters of lentil. *Journal of Agriculture Research* 50: 361–372.

Anand, B., Kumari, M.A., Mallick, M.A. 2016. Phosphate solubilizing microbes: an effective and alternative approach as bio-fertilizers. *International Journal of Pharmacy and Pharmaceutical Sciences* 8(2): 37–40.

Bagyalakshmi, B., Ponmurugan, P., Balamurugan, A. 2012. Impact of different temperature, carbon and nitrogen sources on solubilization efficiency of native potassium solubilizing bacteria from tea (*Camellia sinensis*). *Journal of Biology Research* 3(2): 36–42.

Baharlouei, J., Pazira. E., Khavazi, K., Solhi, M. 2011. Evaluation of inoculation of plant growth-promoting rhizobacteria on cadmium uptake by canola and barley. *2nd International Conference on Environment Science and Technology* 2: 28–32.

Beneduzi, A., Ambrosini, A., L.M.P., Passaglia. 2012. Plant growth-promoting rhizobacteria: their potential as antagonists and biocontrol agents. *Genetics and Molecular Biology* 35(4): 1044–1051.

Bhardwaj, M.W., Ansari, R.K., Sahoo, N., Tuteja. 2014. Biofertilizers function as key player in sustainable agriculture by improving soil fertility, plant tolerance and crop productivity. *Microbial Cell Factories* 13(66): 1–10.

Caraglia, M., Rosa, G.D., Abbruzzese, A., Leonetti, C. 2011. Nanotechnologies: new opportunities for old drugs. The case of amino-bisphosphonates, *Nanomedical Biotherapeutic Diseases* 1: 103e.

Compant, S., Duffy, B., Nowak, J., Clément, C., Barka, E.A. 2005. Use of plant growth-promoting bacteria for biocontrol of plant diseases: principles, mechanisms of action, and future prospects. *Applied Environment and Microbiology* 71: 4951–4959.

Compant, S., Saikkonen, K., Mitter, B., Campisano, A., Blanco, J.M. 2016. Soil, plants and endophytes. *Plant Soil* 405: 1–11.

Damam, M., Kaloori, K., Gaddam, B., Kausar, R. 2016. Plant growth promoting substances (phytohormones) produced by rhizobacterial strains isolated from the rhizosphere of medicinal plants. *International Journal of Pharmacology Science and Reviews* 37(1): 130–136.

Dixshit, A., Shukla, S.K., Mishra, R.K. 2013. *Exploring Nanomaterials with PGPR in Current Agriculture Scenario PGPR with Special Reference to Nanomaterials*. Lab Lambert Academic Publication, Germany, p. 51.

Egamberdieva, D., Lugtenberg, B. 2014. Use of plant growth-promoting rhizobacteria to alleviate salinity stress in plants. PGPR to alleviate salinity stress on plant growth. In M. Miransari (ed.), *Use of Microbes for the Alleviation of Soil Stresses*. Springer, New York, pp. 73–96.

Glick, B.R., Penrose, D.M., Li, J. 1998. A model for the lowering of plant ethylene concentrations by plant growth-promoting bacteria. *Journal of Theoretical Biology* 190(1): 63–68.

Gonzalez, A.L., Larraburu, E.E., Llorente, B.E. 2015. *Azospirillum brasilense* increased salt tolerance of jojoba during in vitro rooting Ind. *Crops Products* 76: 41–48.

Gusain, Y.S., Singh, U.S., Sharma, A.K. 2015. Bacterial mediated amelioration of drought stress in drought tolerant and susceptible cultivars of rice (*Oryza sativa* L.). *African Journal of Biotechnology* 14: 764–773.

Haggag, W.M., Abouziena, H.F., Abd-El-Kreem, F., Habbasha, S. 2015. Agriculture biotechnology for management of multiple biotic and abiotic environmental stresses in crops. *Journal of Chemical Pharmacology* 7(10): 882–889.

Jahanian, A., Chaichi, M.R., Rezaei, K., Rezayazdi, K., Khavazi. K. 2012. The effect of plant growth promoting rhizobacteria (PGPR) on germination and primary growth of artichoke (*Cynaras colymus*). *International Journal of Agriculture Crop and Science* 4: 923–929.

Kamkar, B. 2016. Sustainable development principles for agricultural activities. *Advanced Plant Agriculture Research* 3(5): 1–2.

Kertesz, M.A., Mirleau, P. 2004. The role of soil microbes in plant sulphur nutrition. *Journal of Experimental Botany* 5(5): 1939–1945.

Khan, Z., Rho, H., Firrincieli, A.H., Hung, V., Luna, O., Masciarelli, S.H., Kim Doty, S.L. 2016. Growth enhancement and drought tolerance of hybrid poplar upon inoculation with endophyte consortia. *Current Plant Biology* 1–10.

Kundan, R., Pant, G., Jado, N., Agrawal, P.K. 2015. Plant growth promoting rhizobacteria: mechanism and current prospective. *Journal of Fertilizers and Pesticides* 6: 2.

Liu, W., Wang, Q., Hou, J., Tu, C., Luo, Y., Christie, P. 2016. Whole genome analysis of halotolerant and alkalo tolerant plant growth-promoting rhizobacterium *Klebsiella* sp. D5A. *Scientific Reports* 6: 26710.

Ma, Y., Prasad, M.N.V., Rajkuma, M., Freitas, H. 2011. Plant growth promoting rhizobacteria and endophytes accelerate phytoremediation of metalliferous soils. *Biotechnology Advances* 29: 248–258.

Mahmood, S., Daur, I., Solaimani, S.G.A., Ahmad, S., Madkour, M.H., Yasir, M., Hirt, H., Ali, S., Ali. Z. 2016. Plant growth promoting rhizobacteria and silicon synergistically enhance salinity tolerance of mung bean. *Frontiers in Plant Science* 7: 1–14.

Marschner, H. 1995. *Mineral Nutrition of Higher Plants*. 2nd ed. Academic Press, London.

Mus, F., Crook, M.B., Garcia, K., Costas, A.G., Geddes, B.A., Kouri, E.D., Paramasivan, P. 2016. Symbiotic nitrogen fixation and the challenges to its extension to non-legumes. *Applied Environment and Microbiology* 82(13): 3698–3710.

Nahas, E. (1996). Factors determining rock phosphate solubilization by microorganisms isolated from soil. *World Journal of Microbiology and Biotechnology* 12: 567–572.

Naz I., Ahmad H., Khokhar S.N., Khan K., Shah A.H. 2016. Impact of zinc solubilizing bacteria on zinc contents of wheat. *American Euras Journal of Agriculture and Environmental Science* 16: 449–454.

Pal, K.K., Gardener, B.M. 2006. Biological control of plant pathogens. *Plant Health Instructions* 2: 1117–1142.

Paredes, S.H., Lebeis, S.L. 2016. Giving back to the community: microbial mechanisms of plant-soil interactions. *Functional Ecology* 30(7): 1–10.

Paustian, K., Lehmann, J., Ogle, S., Reay, D., Robertson, P.G., Smith. P. 2016. *Climate-Smart Soils Nature* 532: 49–57.

Pawar, S.T., Bhosale, A.A., Gawade, T.B., Nale T.R. 2016. Isolation, screening and optimization of exopolysaccharide producing bacterium from saline soil. *Journal of Microbiology and Biotechnology Research* 3(3): 24–31.

Perez, Y.M., Charest, C., Dalpe, Y., Seguin, S., Wang, X., Khanizadeh, S. 2016. Effect of inoculation with *Arbuscular mycorrhizal* fungi on selected spring wheat lines. *Sustainable Agriculture Research* 5(4): 24–29.

Prithiviraj, B., Zhou, X., Souleimanov, A., Kahn, W., Smith, D. 2003. A host-specific bacteria-to-plant signal molecule (Nod factor) enhances germination and early growth of diverse crop plants. *Planta* 216(3): 437–445.

Rathore, P. 2015. A review on approaches to develop plant growth promoting rhizobacteria. *International Journal of Recent Science and Research* 5(2): 403–407.

Raza, R., Yousaf, S., Rajer, F.U. 2016a. Plant growth promoting activity of volatile organic compounds produced by bio-control strains. *Science Letters* 4(1): 40–43.

Raza, W., Ling, N., Yang, L., Huang, Q., Shen, Q. 2016b. Response of tomato wilt pathogen *Ralstonia solanacearum* to the volatile organic compounds produced by a biocontrol strain *Bacillus amyloliquefaciens* SQR-9. *Scientific Reports* 6: 248–256.

Roper, M.M., Gupta, V.S. R. (2016). Enhancing non-symbiotic N_2 fixation in agriculture. *Open Agricultural Journal* 10: 7–27.

Rout, M.E., Callaway, R.M. 2012. Interactions between exotic invasive plants and soil microbes in the rhizosphere suggest that everything is not everywhere. *Annals of Botany* 110: 213–222.

Sanlibaba, P., Cakmak, G.A. 2016. Exo-polysaccharides production by lactic acid bacteria. *Applied Microbiology* 2(2): 1–5.

Seshadre, S., Muthukumarasamy, R., Lakshminarasimhan, C., Ignaacimuthu, S. 2002. Solubilization of inorganic phosphates by *Azospirillum halopraeferans*. *Current Science* 79: 565–567.

Shaikh, S., Saraf, M. 2017. Zinc biofortication: strategy to conquer zinc malnutrition through zinc solubilizing PGPR's. *Biomedical Journal of Scientific & Technical Research* 1(1): 224–226.

Shrestha, J. 2016. Review on sustainable agricultural intensification in Nepal. *International Journal of Society of Scientific Research* 4(3): 152–156.
Shrivastava, A. 2014. U.S. Patent Application No. 13/938,176.
Suhag, M. (2016). Potential of biofertilizers to replace chemical fertilizers. *International Advanced Research Journal in Science, Engineering and Technology* 3(5): 163–167.
Vejan, P., Abdullah, R., Khadiran, T., Ismail, S., Nasrulhaq Boyce, A. 2016. Role of plant growth promoting rhizobacteria in agricultural sustainability a review. *Molecules* 21(5): 573.
Verma, M., Mishra, J., Arora, N.K. 2019. Plant growth-promoting rhizobacteria: diversity and applications. In *Environmental Biotechnology: For Sustainable Future*. Springer, Singapore, pp. 129–173.

9

Exploitation of Silver Nanoparticles in Bioremediation

Punabati Heisnam, Abhinash Moirangthem, Yengkhom Disco Singh, Pranab Dutta, Chabungbam Victoria Devi, and B.N. Hazarika
Central Agricultural University

CONTENTS

9.1 Introduction

Since nineteenth century, industrial revolution has brought out certain advanced innovative technologies, resulting in an excessive consumption of resources along with higher disposal of waste into the environment without much attention (Cecchin et al., 2016). Previously, people disposed waste by piling in holes or pit, which creates the need to find new places consistently. So, people considered that using chemical decomposition can degrade the waste materials quickly, but in the meantime, it can have certain adverse effects on the environment. Environmental pollution becomes a major problems that affect biodiversity, ecosystems and human health. The main activities causing air pollution, water pollution and soil pollution include burning of fossil fuels, draining of industrial waste, introduction of chemicals and industrial smoke. Pollutant materials in air such as smoke, mist, fumes, volatile organic compounds and polycyclic aromatic hydrocarbons can lead to many cancerous diseases.

According to World Health Organization (WHO), heavy metals including mercury, cadmium, zinc, nickel, copper, arsenic, chromium and lead and carbon monoxide, nitrogen oxides and sulphur oxides belong to non-degradable compounds and have deleterious effects on human health, wildlife, livestock, crops and ecosystem. Although heavy metals are regarded as naturally occurring essential elements for cell development, when used in higher concentrations, they are very toxic. The accumulation of heavy metals into the soil by the anthropogenic activities do not undergo microbial degradation and hence retain into the soil for a longer period of time (Adriano, 2003). Soil is known to be a non-renewable resource supporting plant growth, but the rising environmental pollution by industrial sector and excessive release

DOI: 10.1201/9781003181224-9

of heavy metals into the soil reduces the plant growth and quality of agricultural products. The introduction of heavy metals into the food chain leads to long-term toxic effects on human health and the environment. Safe drinking water is a must for humans and their health as most of the water in various countries does not meet WHO standards (Khan et al., 2013). Human life, aquatic life and crop production experience hazardous problems due to the poor quality of drinking water. In addition, soil can also be contaminated through various anthropogenic activities, i.e. rapidly expanding industrial areas, mine tailings, randomly used chemical fertilizers, wastewater irrigation and spillage of petrochemicals that produce toxic substances and affect the entire ecosystem even at very small concentrations. Environmental pollution can have many negative impacts on the natural sources (air, water and soil), and this makes hazards to plants, animals and human being. Pollution mainly occurs when the natural or synthetic elements cannot be destroyed and decomposed by natural means or artificially without creating harm to the environment. Besides, the degradation and decomposition process of such heavy metal elements may last from a few days to many years, which sometimes make them more difficult to remediate. In order to mitigate these concerns, bioremediation is a rising knowledge which can be used to completely manage a diverse group of environmental pollutants to achieve sustainable development.

The process of **bioremediation** is to reduce the pollution of contaminated water, soil and air by using microorganisms, plants, hairy root cultures etc. by degrading into less toxic materials for better sustainable environment. The microorganisms such as bacteria, fungi, actinomycetes, yeasts and plants are involved in bioremediation owing to their ability to break down toxic materials into organic matter and the ability to absorb chemical substances(Katarzyna and Christel, 2013). Some of the naturally existing microorganisms contribute amazing characteristics and have the ability to degrade and transform a large number of compounds, viz. toxic metals, radionuclides, polychlorinated biphenyls (PCBs), polycyclic aromatic hydrocarbons (PAHs) and hydrocarbons found in gasoline. In short, microorganisms have the capacity to accelerate the degradation of toxic heavy metals from polluted sites by eliminating, contemplating and recuperating.

The cost of bioremediation is low compared to other traditional methods and sustainable environmental management techniques. So, bioremediation is one of the efficient measure to bring eco-friendly environment by stimulating microorganism to devastate the contaminants. The bioremediation of a contaminated site can be carried out either in situ or ex situ. In in situ methods, the removal of contaminants is done directly on the contaminated site, whereas in ex situ methods, the polluted materials are transferred from the original contaminated site for bioremediation. According to the mode of degradation, there are three categories of bioremediation.

9.1.1 Bioaugmentation

In this method, exogenous microorganisms, i.e. genetically modified microorganisms from wild habitat (autochthonous and allochthonous), are added to the polluted soil and water for reclamation of the contaminated environment. The underlying principle is to improve pollutant degradation rate after supplying the microorganisms along with rapid improvement in the soil or water quality.

9.1.2 Biostimulation

In this process, bioremediation is carried out by stimulating the naturally occurring bacteria in the severely polluted environment. This process involves supplying oxygen and inorganic nutrient substances, viz. phosphorus, nitrogen, acetate and sulphur, to the contaminated soil either in gas or in liquid form to promote the growth of the biostimulant microbial population for the speedy elimination of pollutants. This is more effective, less expensive and eco-friendly when combined with bioaugmentation method.

9.1.3 Intrinsic Bioremediation

In situ bioremediation technique is practised where microorganism present in the environment is introduced in the polluted sites especially soil and water using natural attenuation to degrade contaminants.

Intrinsic bioremediation is usually undertaken in underground petroleum tanks where there is a difficulty of leakage having the chances of the toxic substance to enter into the leakage and pollute the tank. In this process, microorganisms are introduced to remove toxic polluting substances. Although they are less expensive due to the absence of an external force, achieving a complete degradation takes more time.

9.2 Behaviour of Silver Nanoparticles

The focus on nanotechnology has rapidly enhanced during the past few years (Panda et al., 2020). Nanoparticles can be synthesized naturally, chemically and biologically having specific properties for application in the fields of environmental management, agriculture, industry, medicine, etc. The rapid advancement of nanotechnology helps in improving soil, water and air pollution. Many researchers and scientists in the field of nanotechnology have paid increasing attention to the management of environment. The technique of using nanotechnology for removing contaminants is very specific as it can easily remediate all the pollutants without any difficulties.

In these modern material sciences, nanotechnology has become one of the active subjects; however, the metal nanoparticles are of great interest to many researchers because of their unique and exceptional properties that can be applied in a diverse area (catalyst, electronics, medical, etc.) (Keat et al., 2015). Nanomaterials such as zerovalent metals (Ag, Fe, Zn, etc.) and metal oxides (TiO_2, ZnO and iron oxides) have extensively been studied for treating pollutants. Moreover, studies related to environmental management using nanoparticles such as iron, Zn, ZnO and TiO_2 have been conducted by several researchers, but the review of silver nanoparticles (AgNPs) for the remediation of polluted sites has not been initiated. Silver nanoparticles (AgNPs) because of their antimicrobial properties have the ability to destroy both Gram-negative and Gram-positive bacteria such as *Pseudomonas aeruginosa, Escherichia coli* and *Staphylococcus aureus* by inhibiting their growth and thus are applied in many medical aspects as environmental purificants (Keat et al., 2015). In addition, AgNPs also possess antifungal properties that hinder the growth of numerous strains of microbes such as *Candida glabrata, Aspergillus fumigates, Candida albicans* and *Saccharomyces cerevisiae* (Wright et al., 1999). Many researchers focused on the possibility of producing nanoparticles of metals such as silver and gold as they perform different and have new properties along with definite morphology and function (Porcaro et al., 2016). Compared to gold nanoparticles, silver nanoparticles can effectively control and monitor environmental hazards and can be applied in many forms with a much cheaper rate and easy preparation and adaptation.

9.3 Nanomaterials in Environmental Remediation

Currently, the pollution of environment including soil, water and air is a challenge. To overcome this challenge, the development of a number of nanoscale materials is considered a promising solution. Besides the success stories of using nanotechnologies in consumer products and medical sciences, nanoscale materials improve the environmental conditions through direct applications in the polluted site, promoting prevention, detection and removal of contaminants. Nanomaterials because of their small size function as adsorbents and possess some unique properties such as mechanical, sorption, catalysts and optimal, which help in the removal of various gases (CO, NO_2, SO_2, etc.), heavy metals, pollutant chemicals (Fe, Mn, NO_3^- and arsenic), organic pollutants (aliphatic compounds and hydrocarbons) and biological substances (bacteria, viruses, parasites, etc.). Utilizing exact physicochemical and biological transformation properties makes the nanomaterials enhance their products, followed by proceeding in application from pilot-scale to large-scale implementation. Many scientists preferred the application of nanoscale materials in the environment because of their bigger surface area (U.S. EPA. Science Policy Council, 2007). Compared to conventional materials, nanomaterials provide a higher adsorption capacity due to their high surface area and thus help in remediating industrial wastewater. In the ground of environmental remediation, nanomaterials possess different morphological shapes and perform as catalysts for removing chemicals and organic matter from polluted sites (Surajit et al., 2018).

Different forms of nanomaterials including silver (Ag), iron (Fe), silica (Si), carbon nanotubes, calcium alginate, titanium dioxide (TiO_2), zinc (Zn), zinc oxide (ZnO) and polymers have increasingly been used for removing metals from the polluted water as well as soil.. The occurrence of nanoscale materials is grouped into three categories, i.e. natural, incidental and engineered. Naturally occurring nanoscale materials are those that are not created by humans, i.e. made by nature through biochemicals, and are mostly found in fine sand, dust, volcanic ash, etc. Examples are TiO_2, ZnO, carbon nanotubes, etc. Both incidental and engineered nanomaterials are man-made and are designed with specific properties and released into the environment for management practices (U.S. EPA. Science Policy Council, 2007). In this regard, after concerning many reviewers, several examples of applications of nanomaterials for the remediation of air, water and soil are provided in Table 9.1.

The remediation of polluted environment by using nanomaterials seems more sustainable and eco-friendly, but the production of nanomaterials at a lower cost in huge quantity is a big challenge. In bioremediation, the nanomaterials reveal a quantum effect, which makes possible all chemical reactions with less energy activation and also provides the property of detecting many toxic heavy materials by the phenomenon of surface plasmon resonance. According to many research studies, carbon nanotube (CNT) nanomaterials are used for purification of air where the molecules of air adsorbed on the surface of CNTs activate the electrons by changing the shape and macroscopic resistance (Zhang et al., 2010). The high amount of mercury in combusted vapour gases can be removed using silica–titania nanocomposite because titania molecules have a high surface area and photocatalytic properties and nanosilica enhances the absorption of mercury.

Nanomaterials can also help in the management of wastewater and industrial pollutants. The organic pollutants such as lindane (γ-hexachlorocyclohexane) present in wastewater can be removed using FeS nanoparticles. Nowadays, broad studies on water remediation by TiO_2 nanoparticles are also carried out for eliminating the harmful pollutant phenol from wastewater during photocatalytic movement using wet oxidation methods. At present, nanoparticles are generated as minute smart nanocomposite materials such as silver nano-embedded pebbles having multifunctional property and they can completely eliminate ions of many toxic heavy metals, microbial loads, dyes, etc., in 10 minutes from the water and they are designed to be reused.

TABLE 9.1

Nanomaterials in Remediation of Environmental Pollution

S. No.	Type of Nanoparticles	Removal Target	References
1.	Gold coated with chitosan polymer	Cu^{2+} and Zn^{2+} from aqueous solutions	Sugunan et al. (2005)
2.	Carbon nanotubes/Al_2O_3 nanocomposite	Fluoride	Li et al. (2001)
3.	Ag-doped TiO_2	2,4,6-Trichlorophenol	Rengaraj and Li (2006)
4.	Ag-doped TiO_2 nanofibres	Methylene blue dye	Srisitthiratkul et al. (2011)
5.	Multiwalled carbon nanotubes (MWCTs)	Zn^{2+}	Lu and Chiu (2006)
6.	Fe^0 coated with carboxymethyl cellulose polymer matrix	Hexavalent chromium (Cr^{6+}) from aqueous solutions	Yew et al. (2006)
7.	PAMAM dendrimer composite membrane consisting of chitosan and a dendrimer	CO_2 separation from a feed gas mixture of CO_2 and N_2 on porous substrates	Kouketsu et al. (2007)
8.	Silica nanoparticles prepared by mixing salicylic acid and hyper-branched poly(propylene imine)	Removal of polycyclic aromatic hydrocarbons (PAHs) such as pyrene and phenanthrene, and Pb^{2+}, Hg^{2+}, Cd^{2+}, $Cr_2O_7^{2-}$ from aqueous solutions	Diallo et al. (2007)
9.	Cu/Fe/Ag-doped TiO_2	Nitrate (NO_3^-)	Rengaraj and Li (2006)
10.	Poly(methacrylic acid)-grafted chitosan/ bentonite	Th^{4+}	Baybas and Ulusoy (2011)

Source: Fernanda et al. (2018).

9.4 Biological Production of Silver Nanoparticles

Nanoparticles can be produced by physical (e.g. laser ablation, arc discharge method, evaporation–condensation and direct metal sputtering into liquid medium), biological (use of botanicals and microorganisms) and chemical techniques (chemical reduction, UV-initiated photoreduction, photoinduced reduction, irradiation methods, and pyrolysis). However, physical and chemical processes are very expensive, highly toxic, high energy consuming than biological methods. As a result, using microorganisms in bio-based protocols for the synthesis of nanofactories is one of the solutions that hold enormous possibilities to bring environmentally friendly, economically green, and cost-effective. (Prabhu and Poulose, 2012). Comparatively, the biosynthesis of nanoparticles is very simple, hygienic and sustainable.

Moreover, biological methods are also reliable in 'bottom-up approach', which provides an important alternate pathway for the green synthesis of nanomaterials. The microorganisms include bacteria, fungi, enzymes and plant extracts that can be used for the synthesis of metal nanoparticles such as silver. PAHs (polycyclic aromatic hydrocarbons) are an organic contaminant (e.g. phenanthrene, pyrene and naphthalene) mainly caused by unfinished fossil fuel burning, old storage tanks seepage, spilling of oils, domestic waste, etc. The PAHs can remain for a longer period of time in the environment and are unmanageable, so they produce harmful organic pollutants that cause mutant disorders and that are carcinogenic in nature. Culturing of pure bacteria in contaminated paths helps in the degradation and uptake of PAHs.

Some of the biological syntheses of AgNPs are given below.

9.4.1 Bacteria

A variety of different microbes are utilized for the reduction of Ag^+ ions to AgNPs, and these particles formed are found to be spherical. The main point of this biosynthesis is to produce electron from naturally reducing agents (terpenoids, enzymes, flavonoids, oils, carbohydrates, etc.) to convert Ag^+ ions to Ag atoms. But this process is simple compared to chemical methods. A bacterium is known to be one of the very effective reducing agents, which furnish intracellularly and extracellularly organic and inorganic materials. For example, lactobacillus strains can reduce ions having spherical silver particles with an average size of 25–50 nm (Singh et al., 2015).

The mesophilic bacteria (*Bacillus siamensis* and *Bacillus indicus*) and psychrophilic bacteria (*Pseudomonas meridian, Pseudomonas antarctica, Pseudomonas proteolytica*, etc.) can also synthesize AgNPs (Shivaji et al., 2011; Zhang et al., 2014). Some bacteria (*Enterobacter cloacae, Escherichia coli, Klebsiella pneumoniae,* etc.) provide effective synthesis of AgNPs and are mostly controlled by $AgNO_3$ concentration, pH, temperature parameters, etc. (Minaeian et al., 2008). Some studies proved that interaction of bacteria with silver ions promotes the morphology of the synthesized AgNPs (Morones et al., 2005).

9.4.2 Fungi

Fungi are considered as important and attractive reducing agents for the biosynthesis of AgNPs because of their tolerance to heavy metals and ability to internalize. The fungi have the capability to secrete large quantity of protein responsible for producing greater number of nanoparticles. The culture of fungus on agar after transferring into the liquid medium followed by filtration is used for the biosynthesis of AgNPs. Then, the unwanted biomass is removed and silver nitrate is added to the filtrate. Cultured fungus is much better than bacteria as there is a large biomass production without additional procedures for the extraction of filtrate and also presence of mass mycelial is suitable for large-scale production (Velusamy et al., 2016). Additionally, the metabolic activity of fungi can also be influenced by changing the temperature, pH, time and biomass quantity so that the required characteristics and morphology of the nanoparticles can be achieved. According to the reactions, enzymes present in the filtrate lead to the reduction of silver ions to silver elements (Ag^0) at nanometric scale and the occurrence of synthesis is extracellular.

The biosynthesis of nanoparticles through fungi microorganism can be either intracellular or extracellular. In intracellular synthesis, the mycelial medium acts as supplementary for pioneer metal for the synthesis of nanoparticles (Molnar et al., 2018). In case of extracellular synthesis, fungal biomolecules act as supplementary for pioneer metal synthesis, and this method has no proper procedure although they are broadly used (Costa Silva et al., 2017). The development of extracellular nanoparticles was mainly due to the reaction of Ag ions with thermophilic fungus *Humicola* sp. (Syed et al., 2013). The method involved for extracellular synthesis from the fungus *Humicola* sp. is easy and simple and the spherical shape (5 and 25 nm) so form improved dispersing capability of nanoparticles which can be confirmed using TEM micrograph (Chwalibog et al., 2010).

9.4.3 Plant Extracts

The biosynthesis of nanoparticles using plant extracts from roots, stems and leaves mixed with the solution of silver nitrate at a particular pH, temperature and time can be achieved. The plan extract can produce different sizes of AgNPs, i.e. *Carissa carandas* (30–40 nm) (Singh et al., 2021), *Myrmecodia pendans* (10–20 nm) (Zuas et al., 2014), *Citrus maxima* (2.5–5.7 nm) (Sarvamangala et al., 2013), *Rhynchotechum ellipticum* (51–73 nm) (Hazarika et al., 2014) and *Mentha piperita* (90 nm) (Mubarak Ali et al., 2011). AgNPs were synthesized from the plant *Phyllanthus maderaspatensis* within 24 hours and were also of very good quality with a particle size of 59 nm (Annamalai et al., 2014). Besides variation in size, the utilization of different plant extracts produced different shapes of AgNPs. Various shapes of AgNPs such as triangles and hexagon were formed by utilizing of *Potamogeton pectinatus* L. (Abdelhamid et al., 2013). Some studies also reported that AgNPs in the shapes of octagon, octahedron, cube and rhombic dodecahedron were obtained from banana stems (Das and Velusamy, 2013).

The biosynthesis of AgNPs can be carried out from the extracted leaves and callus (tissue culture) of *Sesuvium portulacastrum* L. plants followed by stabilization with alcohol-based polyvinyl (Nabikhan et al., 2010). *Memecylon edule* leaf extract, which contains saponins (water soluble), forms square-shaped AgNPs having a size of 50–90 nm (Elavazhagan and Arunachalam, 2011). Saponins-rich extracts of *Sapindus* species have also been used for the synthesis of nanoparticles (Singh and Sharma, 2019). The leaf extract of *Tephrosia purpurea* containing proteins and flavonoids was used as the major input in the development of AgNPs (Ajitha et al., 2014). After maintaining room temperature, the biosynthesis of AgNPs through *Alternanthera dentata* plant extract is able to produce particles within 10 minutes (Kumar et al., 2014).

9.5 Biogenesis Silver Nanoparticles for Detection of Organic and Inorganic Contaminants

With the evolution of science, nanotechnology is gradually finding its feed in the detection of pollution through various techniques (electrochemical and optical detection, surface plasmon resonance, atomic fluorescent spectroscopy, flame atomic absorption spectroscopy, liquid chromatography, etc.) for cleaning up the polluted site in a sustainable way. Different reduction processes (adsorption, transformation, photocatalysis and catalysis) are involved in the inorganic and organic nanodetection sensor materials for helping in the detection and removal of toxic inorganic and organic pollutant compounds (Surajit et al., 2018). The detected inorganic pollutants are toxic heavy metals, viz. chromium (Cr), cadmium (Cd), arsenic (As), mercury (Hg) and lead (Pb), and the organic pollutants detected are chlorinated phenols, PAHs, PCBs, phenols, pesticides, azo dyes, etc. (Surajit and Ranjan, 2019).

Mostly, all the sensors used here require extensive sample preparation that takes more time, are highly expensive, are complicated and require trained personnel for obtaining measurement from this instrument. Hence, scientists had gone through different experiments of sensors that are easy to handle and cost-effective. The colorimetric sensor was found to be the best option for the fast detection of water contaminants without any dependence on large and complicated instruments, and also in recent years, many colorimetric sensors based on AgNPs have been invented (Paolo et al., 2020).

The biological production and utilization of silver nanoparticles in the field of sensors is an encouraging work in modern nanotechnology. Among the different sensors (electrochemical sensors, biosensors and optimal sensors) reported for the remediation of environment, biosensors responded more positive during sensing (Surajit et al., 2018). The combined detection of different elements through sensors depends mostly on time of reaction, selectivity, signal-to-noise ratio and limitation in detection. Recently, the beneficial effects and strategies of using silver-based materials as an optimal sensor are examined for their selectivity, functionalization, dimensions, etc., in detecting toxic heavy metals and pesticides (Paolo et al., 2020). Some studies reported that the organic pollutant 4-nitrophenol can be detected with the use of biosynthesized AgNPs from wild twig bark (*Acacia nilotica*) because they possess good electrocatalytic properties (Karuppiah et al., 2014). Organic pollutants and pesticides include plant growth regulators, fungicides, herbicides, insecticides and other chemicals. AgNPs can be used in colorimetric sensor to detect pesticide (Paolo et al., 2020). Examples are AgNPs functionalized by cyclen dithiocarbamate for the detection of herbicides, viz. thiram and paraquat, by simple colorimetric method (Roy et al., 2011).

Many toxic inorganic pollutants such as heavy metals, viz. Cr^{6+}, Cd^{2+}, Cr^{3+}, Hg^{2+}, Cu^{2+}, Mn^{2+}, $Pb^{2+}$and Zn^{2+}, are the main cause of human health problems and environmental hazards (Surajit et al., 2018). In the past few years, cost-effective nanoparticles based on colorimetric sensors have been used in the detection of toxic heavy metals. Among the heavy metals, Hg^{2+} is found to be more dangerous. The combined function of ultra-sensitive colorimetric along with Mercaptobenzothiazoleheterocyclic compounds based AgNPs help in the detection of Hg^{2+} (Bhattacharjee et al., 2018). Original AgNPs along with mercury oligonucleotides were also used as a sensor for the detection of Hg^{2+}. Researchers have also developed sensors which are Hg^{2+} sensitive using the exclusive properties of AgNPs (Ahmed et al., 2014).

Another finding reports that functionalized AgNPs can successfully detect the dangerous Cu^{2+} ions with casein peptide leading to change in colour of the solution providing properties of high detection quality of 0.16 μM ranges in the concentration between 0.08 and 1.44 μM (Ghodake et al., 2018). Cu^{2+} is a conversion metal used in both agriculture and industrial sector and is found to be toxic against microorganisms (fungi, viruses, bacteria, etc.). In drinking water, the presence of Cu^{2+} in a high quantity is extremely harmful to human health. Additionally, AgNPs can also be utilized for the detection of Mn^{2+} using colorimeter detector in combination with melamine and 4-mercapto benzoic acid (Surajit et al., 2018). The detection of ions of heavy metals, such as Hg^{2+}, Pb^{2+} and Mn^{2+}, in water medium was done by a green synthesis-mediated metal biosensor, which was established using the reducing agent L-tyrosine (Zhou et al., 2012).

9.6 Behaviour of Silver Nanoparticles in Aquatic Environment

The behaviour of silver in macro-, micro- or nanoform differs in diverse environmental conditions. Here we will concentrate on the nanoform of silver in different conditions. Silver in nanoform or AgNPs have the ability to restrain the development of harmful bacteria and other microorganisms, which is a vital component for the wastewater treatment procedures. But, microbial population of aquatic and terrestrials environment might be in danger due to harmful effect that cannot be mistreated.

As we know, biofilms which are found everywhere in the aquatic ecosystems such as lakes, ponds, piping systems and rivers involve microorganisms in an extracellular polymeric substances medium (which includes proteins, polysaccharides and nucleic acids) (De Faria et al., 2014). In a study, it was found that biofilms formed from wastewater are very much tolerant to the AgNPs treatment. In heterotrophic plate count measurement, the bacteria in a biofilm after 24 hours of application of 200 mg Ag/L AgNPs were reported to be insignificant. Under the treatment of same conditions, removing extracellular polymeric substances which are freely bound leads to the reduction in the capability of wastewater biofilms.

In contrast to this, when treated as a pure planktonic culture, bacteria isolated from the wastewater biofilms were found extremely vulnerable to AgNPs. The results of many research studies show that biofilms play an important role in preventing the antimicrobial effects of AgNPs among the microbial community and EPS (extracellular polymeric substance). In addition to this, decreasing growth rates may lead to the tolerance of many bacteria to AgNPs. Many reports revealed that there is a variation between microorganisms due to the vulnerability to AgNPs. For example, compared to the bacteria of biofilms, Thiotrichales are highly susceptible to AgNPs.

Compared to planktonic cells, nanoparticles level is much more resistant to biofilms (Sheng and Liu, 2011). While treatment with AgNPs, the growth of bacteria was inhibited in small AgNPs concentration after the utilization of *Escherichia coli* (*E. coli* PHL628) in the range of 10–38 mg/L. Nearly quadrupled confrontation to AgNPs concentration was revealed in the culture of biofilms (De Faria et al., 2014; Thuptimdang et al., 2015). Researchers also observed that there is no change in the heterotrophic plate counts when AgNPs are exposed in the sample of biofilm-forming bacteria (Sheng and Liu, 2011).

Previously, the inhibitory effects of AgNPs are strongly influenced by the biofilms in wastewater treatment plants, but in order to comprehend the effects of structures and to improve the detachment influences by biosorption into wastewater biofilms, further complete researches are required. There was a composite and predictable relationship between biofilms and ENP where they totally vary with environmental conditions. Some researchers reported about the harmful effect of metal oxide nanoparticles on beneficial microorganism while processing wastewater treatment owing to reduction in nutrients at different percentage (Vejerano et al., 2015).

Relatively in many previous works, nitrification inhibition process by AgNPs was well explained and carried out even more in a lower concentration (1 mg Ag/L) found to be effective according to the size of particle, and nature of coating. Dissolution of silver is directly related to the toxicity of AgNPs, such as inhibition of enzyme, disruption of membrane, production of energy and expression of gene. The harmfulness of AgNPs can be enhanced by increased ammonia concentration (Mumper et al., 2013) and lowering water concentration (Yang et al., 2013). AgNPs are found to be extra responsive to ammonia-oxidizing bacteria when compared to nitrate-oxidizing bacteria (Yang et al., 2013).

One of the types of wastewater treatment is the activated sludge process, where the recycling of industrial or sewage wastewaters is conducted with bacteria and protozoa under aerobic conditions. Various metal oxide nanoparticles when added into the wastewater found to be dangerous in microbial populations although they are more effective on activated sludge (Hou et al., 2012). During anaerobic sludge treatment, the important phase of metal oxide nanoparticles was chemical speciation, which converts AgNPs to sulphides at 50 mg/kg concentration. These silver sulphides formed can remain constant up to 6 months in biosolids without degradation (Ma et al., 2013). Thus, in anaerobic conditions, the rise in the concentration of AgNPs directly influences the enhancement of sulphides concentration, and this is theorized somewhere else (Blakelock et al., 2016). While processing solids, the establishment of Ag_2S inside the waste sludge had a negative impact and this might be an obstacle for the engineers in the near future for designing or promoting the plants.

9.7 Pros and Cons of Silver Nanoparticles in Nanoremediation

Nanoremediation is a technique to remediate pollutants by using nanomaterials. This technique is widely used to remediate polluted surface water, ground water, soil, sediments and also wastewater. Many research works have been done on nanomaterials for their versatile use. Most of the research proved that nanomaterials are better than conventional techniques in environmental remediation. The small size of nanomaterials allows them to fill the small pores and empty spaces in the contaminated media and degrade the pollutants into less harmful substances.

Although in agriculture and allied sectors, currently the application of nanomaterials has a number of limitations, in the future, it may help reduce costs, increase efficiency and create more environmentally friendly methods. However, their use could have a negative effect on microflora and fauna in the soil.

The examples of nanomaterials that are commonly used in environment remediation are silver, iron, titanium oxide, gold and iron oxide. Out of the different potential nanoparticles, silver nanoparticles are discussed more because of their diversified use. Silver nanomaterials, for example, can treat wastewater polluted with bacteria, fungi and viruses. Meanwhile, gold nanomaterial is effective in removing pesticides from infected water. But the problem that occurs is the release of silver ions into the wastewater, which cannot be broken down at the treatment plant or in nature. Silver ions are toxic to many organisms.

If they act against only the harmful organisms and not the beneficial organisms, then it will be fine. But, if they act against both, then there may arise many problems.

Nanomaterials such as nanofibres, silver nanoparticles and porous materials help in removing hazardous gases, toxic organic substances, inorganic and biological compounds by functioning as adsorbents and catalysts. As mentioned earlier, due to the high surface area, reactivity, sorption capacity and easy dispersibility after application, a nanoparticle proves to have better environmental remediation than conventional technology. But, the application of nanomaterials for waste management has its own limitations, or otherwise, it may result in an increase in 'nano-pollution' that leads to difficulty in detection and the human health and environmental problems may even go unidentified. The widespread use of nanomaterials for the treatment of water and remediation of the environment will be a major concern due to its adverse effects. Many researchers in the meantime concerned about the efficacy of nanoparticles on the environmental problems and biological approaches, and their harmful effects which can cause damage to the DNA and lipid peroxidation. Utilization of nanoparticles for remediating the environmental pollutants will lead to sustainable future.

While utilizing in situ remediation, two reasons must be considered for the reactive product. First, the products of unconsidered materials may become more harmful when compared to parent materials. Second, effect on products effectiveness and remediation rate. Under the conditions of reduction process, nanoiron converts trichloroethylene into dechlorinate and to dichloroethane and vinyl chloride. This process might be objectionable as the vinyl chloride produced is more harmful than trichloroethylene.

Nanomaterials such as Ag can react easily without targeting the compounds. The exposed nanoparticles are likely to react with sediment, soil and materials of underground water when they all gather together. In case of in situ remediation, the dispersing activity in contaminated area is inhibited along with reduction in their effectiveness during remediation (Kaegi et al., 2011). Bioremediation in some conditions may be recommended deliberately at the identical site or with the identical material as nanoremediation.

Commonly, AgNPs present in consumer products are usually drained down and released ultimately into culvert systems and finally reach the wastewater treatment plants. In wastewater treatment, primary screening and pebbles elimination do not totally sieve out AgNPs and additional condensation occurs in wastewater sludge under the treatment of coagulation, whereas in secondary process, growth systems were suspended by allowing the bacteria to undergo organic decomposition within the water. The AgNPs after passing all the processes of treatment are finally deposited in the environment.

The antimicrobial property of AgNPs are mostly depending on size in which the size of 8 nm is found to be more effective than 11–23 nm size (Pal et al., 2007). Also, the effectiveness of antibacterial in AgNPs mostly depends on the shape; that is, shortened triangular is better than rod- and spherical-shaped nanoparticles (Li et al., 2010).

Commonly, in wastewater treatment under flooded condition, sulphur accumulation are found to be higher than other contaminants (Brown, 2017). Throughout the treatment process, AgNPs change into silver ions with a specific complex property and these ions are detached from the wastewater due to the presence of high chloride and sulphide content (Brown, 2017). The release of silver ions depends mostly on the amount and level of precipitate silver, which may enhance along with increasing concentration of oxygen and reduction in pH, and also depends on the type of materials used and how the products have been embedded by AgNPs. Usually, AgNPs are easily released into the atmosphere if they are mixed in liquid products in the form of spray, while the AgNPs combined with solid matrices (paints and textiles) are emitted very slowly. In many countries, recycling of sewage and sludge may result in alarm to the environment by producing toxic materials to the environment, although 90% of AgNPs treatment was adsorbed (Mueller and Nowack, 2008).

A low pH level and occurrence of bleaching may lead to an increase in the dissolution of silver, and this effect causes acidification, leading to high accumulation of Ag in the global marine ecosystem. Oxidation process occurs once the silver enters the environment and it is converted into Ag^+ ions (Su-Juan et al., 2013). In case of marine fish ecosystem, the Ag^+ ions so formed stop the regulating exchange system of Na^+ and Cl^- ions and are accumulated within the fish gill, which causes blood acidosis (Fabrega et al., 2011).

The lack of knowledge about the harmful effects of nanoparticles, information on exposure, bioaccumulation, risk in absorption, etc., are the major concerns, but majority of the funding agencies related to risk research target only agricultural materials or non-food products (cleaning or washing products), providing that this deficiency in knowledge should continue. Due to diverse behaviour and properties of nanoparticles, the environmental hazards and health problems are not easily assessable. The properties of nanomaterials such as size, shape and charge directly influenced the toxicity, absorption, metabolism, excretion and distribution. Therefore, nanomaterials having different shapes and sizes may possess different levels of hazards although they are of the same chemical composition.

9.8 Conclusions and Future Prospects

Since the inception of human race, we are encountering many challenges with the nature. However, till today, we have always been trying to come up with a solution to every problem now and then. The ever-increasing environmental pollution, which causes serious health hazards, becomes the centre of discussion for the entire environmentalists. To combat with the ever-increasing environmental pollution and accumulation of large amounts of pollutants in the environment, nanoparticles have emerged as one of the most promising and potential technologies to bring a remediating solution. The application of nanotechnology can actually help in cleaning the environment. Its unique delivery property and targeted site specificity can act as a sensor to detect pollutants, control the release of pollutants, remediating pollutants when encapsulated with plants or its products as in both in situ and ex situ experiment. Among the various nanoparticles used in the remediation of environment, silver nanoparticles become one of the best candidates despite their toxicity and safety issues. Silver nanoparticles, when mixed with heterogeneous catalysts, can degrade para-nitrophenol. Although the field is well adapted by the researchers, the exact mechanism of bioremediation through the nanoparticles remains unresolved.

In recent years, the impact of nanotechnology in remediating environmental pollution has increased drastically. It has potential for remediating environmental pollutions in developing sensors, in treating with the pollutants and also in making of pollutant free environment.. It provides a platform for transmitting new discoveries to overcome the existing problems on the environment. The toxic metals accumulation in the environment can be resolved by using nanomaterials rapidly. However, to show more suitable technologies for the remediation of the environment, it is crucial to have an amalgamation of multidisciplinary subjects, which range from microelectronics to molecular biology and include more fascinating environmental studies. The efficient utilization of silver nanoparticles in the environment can improve the traditional phytoremediation technologies. Therefore, nanotechnology can make a bridge in between the environment and the materials science.

REFERENCES

Abdelhamid, A.A., Al-Ghobashy, M.A., Fawzy, M., Mohamed, M.B., and Abdel-Mottaleb, M.M.S.A. Phytosynthesis of Au, Ag, and Au-Ag bimetallic nanoparticles using aqueous extract of sago pondweed (*Potamogeton pectinatus* L.). *ACS Sustainable Chemistry and Engineering* 1, no. 12 (2013): 1520–1529.

Adriano, D. C. *Trace Elements in Terrestrial Environments: Biogeochemistry, Bioavailability and Risks of Metals*. Springer, New York (2003), 2nd edition.

Ahmed, M.A., Hasan, N., and Mohiuddin, S. Silver nanoparticles: green synthesis, characterization, and their usage in determination of mercury contamination in seafoods. *International Scholarly Research Notices* (2014): 1–5.

Ajitha, B., Reddy, Y.A.K., and Reddy, P.S. Biogenic nano-scale silver particles by *Tephrosia purpurea* leaf extract and their inborn antimicrobial activity. *Spectrochimica Acta—Part A: Molecular and Biomolecular Spectroscopy* 121 (2014): 164–172.

Annamalai, A., Christina, V.L.P., Christina, V., and Lakshmi, P.T.V. Green synthesis and characterisation of Ag NPs using aqueous extract of *Phyllanthus maderaspatensis* L. *Journal of Experimental Nanoscience* 9, no. 2 (2014): 113–119.

Baybas, D., and Ulusoy, U. The use of polyacrylamide-alumino silicate composites for thorium adsorption. *Applied Clay Science* 51 (2011): 138–146.

Bhattacharjee, Y., Chatterjee, D., and Chakraborty, A. Mercaptobenzo heterocyclic compounds functionalized silver nanoparticle, an ultrasensitive colorimetric probe for Hg (II) detection in water with picomolar precision: a correlation between sensitivity and binding affinity. *Sensors and Actuators B: Chemical* 255 (2018): 210–216

Blakelock, G.C., Xenopoulos, M.A., Norman, B.C., Vincent, J.L., and Frost, P.C. Effects of silver nanoparticles on bacterio plankton in a boreal lake. *Freshwater Biology* 61, no. 12 (2016): 2211–2220.

Brown, J. *Impact of Silver Nanoparticles on Wastewater Treatment. Nanotechnologies for Environmental Remediation: Applications and Implications*. Springer International Publishing, Cham (2017): 255–267

Cecchin, I., Reddy, K. R., Thome, A., Tessaro, E. F., and Schnaid, F. Nanobioremediation: integration of nanoparticles and bioremediation for sustainable remediation of chlorinated organic contaminants in soils. *International Biodeterioration & Biodegradation* 199 (2016): 419–428.

Chwalibog, A., Sawosz, E., Hotowy, A., Szeliga, J., Mitura, S., Mitura, K., Grodzik, M., Orlowski, P., and Sokolowska, A. Visualization of interaction between inorganic nanoparticles and bacteria or fungi. *International Journal of Nanomedicine* 5 (2010): 1085–1094.

Costa Silva, L.P., Oliveira, J.P., Keijok, W.J., Silva, A.R., Aguiar, A.R., Guimaraes, M.C.C., Ferraz, C.M., Araujo, J.V., Tobias, F.L., and Ribeiro, F. Extracellular biosynthesis of silver nanoparticles using the cell-free filtrate of nematophagus fungus *Duddingtonia flagrans*. *International Journal of Nanomedicine* 12 (2017): 6373–6381.

Das, J. and Velusamy, P. Biogenic synthesis of antifungal silver nanoparticles using aqueous stem extract of banana. *Nano Biomedicine and Engineering* 5, no. 1 (2013): 34–38.

De Faria, A.F., De Moraes, A.C.M., and Alves, O.L. Toxicity of nanomaterials to microorganisms: mechanisms, methods, and new perspectives, In: Duran, N., Guterres, S.S., Alves, O.L. (Eds.), *Nanotoxicology*. Springer New York, New York (2014): 363–405.

Diallo, M.S., Falconer, K., Johnson, J.H., and Goddard, W.A., III. Dendritic anion hosts: perchlorate uptake by G5-NH2 poly (propyleneimine) dendrimer in water and model electrolyte solutions. *Environmental Science and Technology* 41 (2007): 6521–6527.

Elavazhagan, T., and Arunachalam, K. D. Memecylonedule leaf extract mediated green synthesis of silver and gold nanoparticles. *International Journal of Nanomedicine* 6 (2011): 1265–1278.

Fabrega, J., Luoma, S., Tyler, C., Galloway, T., and Lead, J. Silver nanoparticles: behaviour and effects in the aquatic environment. *Environment International* 37, no. 2 (2011): 517–531.

Fernanda, D.G., Mohamed, F.A., Daniel, C.W., and Alexis, F. Nanotechnology for environmental remediation: materials and applications. *Molecules* 23, no. 7 (2018).

Ghodake, G.S., Shinde, S.K., Saratale, R.G., Kadam, A.A., Saratale, G.D., Syed, A., Ameen, F., and Kim, D.Y. Colorimetric detection of Cu2+ based on the formation of peptide-copper complexes on silver nanoparticle surfaces. *Beilstein Journal of Nanotechnology* 9 (2018): 1414–1422.

Hazarika, D., Phukan, A., Saikia, E., and Chetia, B. Phytochemical screening and synthesis of silver nanoparticles using leaf extract of *Rhynchotechu mellipticum*. *International Journal of Pharmacy and Pharmaceutical Sciences* 6, no. 1 (2014): 672–674.

Hou, L., Li, K., Ding, Y., Li, Y., Chen, J., Wu, X., and Li, X. Removal of silver nanoparticles in simulated wastewater treatment processes and its impact on COD and NH 4 reduction. *Chemosphere* 87 (2012): 248–252.

Kaegi, R., Voegelin, A., Sinnet, B., Zuleeg, S., Hagendorfer, H., Burkhardt, M., and Siegrist, H. Behavior of metallic silver nanoparticles in a pilot wastewater treatment plant. *Environmental Science Technology* 45, no. 9 (2011): 3902–3908.

Karuppiah, C., Palanisamy, S., Chen, S.M., Emmanuel, R., Ali, M.A., Muthukrishnan, P., and Al-Hemaid, F. M. Green biosynthesis of silver nanoparticles and nanomolar detection of p-nitrophenol. *Journal of Solid-State Electrochemistry* 18 (2014): 1847–1854.

Katarzyna, H., and Christel, B. Application of Microorganisms in Bioremediation of Environment from Heavy Metals. *Environmental Deterioration and Human Health*, Springer (2013): 215–217.

Keat, C. L., Aziz, A., Eid, A.M., and Elmarzugi, N.A. Biosynthesis of nanoparticles and silver nanoparticles. *Bioresources and Bioprocessing* 47, no. 2 (2015).

Khan, N., Hussain, S. T., Saboor A., Jamila, N., and Kim, K. S. Physiochemical investigation of the drinking water sources from Mardan, Khyber Pakhtunkhwa, Pakistan. *International Journal of Physical Sciences* 33, no. 8 (2013): 1661–1671.

Kouketsu, T., Duan, S., Kai, T., Kazama, S., and Yamada, K. PAMAM dendrimer composite membrane for CO2 separation: formation of a chitosan gutter layer. *Journal of Membrance Science* 287 (2007): 51–59.

Kumar, D.A., Palanichamy, V., and Roopan, S.M. Green synthesis of silver nanoparticles using *Alternanthera dentata* leaf extract at room temperature and their antimicrobial activity. *Spectrochimica Acta A: Molecular and Biomolecular Spectroscopy* 127 (2014): 168–171.

Li, X., Lenhart, J., and Walker, H. Dissolution-accompanied aggregation kinetics of silver nanoparticles. *Langmuir* 26, no. 22 (2010): 16690–16698.

Li, Y.H., Wang, S., Cao, A., Zhao, D., Zhang, X., Xu, C., Luan, Z., Ruan, D., Liang, J., Wu, D., and Wei, B. Adsorption of fluoride from water by amorphous alumina supported on carbon nanotubes. *Chemical Physics Letters* 350 (2001): 412–416.

Lu, C., and Chiu, H. Adsorption of zinc (II) from water with purified carbon nanotubes. *Chemical Engineering Science* 61 (2006): 1138–1145.

Ma, R., Levard, C., Judy, J.D., Unrine, J.M., Durenkamp, M., Martin, B., Jefferson, B., and Lowry, G.V. Fate of zinc oxide and silver nanoparticles in a pilot wastewater treatment plant and in processed biosolids. *Environmental Science and Technology* 48 (2013): 104–112.

Minaeian, S., Shahverdi, A.R., Nohi, A.S., and Shahverdi, H.R. Extracellular biosynthesis of silver nanoparticles by some bacteria. *Journal of Sciences* 17, no. 66 (2008): 1–4.

Molnar, Z., Bodai, V., Szakacs, G., Erdelyi, B., Fogarassy, Z., Safran, G., Varga, T., Konya, Z., Toth-Szeles, E., Szucs, R., and Lagzi, I. Green synthesis of gold nanoparticles by thermophilic filamentous fungi. *Scientific Reports* 8 (2018): 3943.

Morones, J.R., Elechiguerra, J.L., Camacho, A., Holt, A., Kouri, J.B., Ramirez, J.T. and Yacaman, M.J. The bactericidal effect of silver nanoparticles. *Nanotechnology* 16, no. 10 (2005): 2346–2353.

Mubarak Ali, D., Thajuddin, N., Jeganathan, K., and Gunasekaran, M. Plant extract mediated synthesis of silver and gold nanoparticles and its antibacterial activity against clinically isolated pathogens. *Colloids and Surfaces B: Biointerfaces* 85, no. 2 (2011): 360–365.

Mueller, N.C. and Nowack, B. Exposure modeling of engineered nanoparticles in the environment. *Environmental Science and Technology* 42, no. 12 (2008): 4447–4453.

Mumper, C.K., Ostermeyer, A., Semprini, L. and Radniecki, T.S. Influence of ammonia on silver nanoparticle dissolution and toxicity to *Nitrosomonas europaea*. *Chemosphere* 93 (2013): 2493–2498.

Nabikhan, A., Kandasamy, K., Raj, A., and Alikunhi, N. M. Synthesis of antimicrobial silver nanoparticles by callus and leaf extracts from saltmarsh plant, *Sesuvium portulacastrum* L. *Colloids and Surfaces B: Biointerfaces* 79, no. 2 (2010): 488–493.

Pal, S., Tak, Y.K., and Song, J.M. Does the antibacterial activity of silver nanoparticles depend on the shape of the nanoparticle? A study of the gram-negative bacterium *Escherichia coli*. *Applied Environmental and Microbiology* 27 (2007): 1712–1720.

Panda, M.K., Panda, S.K., Singh, Y.D., Jit, B.P., Behara, R.K. and Dhal, N.K. Role of nanoparticles and nanomaterials in drug delivery: an overview. *Advances in Pharmaceutical Biotechnology* (2020): 247–265.

Paolo, P., Luca, B., and Venditti, I. Silver nanoparticles as colorimetric sensors for water pollutants: review. *Chemosensors* 8, no. 26 (2020): 1–29.

Porcaro, F., Carlini, L., Ugolini, A., Visaggio, D., Visca, P., Fratoddi, I., Venditti, I., Meneghini, C., Simonelli, L., Marini, C., Olszewski, W., Ramanan, N., Luisetto, I., and Battocchio, C. Synthesis and structural characterization of silver nanoparticles stabilized with 3-mercapto-1-propansulfonate and 1-thioglucose mixed thiols for antibacterial applications. *Materials (Basel)* 12, no. 9 (2016): 1028.

Prabhu, S., and Poulose, E. Silver nanoparticles: mechanism of antimicrobial action, synthesis, medical applications, and toxicity effects. *International Nano Letters* 2, no. 1 (2012): 1–10.

Rengaraj, S. and Li, X.Z. Enhanced photocatalytic activity of TiO2 by doping with Ag for degradation of 2,4,6-trichlorophenol in aqueous suspension. *Journal of Molecular Catalysis A: Chemical* 243 (2006): 60–67.

Roy, B., Bairi, P., and Nandi, A.K. Selective colorimetric sensing of mercury (ii) using turn off–turn on mechanism from riboflavin stabilized silver nanoparticles in aqueous medium. *Analyst* 136, no. 18 (2011): 3605.

Sarvamangala, D., Kondala, K., Murthy, U.S.N., Narasinga Rao, B., Sharma, G.V.R., and Satyanarayana, R. Biogenic synthesis of AGNP's using Pomelo fruit—characterization and antimicrobial activity against Gram+Ve and Gram–Ve bacteria. *International Journal of Pharmaceutical Sciences Review and Research* 19, no. 2 (2013): 30–35.

Sheng, Z., and Liu, Y. Effects of silver nanoparticles on wastewater biofilms. *Water Research* 45 (2011): 6039–6050.

Shivaji, S., Madhu, S., and Singh, S. Extracellular synthesis of antibacterial silver nanoparticles using psychrophilic bacteria. *Process Biochemistry* 46, no. 9 (2011): 1800–1807.

Singh, R. and Sharma, B. Nanoparticles synthesis and nanotechnological applications of Sapindus species. In *Biotechnological Advances, Phytochemical Analysis and Ethnomedical Implications of Sapindus Species*. Springer-Nature, Singapore (2019): 107–110. (ISBN: 978-981-32-9188-1).

Singh, R., Hano, C., Nath, G. and Sharma, B. Green biosynthesis of silver nanoparticles using leaf extract of *Carissa carandas* L. and their antioxidant and antimicrobial activity against human pathogenic. *Biomolecules* 11 (2021): 299. https://doi.org/10.3390/biom11020299.

Singh, R., Shedbalkar, U.U., Wadhwani, S.A., and Chopade, B.A. Bacteriagenic silver nanoparticles: synthesis, mechanism, and applications. *Applied Microbiology and Biotechnology* 99 (2015): 4579–4593.

Srisitthiratkul, C., Pongsorrarith, V., and Intasanta, N. The potential use of nanosilver-decorated titanium dioxide nanofibers for toxin decomposition with antimicrobial and self-cleaning properties. *Applied Surface Science* 257 (2011): 8850–8856.

Sugunan, A., Thanachayanont, C., Dutta, J., and Hilborn, J.G. Heavy-metal ion sensors using chitosan-capped gold nanoparticles. *Science and Technology of Advanced Materials* 6 (2005): 335–340.

Su- juan, Y., Yong-Guang, Y., and Jing-Fu, L. Silver nanoparticles in the environment. *Environmental Science: Processes Impacts* 15, no. 1 (2013): 78–92.

Surajit, D., and Ranjan, H.D. Microbial Diversity in the Genomic Era. Science Direct (2019). https://www.sciencedirect.com/book/9780128148495/microbial-diversity-in-the-genomic-era.

Surajit, D., Chakraborty, J., Chatterjee, S., and Himanshu Kumar, H. Prospects of biosynthesized nanomaterials for the remediation of organic and inorganic environmental contaminants. *Environmental Science Nano* 5 (2018): 2784.

Syed, A., Saraswati, S., Kundu, G.C., and Ahmad, A. Biological synthesis of silver nanoparticles using the fungus *Humicola* sp. and evaluation of their cytoxicity using normal and cancer cell lines. *Spectrochimica Acta—Part A: Molecular and Biomolecular Spectroscopy* 114 (2013): 144–147.

Thuptimdang, P., Limpiyakorn, T., McEvoy, J., Pruss, B.M., and Khan, E. Effect of silver nanoparticles on *Pseudomonas putida* biofilms at different stages of maturity. *Journal of Hazardous Materials* 290 (2015): 127–133.

U. S. EPA. Science Policy Council. Nanotechnology White Paper. U.S. Environmental Protection Agency (2007). Available at: http://www.epa.gov/ncer/nano/publications/whitepaper12022005.pdf.

Vejerano, E.P., Ma, Y., Holder, A.L., Pruden, A., Elankumaran, S., and Marr, L.C. Toxicity of particulate matter from incineration of nanowaste. *Environmental Science: Nano* 2 (2015): 143–154.

Velusamy, P., Kumar, G. V., Jeyanthi, V., Das, J., and Pachaiappan, R. Bio-inspired green nanoparticles: synthesis, mechanism, and antibacterial application. *Toxicology Research* 32 (2016): 95–102.

Wright, J.B., Lam, K., Hansen, D., and Burrell, R.E. 1999. Efficacy of tropical silver against fungal burn wound pathogens. *American Journal of Infection Control* 27, no. 4 (1999): 344–350.

Yang, Y., Wang, J., Xiu, Z., and Alvarez, P.J.J. Impacts of silver nanoparticles on cellular and transcriptional activity of nitrogen-cycling bacteria. *Environmental Toxicology and Chemistry* 32 (2013): 1488–1494.

Yew, S.P., Tang, H.Y., and Sudesh, K. Photocatalytic activity and biodegradation of polyhydroxybutyrate films containing titanium dioxide. *Polymer Degradation and Stability* 91 (2006): 1800–1807.

Zhang, M., Zhang, K., Gusseme, B.D., Verstraete, W., and Field, R. The antibacterial and anti-biofouling performance of biogenic silver nanoparticles by *Lactobacillus fermentum*. *Biofouling* 30, no. 3 (2014): 347–357.

Zhang, X.X., Liu, W.H., Tang, J., and Xiao, P. Study on PD detection in SF6 using multi-wall carbon nanotube films sensor. *IEEE Transactions on Dielectrics and Electrical Insulation* 17, no. 3 (2010): 838–844.

Zhou, Y., Zhao, H., Li, C., He, P., Peng, W., Yuan, L., Zeng, L., and He, Y. Colorimetric detection of Mn2+ using silver nanoparticles cofunctionalized with 4-mercaptobenzoic acid and melamine as a probe. *Talanta* 97 (2012): 331–335.

Zuas, O., Hamim, N., and Sampora, Y. Bio-synthesis of silver nanoparticles using water extract of *Myrmecodia pendan* (*Sarang semut* plant). *Materials Letters* 123 (2014): 156–159.

10

Role of Nano-Biotechnology in Solid Waste Management

Rakesh K. Sahoo and Shikha Varma
Institute of Physics

Saroj Kumar Singh
CSIR-Institute of Minerals and Materials Technology

CONTENTS

10.1 Introduction

As Gilpin (1976) stated, the alteration in the environment caused by a solid, liquid, gaseous or radioactive substance that is discharged or emitted or deposited in the environment in a large volume and a larger area is defined as waste. Allaby (1997) defines waste as 'a substance in any form of matter which

DOI: 10.1201/9781003181224-10

is found no use by the system that generates it and for which a disposal method has to be designed'. A more generalized definition of waste by Hoornweg et al. (1999) is the following: waste is an unconsumed material deliberately cast away for disposal.

Based on the source of origin and toxicity of the waste, the Central Pollution Control Board (Pariatamby, 2014), India, defines municipal solid waste as residential, commercial and treated medical wastes in solid or semisolid form. The hazardous industrial wastes are excluded from the list of municipal solid waste. However, the Ecological Solid Waste Management Act of the Republic of the Philippines (2000) includes non-hazardous institutional and industrial waste, street sweepings, construction debris, agricultural waste and other non-hazardous/non-toxic solid wastes in the list of municipal solid waste. Also, the above institution defines municipal solid waste as the waste generated by local government units, including domestic, commercial, institutional and industrial wastes, and street litters.

Solid waste management is a planned system that takes care of the seven essential aspects, starting from waste generation to final dumping. According to the Ecological Solid Waste Management Act of the Philippines, solid waste management deals with controlling waste generation, storage, collection, transport and final disposal or utilization, taking care of public health, economic viability and environmental sustainability, and being responsive to public attitudes. Additionally, waste utilization is a broad discipline dealing with the engineering and management aspects of modifying the waste hygienically, aesthetically and economically acceptable.

10.1.1 Perspectives of Nano-Biotechnology for Solid Waste Management at the Global Level

As per the report of the Food and Agriculture Organization of the United Nations (FAO), the world population is increasing rapidly, an almost twofold increase between 1960 and 2011 (from 3 billion to 7 billion). This rate is also expected to increase equally by 2050 (FAQ Statistics, 2013). The population growth led to urbanization and industrialization. Consequently, the consumption pattern of the people and hence the proliferation of waste increased at an alarming rate. Several developed and developing courtiers have adopted waste-to-energy technology as a superior choice to tackle the energy demand and the betterment of solid waste management. The developed countries such as the USA, Japan and China have prioritized the WTE conversion in a modernized way for solid waste management (www.eia.gov). In 2015, the USA had 71 WTE plants with a total electricity generating capacity of 2.3 GW, contributing to about 0.4% of entire US electricity generation.

A crowded country like China had around 166 incineration plants in operation till 2013 (Li et al. 2016). In Europe (especially in France), 70 plants are in operation for electricity generation, whereas in Japan, 102 plants are in operation for electricity generation from solid wastes (Bajić et al. 2015). Japan is the top country worldwide, incinerating almost 74% of the waste generated, followed by 54% in Denmark and 50% in Switzerland (Psomopoulos et al. 2009). On the other hand, Italy is the highest electricity-producing country globally (50 kW–1 MW) by anaerobic co-digestion of solid waste (Pantaleo et al. 2013). In Asia, developing countries such as India, Malaysia and Vietnam used anaerobic digestion of organic wastes for energy generation; however, the production scale is very low. Nguyen et al. (2014) reported that around 4% of Vietnam's electricity demand could be fulfilled by using food waste in energy recovery. The potential of waste-to-energy is yet to be explored by the developing countries. It is partially explored in the developed countries where the technology is mature.

10.1.2 Need and Economics of Solid Waste Management

Environmental pollution causing global warming and energy crises are the serious worldwide challenges of this decade. It is the exact time to understand and realize the potential of waste materials as a future renewable source for energy generation and environmental remediation. Thus, solid waste management in an economically viable and environmentally sustainable way is the foremost need of the twentieth century for a clean environment and green energy generation. Many reports have pointed out the importance of recycling over energy recovery from waste materials. On the other hand, Baran et al. (2016) reported the waste-to-energy by thermal conversion as a feasible solution to an effective waste management strategy. However, high operational and maintenance costs in thermal treatment technologies hinder the waste-to-energy technology familiarity in developing countries (Erses-Yay, 2015).

10.2 Waste Generation, Characteristics and Composition

The increase in waste generation worldwide is mainly due to population growth, urbanization and economic development. In the present scenario, the waste generation rate is higher in developed countries than in developing countries because it is directly related to their socio-economics (Ouda et al. 2016). The East Asia and the Pacific, and the North Africa region produce the highest and lowest amounts, around 23% and 6%, of the total waste generated worldwide. According to a World Bank report, the worldwide average waste generation rate per person per day is 0.74 kg, which varies between 0.11 and 4.54 kg. It is also reported that the waste generation rate is directly related to the per capita income of the countries. In this context, the high-income countries generate 34% of the world's waste (approximately 683 million tons). It is expected that the per capita waste generation in high-, low- and middle-income countries will increase about 19, 40 and 40 percentages by 2050. It is also speculated that in low-income countries, the per capita waste generation rate will be much higher (three times by 2050) than in high-income countries (Kumar and Samadder, 2017). Globalization mainly plays a vital role. The lifestyle, food habits, consumption patterns and living standards of developing countries are changing rapidly (Khan et al. 2016) (Figure 10.1).

The waste composition differs across income levels, reflecting varied patterns of consumption. High-income countries generate relatively less food and green waste, at 32% of total waste, and generate more dry waste that could be recycled, including plastic, paper, cardboard, metal and glass, which account for 51% of waste. Middle- and low-income countries generate 53% and 57% food and green waste, respectively, with the fraction of organic waste increasing as economic development levels decrease. In low-income countries, materials that could be recycled account for only 20% of the waste stream. Across regions, there is not much variety within waste streams

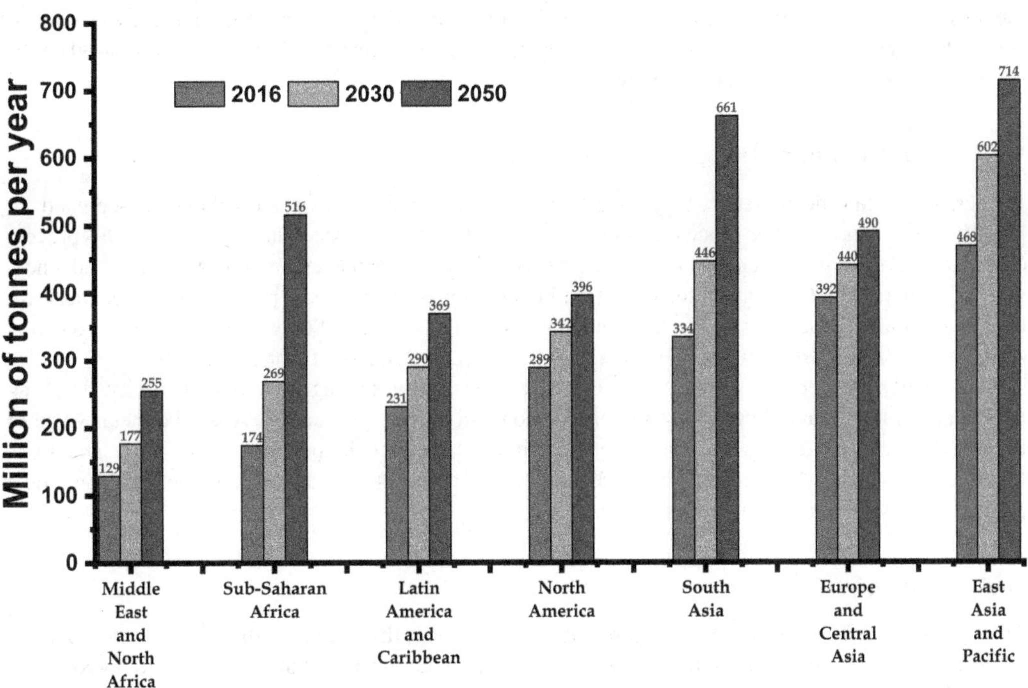

FIGURE 10.1 Projected waste generation, by region (millions of tonnes/year). (Adopted from the World Bank IBRD, *IDA Data Base on Solid Waste Management*, 2019.)

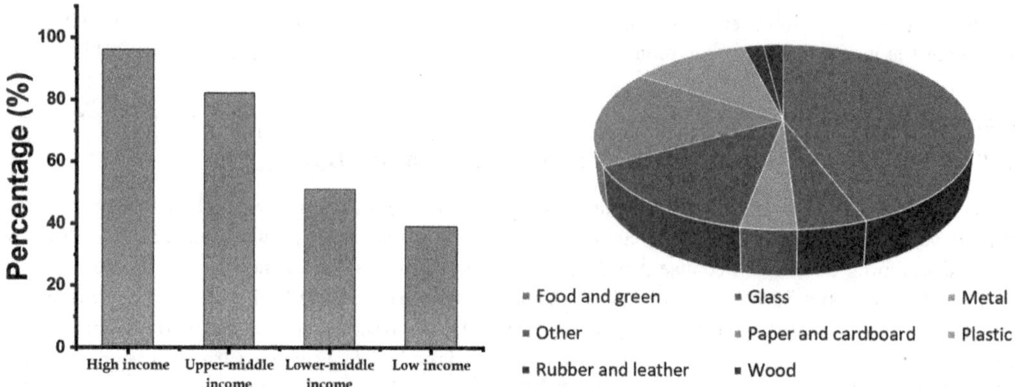

FIGURE 10.2 Waste collection rates, by income level (%), and global waste composition (%). (The World Bank IBRD, *IDA Data Base on Solid Waste Management*, 2019.).

beyond those aligned with income. On average, all regions generate about 50% or more organic waste, except for Europe, Central Asia and North America, generating higher portions of dry waste (Figure 10.2).

10.3 Different Technological Options

10.3.1 Thermal Conversion Technologies

Thermal technology is a major solid waste treatment technology adopted worldwide. This technology has received special attention in the treatment of dry and non-biodegradable wastes and their conversion to usable energy, fuel oil or gas with fewer emissions. In this process, the waste materials are treated in the presence/absence of an oxygen environment with thermal energy or heat supply. The thermal conversion technology is classified into three major categories, which are discussed in detail below based on the heating environment and operating temperature.

10.3.1.1 Incineration

Incineration is an efficient technology adopted worldwide to reduce the waste volume and demand for landfill space. This attractive process can reduce almost 95% of the waste volume. Typically, this process eliminates the methane gas emission as compared to other waste management processes. In the advancement of modern pollution control technologies, incineration is considered a prime process for waste-to-energy generation, especially in developed countries. As per the United States Environmental Protection Agency (US EPA), waste incineration generates clean and green energy (Leme et al. 2014).

Additionally, incineration of waste has the advantages of generating heat and electricity. The slag of the incineration plant is used in road construction and cement production. Also, the metal value of the slag is recovered (ferrous and non-ferrous) from the slag beneficiation (Morf et al. 2013). Due to its large-scale waste digestion and electricity generation capacity, it is considered a reliable and economical process for developing countries worldwide.

10.3.1.2 Pyrolysis

The significant difference between incineration and pyrolysis is the change in their heating environment and heating rate and temperature range. Pyrolysis is carried out in an inert atmosphere, and based on the required product, the temperature range is selected. For example, if the product is wax, oil or char, the wastes are treated in the temperature range of 400°C–800°C, whereas for gaseous output, the temperature is selected above 800°C. Additionally, the quality and quantity of the product typically depend upon the heating rate, process temperature, residence time and composition of the input wastes (Kalyani and Pandey, 2014). However, this process has not achieved a commercial scale. Several reports are there to treat a specific type of waste using pyrolysis; however, they addressed the change in the process parameters. This process has particular importance particularly in recycling the scrap tyre to recover wire, oil, carbon black and gaseous products (Barzallo-Bravo et al. 2019). It is reported that the gas produced by the pyrolysis of the wastes has a conversion efficiency greater than 25% for energy recovery using a gas turbine (Baggio et al. 2008).

10.3.1.3 Gasification

Gasification is another thermal conversion technology for waste treatment and is operational in a controlled oxygen environment at high temperature. In general, in this process, the organic compounds are converted into syngas, which is further combusted to produce energy. In this process, a specific type of waste (wood, coal, etc.) is treated in a homogeneous flow condition, and this is a normal process in the coal industry (Arafat et al. 2015). Recently, this process has received particular attention across the globe due to its less harm to the environment during the waste-to-energy conversion. Modern gasification plants have an enclosure that effectively minimizes the release of contaminants and releases less carbon dioxide than incineration plants of the same capacity (Ma et al. 2019). Asian countries, particularly Japan, have taken the lead in gasification technology to produce energy from the wastes in almost 80 plants, followed by Europe, Africa and America (Ouda et al. 2016). The gasification processes are better than incineration processes based on environmental emission, waste-to-energy conversion efficiency and flue gas rate used during cleaning and pyrolysis. However, the commercial scalability and adaptability of the above two processes (pyrolysis and gasification) in developing countries are yet to be established.

10.3.2 Biological Conversion Technology

Wastes with high organic content/bio-digestible compounds and moisture contents are highly preferable for biological conversion technology (Barzallo-Bravo et al. 2019; Kainthola et al. 2019). This technology is a very environmentally friendly and cleaner way of energy generation and recovery through anaerobic digestion.

10.3.2.1 Anaerobic Digestion

In this process, the organic fraction of the wastes is separated and treated with methanogenic bacteria in a closed environment without oxygen. The organic matter undergoes microbial degradation producing methane-rich biogas and effluent. This biogas is rich in methane (50%–70%) and contains other gaseous contents such as carbon dioxide, water vapour and ammonia (Surendra et al. 2014). The yield and quality of the output gas directly depend on the process parameters and composition of waste. Anaerobic digestion follows four-step biological processes starting with (i) hydrolysis, (ii) fermentation, (iii) acetogenesis and (iii) methanogenesis (Wang, 2014). In the hydrolysis step, the complex organic content in the waste is converted to simple water-soluble compounds such as sugar, amino acids and fatty acids. In the fermentation step, these soluble organic compounds are digested to form simple acid radicals, H_2 and CO_2 as by-products. In methanogenesis, methane formation takes place. A simple schematic of these four steps is presented in Figure 10.3 as presented by Wang in 2014 (Wang, 2014). Murphy and McKeogh (2004) reported that if 1 tonne of waste with 60% organic and 40% moisture content is anaerobically digested, it can generate around 150 kg of methane, a substitute for natural gas in domestic and industrial uses.

Recently, anaerobic digestion has been considered an economical and environmentally sustainable technology for waste treatment and waste-to-energy recovery (Wainaina et al. 2020). Earlier, the process was used for the same purpose. Still, due to the high nitrogen content in the waste, the energy recovery is low and ammonia formation is very high. On the other hand, the by-products from this process were used as a fertilizer. The soil conditioner contains a high amount of nitrogen and unwanted by-products,

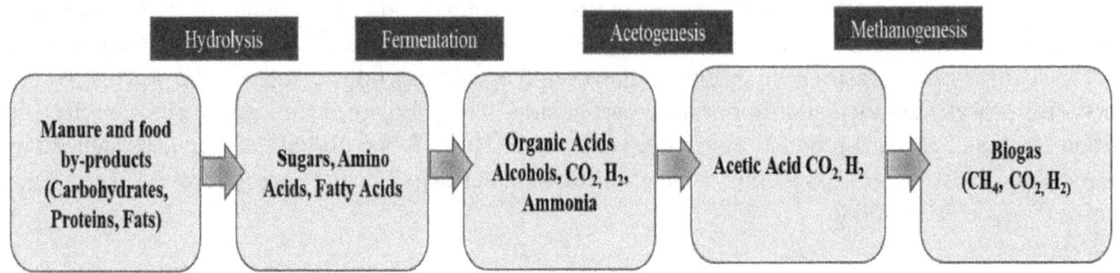

FIGURE 10.3 Schematic showing the steps of the anaerobic digestion of waste-to-energy process. (Adopted from Wang, J., Front. Energy Res., 2, 2014.)

which reduces the soil fertility value. In 2014, European legislation prohibited using anaerobic digested solid as a fertilizer (Browne et al. 2014). In recent years, several process modifications have been carried out to enhance the yield and quality of the product gas. The wastes are mixed with low nitrogen content waste with a suitable inoculum (Gómez et al. 2006), to increase the digestion rate and reduce the nitrogen content in the substrate. Additionally, in the purification step, unwanted gases such as CO_2, ammonia and trace gases are removed from the product to increase the yield and quality of the product to be used as a transportation fuel called biomethane.

10.3.2.2 Landfilling

Generally, this process is carried out in two different ways. First, sanitary landfills result in the controlled release of waste into the land, and the other is unsanitary landfilling. The latter is primarily adopted in developing countries to manage the increasing burden of waste. This type of waste management creates several serious environmental pollution issues (Cherubini et al. 2009; Emery et al. 2007), such as groundwater contamination and land degradation. As reported by Muller et al., the landfill leachates containing recalcitrant compounds are a major contaminant for the surface and groundwater aquifers. However, this type of waste management is highly discouraged in developed countries. Several countries across the globe have considered landfilling as the last option for managing solid waste where there is a limited land area available for dumping wastes (Talyan et al. 2008).

10.4 Nanotechnology and Waste Management

Solid wastes, especially from the industry and urban areas, contain several types of organic and inorganic pollutants that cannot be removed using existing technologies used in solid waste management. Nanotechnology is becoming a modern and alternate technology to treat solid wastes (Mitter and Hussey, 2019). Some of the cases where nanotechnology is used in effective solid waste management are discussed below.

10.4.1 Nano-Tags in Solid Waste Management

Tags made with novel nanomaterials can be incorporated into/on the product during the manufacturing process to track it throughout its lifespan. These tags are unique in size and physical appearance (fine size invisible to the naked eye). Still, the manufacture can easily identify them throughout their lifecycle (Pradeep and Sajanlal, 2011). At the end of life, when the product is thrown into the waste stream, it can be appropriately managed by sorting, separating and sending it to appropriate recycling and reuse based on the tags used by the chemical and product manufacturers. This smart identification process will maximize the recycling and recovery rate of specific wastes such as e-wastes.

10.4.2 Nano-Filters in Waste Management

The significant potential environmental impacts related to landfill leachate are pollution of groundwater and surface waters. Landfill leachate contains pollutants categorized into four groups (dissolved organic matter, inorganic macro-components, heavy metals and xenobiotic organic compounds). Waste and landfill leachates have various toxic and non-biodegradable materials such as heavy metals, dissolved organic matter, inorganic macro-components, arsenic and xenobiotic compounds, which pollute the environment and create several health issues. Common waste management processes are not much standard to eliminate the above contaminants from waste and landfill leachates. Modern technology using nano-filters can reduce these contaminants to a considerable extent in an economical manner. By using nano-filters, the ammonia and COD in the landfill leachates can be reduced to 50% and 60%–70%, respectively, compared to without using a filter. The heavy metals, arsenic and toxic metals such as Ar, Se, Cd, Cr and U were filtered out considerably by the nano-filters in the treatment of wastewater (Li et al. 2016).

Another major problem in the filtration process is the fouling of the film. Conventionally, to avoid this problem, the membrane is cleaned or replaced, which is very costly. However, the use of a nano-filter which has inherent antifouling properties is more capable and economical to filter the waste for a more extended period. Additionally, the carbon-based nanoparticles, i.e. fullerene, have intrinsic antibacterial and antifouling characteristics. Thus, the nanomaterials consisting of fullerene are coated on the surface of the transmission pipe and filtering components to minimize the fouling and clogging in the pipeline (Anand et al. 2019). Recently, nanoparticles have been used to prepare high-performance green concrete from solid waste materials to reduce the usage of natural resources and greenhouse emission gases (Vishwakarma and Ramachandran, 2018).

10.4.3 Nano-Photocatalysts and Nano-Porous Catalysts

Nanoparticles with a high surface-to-volume ratio in the presence of water, oxygen and sunlight/UV light decompose very harmful compounds into less toxic simple compounds. However, these particles as individuals rarely participate in chemical reactions such as oxidation or reduction. Several efficient and low-cost catalysts are used in solid waste treatment, especially in the treatment of landfill leachates.

i. A layer of photocatalyst particles is coated on the fixed membrane (Bosc et al. 2005).
ii. A silica nanostructure decorated with photocatalyst and titanium dioxide nanoparticles is used to treat leachates (Tang et al. 2018).
iii. The hydrophilic nanoparticles such as titanium dioxide (Hu et al. 2016), zinc oxide (Le et al. 2019; Das et al. 2019) and iron oxide (Safari et al. 2019) are used to absorb heavy metals from the landfill leachate and contaminated wastewater.
iv. In the gasification of wastes, porous nano-crystals such as Fe (III)/CaO (Zhang et al. 2014), acidic zeolite (Aho et al. 2007), ZrO_2 and TiO_2 (Lu et al. 2010) are used to catalyse the chemical reaction and the decomposition of syngas generated in the gasification process to form ethanol. Syngas contains carbon dioxide, carbon monoxide, hydrogen and methane, and the carbon monoxide decomposes to form ethanol.

10.5 Nano-Biotechnology in Solid Waste Management

Using nano-biotechnology for deriving biomaterials from bio-waste is highly challenging and also the need of future biomedical device design which can replace part or function of the human body. Some of the well-reported biomaterials derived from bio-waste having potential for multifunctional applications are discussed. The biological options are pretty open to the treatment of solid waste.

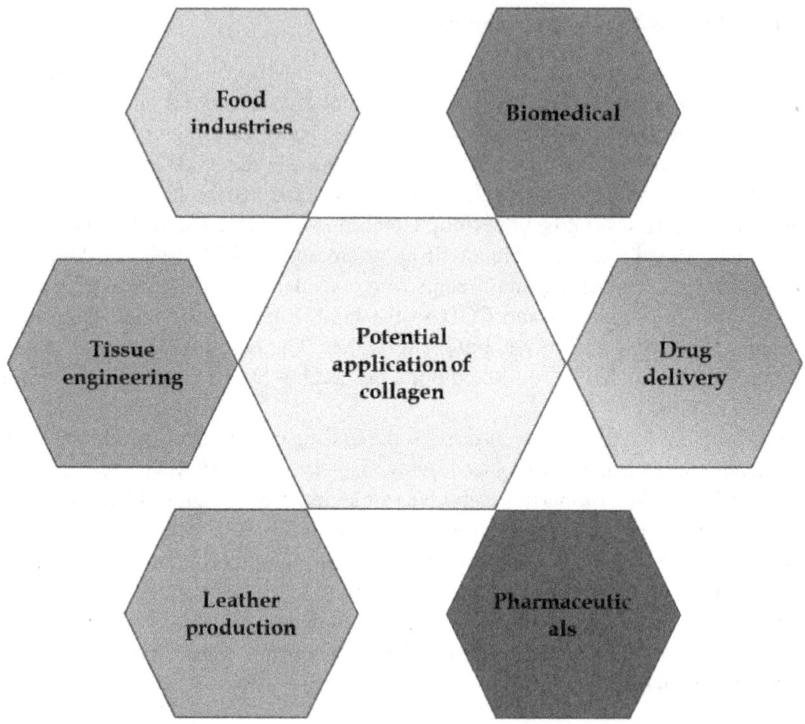

FIGURE 10.4 Applications of collagen. (Adopted from Xu, C., Nasrollahzadeh, M., Sajjadi, M., Maham, M., Luque, R., and Puente-Santiago, A.R., *Renew.* Sustain. Energy Rev., 112, 195–252, 2019a.)

10.5.1 Collagen and Collagen-Based Biopolymers

Collagen is the most abundant fibrous protein found in animal and human bodies. Bio-wastes, especially fish residues, are a reliable and eco-friendly source of collagen. Several routes are adopted for the synthesis of collagen from fish residues (Zhang et al. 2020). Recently, a process based on extrusion and hydro-extraction has been adopted for the extraction of collagen from tilapia fish scale (Huang et al. 2016). In this process, type I collagens were produced with a yield of 16 g protein/100 g crude protein containing tilapia fish scale basis, with minimal waste products. Since collagen has some inherent bioactive properties such as biodegradability, bio-compatibility, non-immunogenicity and low antigenicity, it can be extensively used in several biomaterials production (as shown in Figure 10.4) (Xu et al. 2019a).

10.5.2 Chitin or Chitosan-Derived Biomaterials

Chitin is extracted from the crustacean waste of the fishing industry, mostly found in bio-wastes such as shrimp, crab, lobster, prawn and krill shells (Yang et al. 2019). These wastes contain chitin in the range of 20%–30% by weight. Chitosan is prepared by the deacetylation or alkaline hydrolysis of the acetamide group present in chitin (Pacheco et al. 2011). Atom transfer radical polymerization and graft copolymerization techniques have been used to modify and attach surface functional groups to the chitosan and its derivatives (Thakur and Thakur 2014). Grafting with copolymers increases the functionality of chitosan and chitosan-derived materials (Rajeswari et al. 2016). Chitosan and its derivatives are found in several eco-friendly applications, as shown in Figure 10.5. In wastewater treatment, chitosan is used as a coagulant or flocculent to remove charged particles. To adsorb aqueous nitrate ions, the complexion of Pb(II) and Cd(II), polyvinyl alcohol–chitosan composite, goethite/chitosan composite and carboxymethyl chitosan are used. Nano-chitosan demonstrates superior drug delivery capacity; for example,

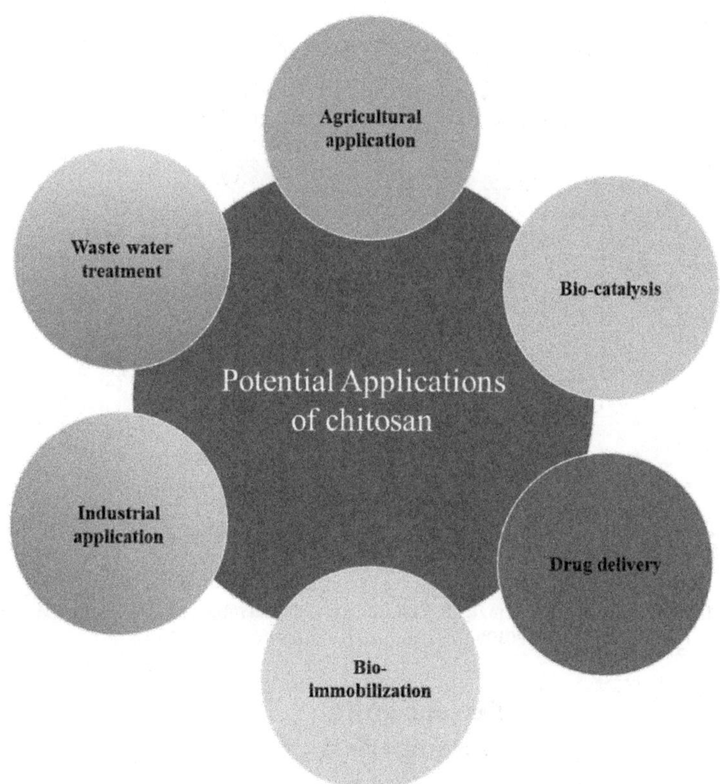

FIGURE 10.5 Applications of chitosan. (Adopted from Xu, C., Nasrollahzadeh, M., Sajjadi, M., Maham, M., Luque, R., and Puente-Santiago, A.R., *Renew. Sustain. Energy Rev.*, 112, 195–252, 2019a.)

catechin-loaded chitosan nanoparticles are highly efficient for the controlled the release of polyphenol in the gastrointestinal tract (Thakur and Thakur 2014; Rajeswari et al. 2016).

10.5.3 Hydroxyapatite (HA)-Derived Biomaterials

Bio-wastes such as eggshells, seashells, animal bones and algae are processed to extract hydroxyapatite biomaterials and have the potential for several biomedical applications (Figure 10.6).

i. Hydroxyapatite derived from bio-waste is highly efficient in removing selenium from water compared to the commercial HA (fourfold higher) (Kongsri et al. 2013).
ii. Due to its superior cell adhesion behaviour, it can be used as a bone scaffold and in tissue regeneration (Pon-On et al. 2016).
iii. HA derived from fish bones of rainbow trout, cod and salmon contain more osteoblasts than commercial HA (Sunil and Jagannatham 2016).
iv. HA prepared from eggshells and seashells is an efficient catalyst for chemical reactions (Xu et al. 2019b).

10.5.4 Bio-Plastics

Day by day, plastic is becoming the biggest threat to the environment and mankind due to its non-degradability. Most plastics synthesized from biomaterials are degradable, with minor exceptions to cellulose acetate that does not decompose in the environment and bio-PET from bio-based ethylene glycol (as

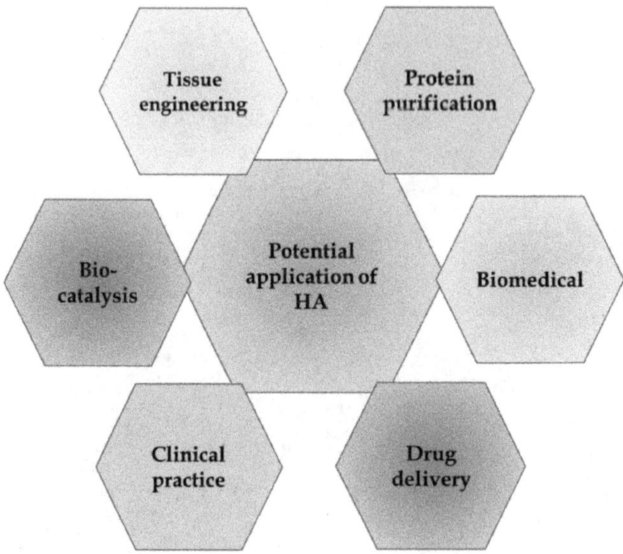

FIGURE 10.6 Applications of hydroxyapatite. (Adopted from Xu, C., Nasrollahzade, M., Selva, M., Issaabadib, Z., and Luque, R., *Chem.* Soc. Rev. 48, 4791–822, 2019b.)

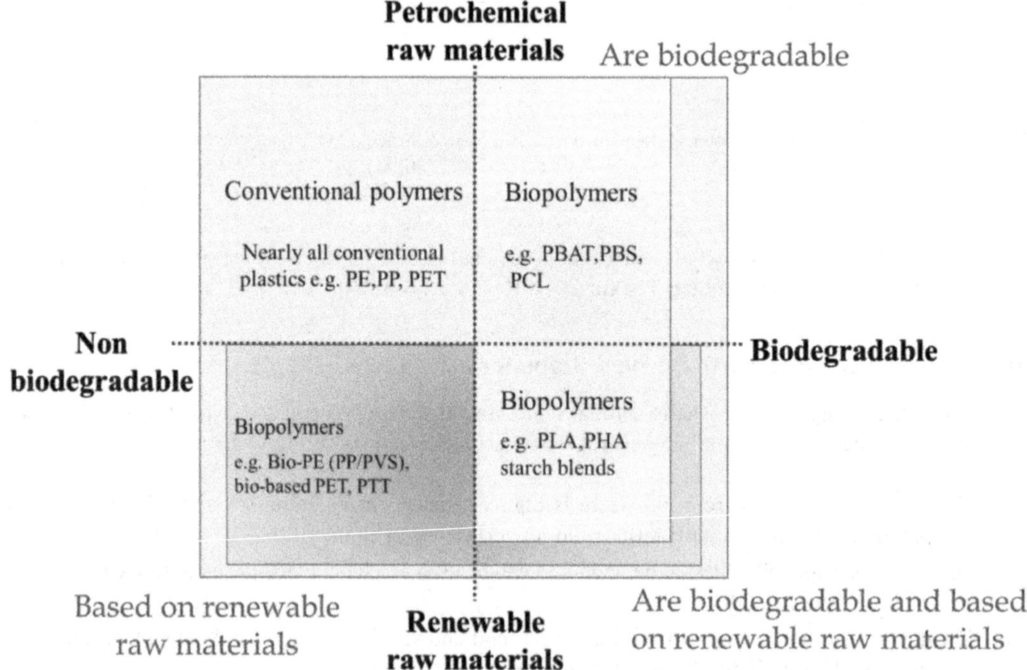

FIGURE 10.7 Different type of plastics. (Adopted from Xu, C., Nasrollahzade, M., Selva, M., Issaabadib, Z., and Luque, R., *Chem. Soc. Rev.* 48, 4791–822, 2019b.)

shown in Figure 10.7) (Xu et al. 2019b). Several bio-wastes are utilized in the synthesis of bio-plastics. Vasconcelos et al. (2019) used myofibrillar proteins extracted from the gilded fish residue to synthesize bio-plastic films. Bayer et al. (2014) reused edible vegetable wastes such as rice hulls, cocoa pod husks, wastes of parsley and spinach stems to synthesize bio-plastic in the presence of trifluoroacetic acid.

Perotto et al. (2018) reported the synthesis of bio-plastic from the powder of vegetable wastes such as carrots, radicchio, parsley and cauliflower. Nano-biotechnological options for increasing the preparation of biodegradable plastics from ca. 2.05 million tonnes in 2017 to 2.44 million tonnes by 2022 have been targeted to be explored worldwide.

10.6 Thermal Plasma Processing in Solid Waste Management

Thermal plasma processing has received great attention in solid waste management, especially in municipal and medical waste management, producing minimal air pollutants with inert slag as the residue. The organic part in the above type of wastes is converted into syngas (consisting of H_2 and CO), and the inorganic part is vitrified into inert slag. The significant advantages of this process are the following:

i. The high temperature, high enthalpy and faster reaction time provide high throughput in a small reactor.
ii. As compared to the traditional combustion process, the volume of gas generated is minimal and the required oxidant for gasification is also very less.
iii. High heat flux densities reduce the reaction time to reach steady state, and rapid quenching helps in vitrifying the residue.
iv. The process is versatile in treating solid, liquid and gaseous wastes with a three-step process inside the reactor, such as pyrolysis without oxygen, gasification with oxygen and rapid quenching in slag viterification.

We are using our indigenously developed thermal plasma reactors to treat and convert several solid wastes into value-added products.

10.6.1 Bio-Wastes to Carbon/Carbon Hybrid

Bio-wastes such as rice husk (Singh et al. 2002), banana peel, banana peduncle biomass, effluent sludge of phosphate fertilizer industries (Singh et al. 2002) and polystyrene (Sahoo et al. 2019) are converted to carbon hybrid and carbon-based nanomaterials.

10.6.2 Industrial Wastes to Value-Added Products

The thermal plasma reactor was used to treat the industrial wastes such as aluminium dross, mine waste to mullite (Pani et al. 2015), magnesium carbonate to magnesium aluminate (Dash et al. 2017), natural sillimanite rock to fused mullite (Mekap et al. 2019), boric powder to boron carbide – a high-value material (Mishra et al. 2015). In a particular case, the low-quality rejected gemstones were processed by our group to a high-value material (Sahoo et al. 2014, 2015, 2016a,b, 2017; Rout et al. 2017; Rao et al. 2016). A new process was developed using thermal plasma to treat the low-quality gemstones on a large scale with a very minimal time (Sahoo et al. 2017).

10.7 Conclusions and Future Perspectives

This chapter comprehensively summarized the present scenario of nano-biotechnology in solid waste management. An attempt was made to summarize the current scenario in solid waste management worldwide. Additionally, the two major aspects of solid waste management, waste-to-energy and recycling, were addressed in detail. After reviewing all the literature, it is observed that waste recycling is a preferred option in solid waste management than energy recovery. Additionally, several important aspects related to this study are summarized below.

i. It is noted that countries, especially developed countries, are paying attention to both energy recovery and recycling of waste. The developed countries with a high rate of energy recovery have a considerable rate of recycling, whereas in developing countries, the recycling rate is very low due to the lack of technological options; landfilling is the only option.
ii. Anaerobic digestion of organic wastes, incineration of mixed wastes, pyrolysis, gasification of a specific type of waste, such as wood, tyre, e-waste and plastics, and landfilling of inert wastes are considered the preferred waste-to-energy recovery options.
iii. Managing plastic wastes is a very crucial challenge in solid waste management. Nano-biotechnological options for increasing the preparation of biodegradable plastics from ca. 2.05 million tonnes in 2017 to 2.44 million tonnes by 2022 have been targeted to be explored worldwide.

The study found that the nano-biotechnological options are yet to be explored on a bigger scale for the smart management of solid waste in the future. The developed countries where the technology is mature prioritize the use of nano-biotechnological options in solid waste management. However, the establishment and commercialization of this technology involves tackling the severe health and environmental issues of solid waste management in the long run.

REFERENCES

Aho, A., Kumar, N., Eränen, K., Salmi, T., Hupa, M., and Murzin, D.Y. 2007. Catalytic pyrolysis of biomass in a fluidized bed reactor: Influence of the acidity of h-beta zeolite. *Process Safety and Environmental Protection*. 85:473–80.

Allaby, M. 1997. *Dictionary of Environment*. Macmillan Press Ltd. 512.

Anand, A., Unnikrishnan, B., Wei, S.-C., Chou, C.P., Zhang, L.-Z., and Huang, C.-C. 2019. Graphene oxide and carbon dots as broad-spectrum antimicrobial agents – a minireview. *Nanoscale Horizons*. 4:117–37.

Arafat, H.A., Jijakli, K., and Ahsan, A. 2015. Environmental performance and energy recovery potential of five processes for municipal solid waste treatment. *Journal of Cleaner Production*. 105:233–40.

Baggio, P., Baratieri, M., Gasparella, A., and Longo, G.A. 2008. Energy and environmental analysis of an innovative system based on municipal solid waste (MSW) pyrolysis and combined cycle. *Applied Thermal Engineering*. 28:136–44.

Bajić, B.Ž., Dodić, S.N., Vučurović, D.G., Dodić, J.M., and Grahovac, J.A. 2015. Waste-to-energy status in Serbia. *Renewable and Sustainable Energy Reviews*. 50: 1437–44.

Baran, B., Mamis, M.S., and Alagoz, B.B. 2016. Utilization of energy from waste potential in Turkey as distributed secondary renewable energy source. *Renewable Energy*. 90:493–500.

Barzallo-Bravo, L.A., Carrera-Villacrés, D., Vargas-Verdesoto, R.E., Ponce-Loaiza, L.K., Correoso, M., and Gavilanes-Quishpi, Á.P. 2019. Bio-digestion and post-treatment of effluents by bio-fermentation, an opportunity for energy uses and generation of organic fertilizers from bovine manure. *International Journal of Recycling of Organic Waste in Agriculture*. 8: 431–8.

Bayer, I.S., Guzman-Puyol, S., Heredia-Guerrero, J.A., Ceseracciu, L., Pignatelli, F., and Ruffilli, R., 2014. Direct transformation of edible vegetable waste into bioplastics. *Macromolecules*. 47:5135–43.

Bosc, F., Ayral, A., and Guizard, C. 2005. Mesoporous anatase coatings for coupling membrane separation and photocatalyzed reactions. *Journal of Membrane Science*. 265:13–9.

Browne, J.D., Allen, E., and Murphy, J.D. 2014. Assessing the variability in biomethane production from the organic fraction of municipal solid waste in batch and continuous operation. *Applied Energy*. 128:307–14.

Cherubini, F., Bargigli, S., and Ulgiati, S. 2009. Life cycle assessment (LCA) of waste management strategies: Landfilling, sorting plant and incineration. *Energy*. 34:2116–23.

Das, A., Sahoo, R.K., Kumar, Mishra, D., Singh, S.K., Mane, R.S., and Kim, K.H. 2019. Thermal plasma-inspired synthesis of ZnO1−XMnx dilute magnetic semiconductors for enhanced visible light photocatalysis. *Applied Surface Science*. 467–468:1059–69.

Dash, S., Sahoo, R.K., Das. A., Bajpai, S., Debasish, D., and Singh, S.K. 2017. Synthesis of MgAl2O4 spinel by thermal plasma and its synergetic structural study. *Journal of Alloys and Compounds*. 726:1186–94.

Emery, A., Davies, A., Griffiths, A., and Williams, K. 2007. Environmental and economic modelling: A case study of municipal solid waste management scenarios in Wales. *Resources, Conservation and Recycling*. 49:244–63.

Erses-Yay, A.S. 2015. Application of life cycle assessment (LCA) for municipal solid waste management: A case study of Sakarya. *Journal of Cleaner Production*. 94:284–93.

FAO Statistics. 2013. Food and Agriculture Organisation of the United Nations. http://faostat3.fao.org/faostat-gateway/go/to/home/E.

Gilpin, A. 1976. *Dictionary of Environmental Terms*. London: Routledge and Kegan Paul Ltd. 169.

Gómez, X., Cuetos, M.J., Cara, J., Morán, A., and García, A.I. 2006. Anaerobic co-digestion of primary sludge and the fruit and vegetable fraction of the municipal solid wastes: Conditions for mixing and evaluation of the organic loading rate. *Renewable Energy*. 31:2017–24.

Hoornweg, D., Thomas, L., and Otten, L. 1999. Composting and its- applicability in developing countries. *Urban Waste Management Working Paper Series 8*. Washington, DC: World Bank.

Hu, L., Zeng, G., Chen, G., Dong, H., Liu, Y., and Wan, J., 2016. Treatment of landfill leachate using immobilized *Phanerochaete chrysosporium* loaded with nitrogen-doped TiO_2 nanoparticles. *Journal of Hazardous Materials*. 301:106–18.

Huang, Chun-Yung, Kuo, Jen-Min, Wu, Shu-Jing, Tsai, Hsing-Tsung. 2016.Isolation and characterization of fish scale collagen from tilapia (Oreochromis sp.) by a novel extrusion–hydro-extraction process.Food Chemistry.190:997-1006.

Kainthola, J., Kalamdhad, A.S., and Goud, V.V. 2019. A review on enhanced biogas production from anaerobic digestion of lignocellulosic biomass by different enhancement techniques. *Process Biochemistry*. 84:81–90.

Kalyani, K.A., and Pandey, K.K. 2014. Waste to energy status in India: A short review. *Renewable and Sustainable Energy Reviews*. 31:113–20.

Khan, D., Kumar, A., and Samadder, S.R. 2016. Impact of socioeconomic status on municipal solid waste generation rate. *Waste Management*. 49:15–25.

Kongsri, S., Janpradit, K., Buapa, K., Techawongstien, S., and Chanthai, S. 2013. Nanocrystalline hydroxyapatite from fish scale waste: Preparation, characterization and application for selenium adsorption in aqueous solution. *Chemical Engineering Journal*. 215–216:522–32.

Kumar, A., and Samadder, S.R. 2017. An empirical model for prediction of household solid waste generation rate – A case study of Dhanbad, India. *Waste Management*. 68:3–15.

Le, A.T., Pung, S.-Y., Sreekantan, S., Matsuda, A., and Huynh, D.P. 2019. Mechanisms of removal of heavy metal ions by ZnO particles. *Heliyon*. 5:e01440.

Leme, M.M.V., Rocha, M.H., Lora, E.E.S., Venturini, O.J., Lopes, B.M., and Ferreira, C.H. 2014. Techno-economic analysis and environmental impact assessment of energy recovery from Municipal Solid Waste (MSW) in Brazil. *Resources, Conservation and Recycling*. 87:8–20.

Li, J., Zhao, L., Qin, L., Tian, X., Wang, A., and Zhou, Y, 2016. Removal of refractory organics in nanofiltration concentrates of municipal solid waste leachate treatment plants by combined Fenton oxidative-coagulation with photo – Fenton processes. *Chemosphere*. 146:442–9.

Lu, Q., Zhang, Y., Tang, Z., Li, W.-Z, and Zhu, X.-F. 2010. Catalytic upgrading of biomass fast pyrolysis vapors with titania and zirconia/titania based catalysts. *Fuel*. 89:2096–103.

Li, J., Zhao, L., Qin, L., Tian, X., Wang, A., Zhou, Y., et al. 2016. Removal of refractory organics in nanofiltration concentrates of municipal solid waste leachate treatment plants by combined Fenton oxidative-coagulation with photo – Fenton processes. *Chemosphere*. 146:442–449.

Ma, C., Liu, X., Min, J., Li J., Gong, J., and Wen, X., 2019. Sustainable recycling of waste polystyrene into hierarchical porous carbon nanosheets with potential applications in supercapacitors. *Nanotechnology*. 31:035402.

Mekap, A., Sahoo, R.K., Das, A., Debasish, D., Bajpai, S., and Singh, S.K. 2019. Two-step plasma mediated synthesis of mullite and sillimanite powder and their suspensive spray coating on stainless steel. *Surface and Coatings Technology*. 372:103–10.

Mishra, A., Sahoo, R.K., Singh, S.K., and Mishra, B.K. 2015. Synthesis of low carbon boron carbide powder using a minimal time processing route: Thermal plasma. *Journal of Asian Ceramic Societies*. 3: 373–6.

Mitter, N., and Hussey, K. 2019. Moving policy and regulation forward for nanotechnology applications in agriculture. *Nature Nanotechnology*. 14:508–10.

Morf, L.S., Gloor, R., Haag, O., Haupt, M., Skutan, S., and Lorenzo, F.D., 2013. Precious metals and rare earth elements in municipal solid waste – Sources and fate in a Swiss incineration plant. *Waste Management*. 33:634–44.

Murphy, J. D., and McKeogh, E. 2004. Technical, economic and environmental analysis of energy production from municipal solid waste. *Renewable Energy* 29(7):1043–1057.

Nguyen, H.H., Heaven, S., and Banks, C. 2014. Energy potential from the anaerobic digestion of food waste in municipal solid waste stream of urban areas in Vietnam. *International Journal of Energy and Environmental Engineering*. 5:365–74.

Ouda, O.K.M., Raza, S.A., Nizami, A.S., Rehan, M., Al-Waked, R., and Korres, N.E. 2016. Waste to energy potential: A case study of Saudi Arabia. *Renewable and Sustainable Energy Reviews*.61:328–40.

Pacheco, N., Garnica-Gonzalez, M., Gimeno, M., Bárzana, E., Trombotto, S., and David, L., 2011. Structural characterization of chitin and chitosan obtained by biological and chemical methods. *Biomacromolecules*. 12:3285–90.

Pani, S., Sahoo, R., Dash, N., Singh, S., and Kumar, B. 2015. Cost effective and minimal time synthesis of mullite from a mine waste by thermal plasma process. *Advanced Materials Letters*. 6:316–24.

Pantaleo, A., Gennaro, B.D., and Shah, N. 2013. Assessment of optimal size of anaerobic co-digestion plants: An application to cattle farms in the province of Bari (Italy). *Renewable and Sustainable Energy Reviews*. 20:57–70.

Pariatamby, A. 2014. *Masaru Tanaka, Municipal Solid Waste Management in Asia and the Pacific Islands Challenges and Strategic Solutions*. Singapore: Springer-Verlag.

Perotto, G., Ceseracciu, L., Simonutti, R., Paul, U.C., Guzman-Puyol, S., Tran, T.-N., et al. 2018. Bioplastics from vegetable waste via an eco-friendly water-based process. *Green Chemistry*. 20:894–902.

Pon-On, W, Suntornsaratoon P, Charoenphandhu N, Thongbunchoo J, Krishnamra N, and Tang, I.M. 2016. Hydroxyapatite from fish scale for potential use as bone scaffold or regenerative material. *Materials Science and Engineering: C*. 62:183–9.

Pradeep, T., and Sajanlal, P.R. 2011. US Patent Publication No 2011/0043331 A1.

Psomopoulos, C.S., Bourka, A., and Themelis, N.J. 2009. Waste-to-energy: A review of the status and benefits in USA. *Waste Management*. 29:1718–24.

Rajeswari, A., Amalraj, A., and Pius, A. 2016. Adsorption studies for the removal of nitrate using chitosan/PEG and chitosan/PVA polymer composites. *Journal of Water Process Engineering*. 9:123–34.

Rao, K.S., Sahoo, R.K., Dash, T., Magudapathy, P., Panigrahi, B.K., and Nayak, B.B., 2016. N and Cr ion implantation of natural ruby surfaces and their characterization. *Nuclear Instruments and Methods in Physics Research Section B: Beam Interactions with Materials and Atoms*. 373:70–5.

Rout, P.P., Sahoo, R.K., Singh, S.K., and Mishra, B.K. 2017. Spectroscopic investigation and colour change of natural topaz exposed to PbO and CrO3 vapour. *Vibrational Spectroscopy*. 88:1–8.

Safari, S., Alam, M.S., von Gunten, K., Samborsky, S., and Alessi, D.S. 2019. Inhibition of naphthalene leaching from municipal carbonaceous waste by a magnetic organophilic clay. *Journal of Hazardous Materials*. 368:578–83.

Sahoo, R.K., Mohapatra, B.K., Singh, S.K., and Mishra, B.K. 2014. Influence of temperature on surface colouration of the lead oxide treated natural gem ruby. *Advanced Science Letters*. 20:622–5.

Sahoo, R.K., Mohapatra, B.K., Singh, S.K., and Mishra, B.K. 2015. Aesthetic value improvement of the ruby stone using heat treatment and its synergetic surface study. *Applied Surface Science*. 329:23–31.

Sahoo, R.K., Mohapatra, B.K., Singh, S.K., and Mishra, B.K. 2016a. Aesthetic value addition and surface study of Odisha ruby stones by heat treatment with different metal oxide additives. *Advanced Science Letters*. 22:336–40.

Sahoo, R.K., Rout, P.P., Singh, S.K., Mishra, B.K., and Mohapatra, B.K. 2017. Synergetic surface and chemical durability study of the aesthetically enhanced natural quartz by heat treatment. *Metallurgical and Materials Transactions A*. 2017;48:1111–20.

Sahoo, R.K., Singh, S.K., and Mishra, B.K. 2016b. Surface and bulk 3D analysis of natural and processed ruby using electron probe micro analyzer and X-ray micro CT scan. *Journal of Electron Spectroscopy and Related Phenomena*. 211:55–63.

Sahoo, R.K., Singh, S.K., Yun, J.M., Kwon, S.H., and Kim, K.H. 2019. Sb_2S_3 nanoparticles anchored or encapsulated by the sulfur-doped carbon sheet for high-performance supercapacitors. *ACS Applied Materials & Interfaces*. 11:33966–77.

Singh, S.K., Mohanty, B.C., and Basu, S. 2002. Synthesis of SiC from rice husk in a plasma reactor. *Bulletin of Materials Science*. 25:561–3.

Sunil, B.R., and Jagannatham, M. 2016. Producing hydroxyapatite from fish bones by heat treatment. *Materials Letters*. 185:411–4.

Surendra, K.C., Takara, D., Hashimoto, A.G., and Khanal, S.K. 2014. Biogas as a sustainable energy source for developing countries: Opportunities and challenges. *Renewable and Sustainable Energy Reviews*. 31:846–59.

Talyan, V., Dahiya, R.P., and Sreekrishnan, T.R. 2008. State of municipal solid waste management in Delhi, the capital of India. *Waste Management*. 28:1276–87.

Tang, X., Feng, Q., Liu, K., Luo, X., Huang, J., and Li, Z. 2018. A simple and innovative route to remarkably enhance the photocatalytic performance of TiO2: Using micro-meso porous silica nanofibers as carrier to support highly-dispersed TiO2 nanoparticles. *Microporous and Mesoporous Materials*. 258:251–61.

Thakur, V.K., and Thakur, M.K. 2014. Recent advances in graft copolymerization and applications of chitosan: A review. *ACS Sustainable Chemistry & Engineering*. 2:2637–52.

Vasconcelos da Silva Pereira, G., Vasconcelos da Silva Pereira, G., Furtado de Araujo, E., Maria Paixão Xavier Neves, E., Regina Sarkis Peixoto Joele, M., and de Fátima Henriques Lourenço, L. 2019. Optimized process to produce biodegradable films with myofibrillar proteins from fish byproducts. *Food Packaging and Shelf Life*. 21:100364.

Vishwakarma, V. and Ramachandran, D., 2018. Green Concrete mix using solid waste and nanoparticles as alternatives – A review. *Construction and Building Materials*. 162:96–103.

Wainaina, S., Awasthi, M.K., Sarsaiya, S., Chen, H., Singh, E., and Kumar, A., 2020. Resource recovery and circular economy from organic solid waste using aerobic and anaerobic digestion technologies. *Bioresource Technology*. 301:122778.

Wang, J. 2014. Decentralized biogas technology of anaerobic digestion and farm ecosystem: Opportunities and challenges. *Frontiers in Energy Research*. 2:1–12.

The World Bank IBRD. IDA Data Base on Solid Waste Management. 2019. https://www.worldbank.org/en/topic/urbandevelopment/brief/solid-waste-management.

Xu, C., Nasrollahzadeh, M., Sajjadi, M., Maham, M., Luque, R., and Puente-Santiago, A.R. 2019a. Benign-by-design nature-inspired nanosystems in biofuels production and catalytic applications. *Renewable and Sustainable Energy Reviews*. 112:195–252.

Xu, C., Nasrollahzade, M., Selva, M., Issaabadib, Z., and Luque, R. 2019b. Waste-to-wealth: Biowaste valorization into valuable bio(nano)materials. *Chemical Society Reviews*. 48:4791–822.

Yang, H., Gözaydın, G., Nasaruddin, R.R., Har, J.R.G., Chen, X., and Wang, X. 2019. Toward the shell biorefinery: Processing crustacean shell waste using hot water and carbonic acid. *ACS Sustainable Chemistry & Engineering*. 7:5532–42.

Zhang, X., Sun, L, Chen, L, Xie, X, Zhao, B, and Si, H, 2014. Comparison of catalytic upgrading of biomass fast pyrolysis vapors over CaO and Fe(III)/CaO catalysts. *Journal of Analytical and Applied Pyrolysis*. 108:35–40.

Zhang, X., Xu, S., Shen, L., and Li, G. 2020. Factors affecting thermal stability of collagen from the aspects of extraction, processing and modification. *Journal of Leather Science and Engineering*. 2:19.

11

A Sustainable Approach to the Degradation and Detoxification of Textile Industry Wastewater for Environmental Safety

Roop Kishor, Arpita Singh, Nandkishor More, and Ram Naresh Bharagava
Babasaheb Bhimrao Ambedkar University (A Central University)

CONTENTS

11.1 Introduction

Textile industries (TIs) use large volumes of water and a wide range of recalcitrant chemicals in the different stages of production of textiles (Haq et al., 2018; Zhuang et al., 2020; Kishor et al., 2021a). For example, 1.6 million L of water is used for the production of 8,000 kg of fabric per day. Globally, more than 10,000 tons of textile dyes and ~8,000 chemicals are used in TIs (Lade et al., 2016; Kishor et al., 2021a). Approximately 20% of wastewater is discharged from the dyeing and finishing stages, which causes severe threats to public health and environment. Textile industry wastewater (TIWW) is characterized by its high temperature, pH, BOD (biological oxygen demand), COD (chemical oxygen demand), TDS (total dissolved solids), TSS (total suspended solids), DS (dissolved solids) and residual dyes, making it more complex and highly toxic (Chandanshive et al., 2020; Kishor et al., 2021b). Several natural fibres such as jute, cotton, silk and wool and a variety of synthetic fibres such as polyamide, polyester, viscose, nylon and acrylic are used by TIs (Sun et al., 2020).

Dyes are synthetic organic compounds used to add colour to substrates such as cloth, paper, leather and other objects. Dyes are aromatic, heterocyclic and recalcitrant and possess different chromophore groups such as azo (–N=N–), nitro (–N=O), carbonyl (–C=O) and quinoid groups and auxochrome groups such

DOI: 10.1201/9781003181224-11

as hydroxyl (–OH), carboxyl (–COOH), amine (–NH_2) and sulphonate (–SO_3H) groups (Saxena et al., 2020; Kishor et al., 2021a). The chemical structure of different dyes is shown in Figure 11.1.

Dyes are causing carcinogenic, mutagenic, genotoxic and allergenic threats to all forms of life (Chandanshive et al., 2017; Haq et al., 2018; Kishor et al., 2021c). TIs also use a variety of toxic chemicals such as phthalates, surfactants, pentachlorophenol, phenols, chlorobenzenes, binders, dioxins and toxic metals (Herrera-González et al., 2019; Kishor et al., 2021a). These pollutants are recalcitrant and persist in water and soil for a long period of time and cause severe threats to public health and decrease soil fertility and photosynthetic activity of aquatic plants, leading to the development of axenic conditions for both aquatic fauna and flora (Haq et al., 2018; Sun et al., 2020; Kishor et al., 2020). These chemicals accumulate in living tissues through food chain, posing severe effects on human and animal's health (Leo et al., 2019; Zhuang et al., 2020; Kishor et al., 2021d). TIWW is also toxic to plants (Haq et al., 2018; Kishor et al., 2018). Therefore, adequate treatment of TIWW is urgently required before its final disposal into the environment (Figure 11.2).

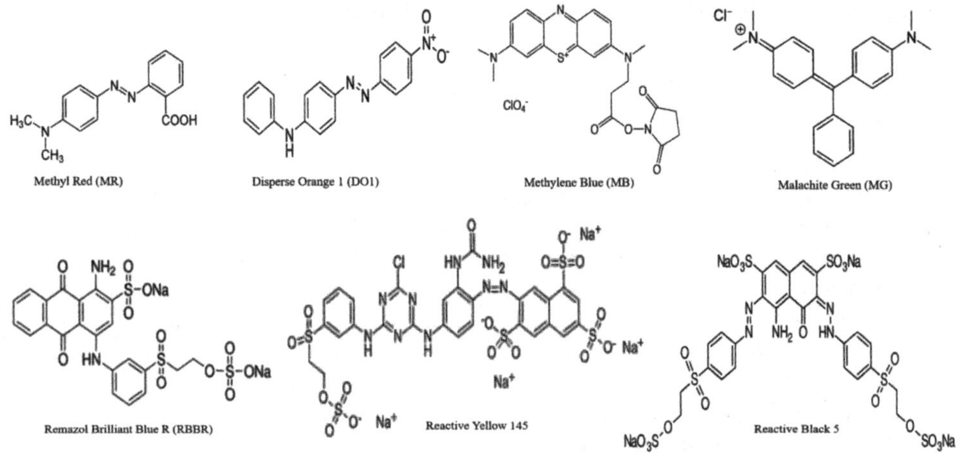

FIGURE 11.1 The chemical structure of different dyes used in TIs.

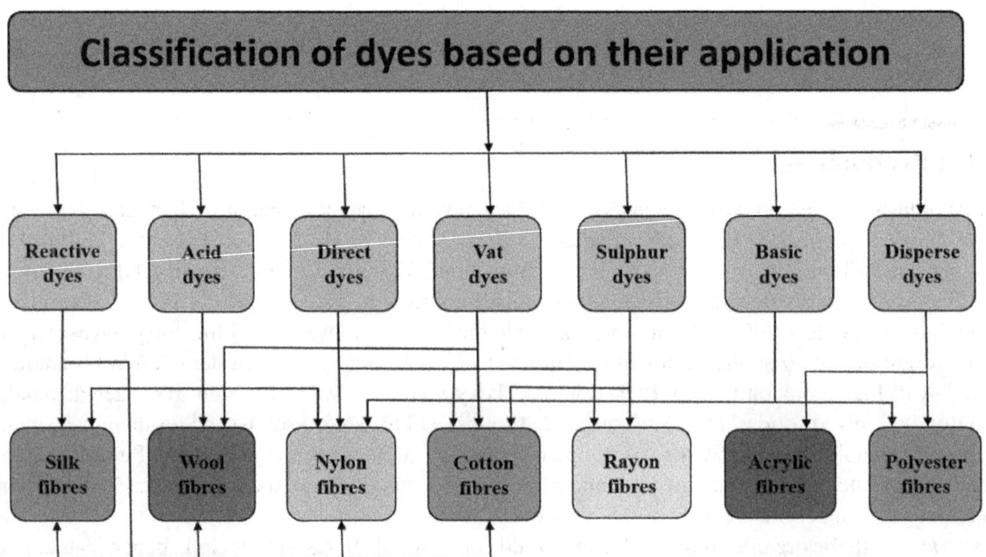

FIGURE 11.2 Different classes of dyes used based on the fabric.

Different physical, chemical, filtration and advanced oxidation processes (AOPs) have been applied in the degradation of wastewaters. But, these methods require high cost, expensive/high amount of chemicals and complicated procedures and produce large volumes of toxic sludge (Khan and Malik 2017; Kishor et al., 2018; Sun et al., 2020). Further, biological treatment is a green, eco-friendly and cost-effective method for the remediation of TIWW (Saxena et al., 2020; Kishor et al., 2021b). The biological treatment may be carried out under aerobic, anaerobic or facultative anaerobic conditions by bacteria, fungi, yeast, algae and plants (Khan and Malik 2017; Bharagava et al., 2018; Kishor et al., 2020). The biological agents convert pollutants into water and carbon dioxide. Further, they are also able to reduce BOD, COD, TDS, TSS, TOC and turbidity and to detoxify various metals from TIWW (Chandanshive et al., 2020; Zhuang et al., 2020).

Thus, this chapter discusses in detail the knowledge about different textile processing steps, wastewater generation and its characteristics and environmental health hazards. It also deals with toxicity in environment and living organisms, treatment methods, key issues, challenges and future perspectives.

11.2 Overview of Textile Industries

TI is one of the largest and most effective industries, which affects every nation in world either directly or indirectly and is estimated to be ~USD 920 billion. In India, ~3,400 TIs are available, which share ~5% of global, ~5% of global export, 5% of global textile and apparel trade. TIs contribute ~4% of the country's national gross domestic product (GDP), 27% of the country's export income and ~14% of the overall Index of Industrial Production (IIP). The textile market is estimated at US$ 75 billion in 2002–2021. It is also expected to recover and grow at 10% CAGR from 2019 to 2020 to reach US$ 190 by 2025–2026.

In addition, India is the largest exporter and producer of textiles after China and the fourth largest producer and exporter of apparel after China, Bangladesh and Vietnam.

11.3 Wastewater Generation, Characteristics and Its Toxicity in the Environment and Human Health

TIs discharge large volumes of wastewater into the environment. For example, ~200 L of groundwater is used for the production of 1 kg textile products (Kishor et al., 2020). TIs use ~1.6 million L of freshwater to produce 8,000 kg of textile fabric per day. Different textile production steps such as sizing, scouring, bleaching, dyeing, printing, washing and finishing are involved in the production of textiles products. Approximately 20% wastewater is released from the finishing and dyeing stages (Sun et al., 2020; Kishor et al., 2021a). TIWW is characterized by its dark colour, high temperature, pH, BOD, COD, TDS, TSS, chlorides, sulphates, phosphates, total nitrogen, total solids and different metals (Khan and Malik 2017; Haq et al., 2018; Kishor et al., 2021b). An overview of TIs including the various steps, the chemicals used and their application and their toxic effects on the environment and living beings is shown in Table 11.1.

TIWW has alkalis, binders, dispersants, dyestuffs, pentachlorophenol, chlorobenzenes, phenols, mordants, surfactants, reducing agents and dioxins (Kishor et al., 2021a). Methylene blue, Congo red, reactive blue, methyl red, methyl orange, Remazol Brilliant Blue R and azure B dye are reported in TIWW (Kishor et al., 2018; Bharagava et al., 2018; Haq et al., 2018). TIWW also has metals (Cr, Pb, Cu, Sb, Cd, As, Ni and Zn) (Khan and Malik, 2017; Tara et al., 2019; Kishor et al., 2021a). TIWW has serious effects on water/soil and public health. It causes mutagenic, carcinogenic and allergenic threats to all forms of life (Chandanshive et al., 2018; Kishor et al., 2021c). TIWW causes colouration of water bodies, leading to the reduction in the penetration power of sunlight, photosynthetic activity and dissolved oxygen content, which affects the normal life of aquatic fauna and flora (Khan and Malik 2017; Haq et al., 2018; Kishor et al., 2021b).

TABLE 11.1

Different Chemicals Used in Various Stages of TIs, Their Applications, Characteristics and Toxic Threats to the Environment and Living Beings

Stage	Chemicals	Application	Wastewater Characteristics	Threats to Human Health and Environment
Sizing	Polyvinyl alcohol, polycyclic acids, carboxymethyl cellulose and polyacrylamide	To provide high potency to fibre	High BOD and COD	Toxic and alter the soil properties
Desizing	Mineral acids and oxidative agents (amylase enzyme)	To remove unwanted sizing materials (starch) and also improve the absorbency of fabric	High COD and BOD	Cause bloating, diarrhoea and irritation of eyes and skin
Scouring	Aqueous sodium hydroxide, glycerol, ethers, anti-static, sodium silicate and soap	To eliminate the wax fats, pectin and lubrication oil	Dark colour, high BOD, pH and temperature (70°C–80°C)	Harmful to liver and kidney
Bleaching	Sodium hypochlorite, sodium chlorite, hydrogen peroxide surfactants, peracetic acid and sodium silicate	To remove colour and improve the whiteness of fibres	High alkalinity, high pH and TDS	Toxic to human health and cause severe irritation of respiratory tract and eyes
Mercerizing	Sodium hydroxide (18%–24%)	To enhance the chemical and physical properties of fibre, such as dye absorbency, tensile strength, lustre, dimensional stability and fabric hygroscopicity	High pH, SS and BOD	Causes severe impacts in living beings and soil/water
Dyeing	Auxiliary chemicals and different dyes depending on fibres	To improve the attachment of dye molecules and provide colour to fibres	Highly dark colour, high toxicity, high BOD, COD, TDS, TSS and TOC	Cause carcinogenic and mutagenic effects in living beings and the environment
Printing	Urea, pentachlorophenol pastes, formaldehyde phthalates, dyes, starches, gums, solvents, acids, metals and reducing agents	To provide the multi-colour design on fibres	Strong colour, High pH, BOD, SS, COD and high toxicity	Highly toxic to living beings and the environment
Washing	Alkylphenol ethoxylates, sodium stearate, alkyl aryl sulphonates and sodium palmitate	To remove unfixed materials	High BOD, COD, TDS, SS, TSS and intensive colour	Have serious effects on soil/ water and human health
Finishing	Chlorophenols, formaldehyde, perfluorinated chemicals, resins, waxes, acetate and chlorinated compounds	To improve and maintain the specific properties of fibres such as stain-proofing, softening, waterproofing, flame retardancy and antimicrobial activity	High toxicity, low alkalinity and BOD	Highly toxic to public health and cause aquatic and soil pollution

TIWW is reported to cause haemorrhage, mucus membrane, nausea and ulceration of skin and also cause spleen and urinary bladder cancers in humans and deformities in mammalian cells (Saxena et al., 2020; Kishor et al., 2021a). TIWW is well reported to affect the plant growth parameters such as seed germination and seedling growth as well as to decrease microbial activity/diversity. Azo dyes are reported to cause cancers in organs of living beings, such as bladder, spleen and liver; normal aberrations in model organisms; and chromosomal deformities in mammalian cells (Kishor et al., 2021a).

11.4 Different Sustainable Approaches to the Degradation and Detoxification of Textile Industry Wastewater

A green, eco-friendly and cost-effective method used for the degradation of textile wastewaters is called a sustainable approach (Chandanshive et al., 2020; Kishor et al., 2021d). The biological method uses metabolic activities of bacteria, algae, fungi, yeasts and plants to degrade or transform pollutants into a simple and non-toxic form. The advantages and disadvantages of different sustainable treatment approaches used in the treatment of TIWW are shown in Table 11.2.

11.4.1 Bacterial Treatment (BT)

BT is an effective method for the treatment of industrial wastewater. Different bacterial strains such as *Arthrobacter, Halomonas, Shewanella, Pseudomonas, Klebsiella, Enterococcus, Aeromonas, Bacillus* and *Lysinibacillus* are well reported in the degradation of textile wastewater (Chen et al., 2018; Bharagava et al., 2018; Kishor et al., 2021c), as shown in Table 11.3.

TABLE 11.2

Advantages and Disadvantages of Different Sustainable Treatment Approaches Used in the Treatment of TIWW

Treatment technology	Advantages	Disadvantages
Bacteria	Cost-effective, green, environmentally friendly, effective for azo dyes and toxic pollutants, sludge can be recycled and biogas production	Failure of long-term treatment due to exhaustion and disposal of bacterial biomass, time-consuming, sensitive against toxic or recalcitrant chemicals and generation of aromatic colourless amine as a metabolite
Fungi/yeasts	Eco-friendly, inexpensive, minimal site disruption, greater public acceptance, strong ability to decolourize and detoxify dye-contaminated effluents	Slow growth rate and require large space, depend on environmental parameters and produce toxic metabolites
Algae/microalgae	Eco-friendly, cost-effective, no additional requirements of nutrients, potential biosorbents, improved CO_2 balance and reduced greenhouse gas (GHG) emissions, low energy demand for oxygen supply and production of green fuels such as biogas, bioethanol and bio-oils	Require large land area and longer retention time, light-dependent process, slow process and difficulty in transformation of metals
Enzymes	Eco-friendly, possibility of complete remediation of pollutants, enhanced treatment efficiency in short time, suitable for a specific compound, low resource consumption, reduced effluents load and reduced CO_2 emission	Enzyme isolation and purification is tedious, requirement of optimum conditions, long treatment time, sensitive to temperature and pH, inactive against toxic compounds and not applicable at large scale due to viability issues
Plants	Eco-friendly, solar-driven, cost-effective, negligible nutrients used, possible recovery and reuse of valuable metals, limit soil erosion, less waste to dispose of, applicable for laboratory, pilot and field studies	Slow process, depend on the depth of contamination, difficulty to achieve acceptable levels of decontamination, toxic metals leach into groundwater, the possibility of food chain contamination

TABLES 11.3

Different Microbial Agents Used for the Treatment of Textile Industry Wastewater by Various Researchers

Strain Name	Wastewater/Dye	Optimized Conditions (pH, Temp., Dye Conc., Agitation and Time)	Treatment Efficiency (%)	References
Bacterial Culture				
Bacillus cohnii RKS9	Textile wastewater and Congo red dye	pH 7.2, 32°C, 100 mg/L, static and 100 rpm, 48 and 12 hours	93.2 and 99	Kishor et al. (2021b)
Pseudomonas putida	Textile wastewater	pH 7, 35°C, 80 rpm, aerobic and 90 hours	Colour (87) and COD (69)	Sen et al. (2019)
Halomonas sp.	Azo dye	30°C and 24 hours	-	Herrera-González et al. (2019)
Pseudomonas aeruginosa Gb30	Reactive Black 5 and cadmium	pH 8, 37°C, 50 and 0.629 mg/L, static and 24 hours	35 and 44	Louati et al. (2020)
Bacillus sp. KM201428	Reactive Black 5	3.9 mg/L, pH 9, 25°C and 120 hours	97	Wanyonyi et al. (2019)
Aeromonas hydrophila	Crystal violet	pH 7, 35°C, 50 mg/L, static and 8 hours	99	Bharagava et al. (2018)
Arthrobacter soli BS5	Reactive Black 5	pH 5–9, 37°C, 50 mg/L and 120 hours	98	Khan and Malik (2017)
Proteus mirabilis LAG	Reactive Blue 13	pH 7, 35°C, static and 5 hours	84	Holkar et al. (2016)
Lysinibacillus sp. RGS	Reactive Orange 4	pH 6.6, 30°C, 50 mg/L, static and 5 hours	TOC (93) and COD (90)	Saratale et al. (2015)
Fungal/Yeast Culture				
Aspergillus strain	Azo dyes mixture	pH 7, 35°C, 100 mg/L, static and 210 minutes	86	Ameen et al. (2021)
Trametes versicolor	Reactive Blue 19	pH 4, 50°C, 200 mg/L, shaking (120 rpm) and 210 minutes	85	Dauda and Erkurt (2019)
Oudemansiella canarii	Congo red	pH 5.5, 30°C, 50 mg/L and 24 hours	80	Iark et al. (2019)
Pichia Kudriavzevii CR-Y103	Reactive orange	pH 6, 30°C, 50 mg/L, shaking and 24 hours	100	Rosu et al. (2018)
Aspergillus bombycis	Reactive Red 31	pH 6, 35°C, static and 12 hours	94	Khan and Fulekar (2017)
Pichia occidentalis G1	Acid red B	pH 5, 30°C, 50 mg/L, shaking and 16 hours	98	Song et al. (2017)
Diaporthe sp.	Methyl violet	100 mg/L and 24 hours	84.87	Ting et al. (2016)
Pichia pastoris strain SMD1168H	Indigo carmine	pH 10, 40°C, 100 rpm and 1 hour	100	Wang et al. (2015)
Algal/Microalgal Culture				
Spirulina platensis	Indigo blue	25 mg/L	>90	Robledo-Padilla et al. (2020)
Scenedesmus	Methylene blue	pH 9, 30°C, 200 mg/L and 120 rpm	87.69	Afshariani and Roosta (2019)
Chlorella vulgaris	Dye effluent	pH 8, 30°C and 10 days	100	Devaraja et al. (2017)
Chlorella vulgaris PSBDU06	Indigo blue	pH 5 and 24 hours	49.03	Revathi et al. (2017)
Chlorella pyrenoidosa NCIM 2738	Reactive red	pH 3, 25°C, 50 mg/L and 30 minutes	COD (82.73) and BOD (56.44)	Sinha et al. (2016)

Bacteria are easy to be cultured, are fast growing, can grow on different substances and are able to adapt variable environmental conditions compared to other microbes (Kishor et al., 2021c). They possess degradative (ligninolytic) enzymes that play a major role in the mineralization of pollutants (Bharagava et al., 2018; Haq et al., 2018). For example, *Pseudomonas* sp. Gb30 was able to remove Reactive Black 5 dye and cadmium from industrial wastewaters within 24 hours (Louati et al. 2020). Similarly, *Bacillus cohnii* is also able to remove Congo red dye and reduce the toxicity of TIWW (Kishor et al., 2021b).

11.4.2 Fungal/Yeast Treatment

Several fungal/yeast species are well reported for the degradation of TIWW, as shown in Table 11.3. Different fungal species produce degradative enzymes, which degrade and decolourize textile dyes present in industrial wastewater (Sen et al., 2016; Haq et al., 2018). Degradative enzymes such as lignin peroxidase (LiP), manganese peroxidase (MnP) azoreductase and laccase catalyse or convert recalcitrant pollutants into non-toxic and mineralized compounds (Sen et al., 2016; Kishor et al., 2021a). For example, the *Aspergillus* strain has capabilities to degrade acid blue, Disperse Red 1 and Congo red dyes (Ameen et al., 2021).

Trametes versicolor was also able to decolourize Reactive Blue 19 dye (Dauda and Erkurt 2019). Different fungal strains such as *Coriolopsis, Penicillium simplicissimum* and *Pleurotus eryngii* are shown to decolourize and remove dyes, BOD, COD and TDS (Sen et al., 2019; Kishor et al., 2021a). The main problems of using fungi alone is the long treatment duration and inefficacy against toxic compounds (Kishor et al., 2021b).

11.4.3 Algal Treatment

Algae are photosynthetic, eukaryotic organisms found in freshwater and saltwater (Sinha et al., 2016; Afshariani and Roosta 2019). Algae involve mainly three mechanisms to remove pollutants: first, algae utilize chromophore groups for their growth; second, they convert chromophore groups into non-chromophore groups; and finally, these compounds are absorbed into their own biomass (Sinha et al., 2016; Kishor et al., 2018). Many algae species such as *Chlorella pyrenoidosa, Spirogyra rhizopus, Desmodesmus, Cosmarium, Nostoc muscorum, Pithophora* and *Sargassum* are shown to degrade and decolourize many textile dyes (Sinha et al., 2016; Kishor et al., 2020, 2021a), as shown in Table 11.3. Algal treatment has many intrinsic advantages: it is a green, eco-friendly and cost-effective method; uses potential biosorbents, requires no additional nutrients; improves CO_2 balance; reduces greenhouse gas (GHG) emissions; and produces biogas, bioethanol and bio-oils.

11.4.4 Enzymatic Treatment

Enzyme treatment is the most effective method for the degradation of recalcitrant pollutants. A variety of microorganisms are capable of producing different enzymes during the degradation of toxic chemical pollutants, as listed in Table 11.4. Enzymes such as azoreductase, LiP, MnP and laccase are able to degrade/convert many pollutants into non-toxic and simple compounds (Haq et al., 2018; Saxena et al., 2020; Kishor et al., 2021a).

For example, azoreductase enzyme breaks down azo groups present in azo dyes and produces colourless aromatic amine. *Halomonas* sp. GT and *Bjerkandera adusta* CCBAS 930 are capable of degrading dyes in wastewater by producing azoreductase enzyme (Tian et al., 2018; Rybczyńska-Tkaczyk et al., 2020). Navada and Kulal (2019) reported that laccase produced by *Phomopsis* sp. can be used as a mediator during the degradation of textile wastewater. There are several advantages such as energy saving, low resource consumption, reduced effluents load, reduced CO_2 emission, enhanced treatment efficiency and suitability for specific pollutants.

11.4.5 Mixed/Consortium Treatment

For the effective treatment of TIWW, the development and utilization of microbial consortium becomes an emerging, eco-friendly, inexpensive, faster and effective approach to the adequate treatment of

TABLE 11.4
Many Enzymes Used by Various Researchers in the Treatment of Textile Industry Wastewater/Dyes

Enzymes	Strains	Textile Wastewater/Dyes	Treatment Efficiency and Decolourization (%)	Time (hour/ day)	References
Manganese peroxidase, lignin peroxidase, azoreductase and laccase	*Bacillus cohnii* RKS9	Textile wastewater and Congo red (CR)	93.2 and 99	48 and 12 hours	Kishor et al. (2021)
Manganese peroxidase, lignin peroxidase, azoreductase, laccase and horseradish-type peroxidase	*Bjerkandera adusta* CCBAS 930 (fungi)	Alizarin Blue Black B (ABBB) and Acid Blue 129 (AB129)	89.22%	20 days	Rybczyńska-Tkaczyk et al. (2020)
Azoreductase	*Shewanella* sp. ST2, *Oceanimonas* sp. ST3, *Enterococcus* sp. ST5 and *Clostridium bifermentans* sp. ST12 (bacteria)	Acid Orange 7 (AO7) and Reactive Black 5 (RB5)	90%	48 hours	Zhuang et al. (2020)
Laccase	*Trametes versicolor* (fungi)	Reactive Blue 19 (RB19)	85%	210 minutes	Dauda and Erkurt (2019)
Laccase	*Pleurotus ostreatus* HAUCC 162 (fungi)	Remazol Brilliant Blue R, malachite green and methyl orange and bromophenol	91.5%	24 hours	Zhuo et al. (2019)
Azoreductase	*Halomonas* sp. GT (bacteria)	Acid brilliant blue GR	100%	96 hours	Tian et al. (2018)
Laccase	*Phomopsis* sp. (fungi)	Textile wastewater	Colour (99%), COD (67%) and BOD (47%)	2.5 hours	Navada and Kulal (2019)
Lignin peroxidase	*Serratia liquefaciens* (bacteria)	Azure B (AB)	90%	48 hours	Haq et al. (2018)
Laccase, veratryl alcohol oxidase, lignin peroxidase, tyrosinase and azoreductase	*Portulaca grandiflora* and *Gaillardia grandiflora* (plant)	Textile wastewater	73%	30 days	Chandanshive et al. (2018)
Laccase	*Pleurotus pulmonarius* (fungi)	Malachite green (MG)	68.6%	36 hours	Leo et al. (2019)
Lignin peroxidase and laccase	*Aeromonas hydrophila* (bacteria)	Crystal violet (CV)	99 %	8 hours	Bharagava et al. (2018)
Laccase, manganese peroxidase, lignin peroxidase and azoreductase	*Brevibacillus aydinogluensis* PDF25, *Geobacillus thermoleovorans* NP1, *Anoxybacillus flavithermus* 52-1A, *Bacillus thermoamylovorans* DKP and *Bacillus circulans* BWL1061 (bacteria)	Direct black G (DBG)	97%	8 hours	Chen et al. (2018)

(Continued)

TABLE 11.4 (*Continued*)

Many Enzymes Used by Various Researchers in the Treatment of Textile Industry Wastewater/Dyes

Enzymes	Strains	Textile Wastewater/Dyes	Treatment Efficiency and Decolourization (%)	Time (hour/ day)	References
Azoreductase, NADH-DCIP reductase and laccase	*Bacillus circulans* BWL1061 (bacteria)	Methyl orange (MO)	99.22%	4 hours	Liu et al. (2017)
Laccase	*Ganoderma lucidum* BCRC 3612 (fungi)	Acid Orange 7 (AO7)	>90%	14 days	Lai et al. (2017)
Laccase, veratryl alcohol oxidase, lignin peroxidase and azoreductase	*Typha angustifolia* and *Paspalum scrobiculatum* (plant)	Congo red (CR) (100 mg/L) and textile industry effluent	Colour (94%), COD (70%), BOD (7%5), TDS (57%) and TSS (47%)	48 and 96 hours	Chandanshive et al. (2017)
Laccase and peroxidase	*Bacillus aryabhattai* DC100 (bacteria)	Coomassie brilliant blue G-250 (CBBG250) and indigo carmine (IC)	100%	72 hours	Paz et al. (2017)
Azoreductase	*Alcaligenes* sp. (bacteria)	Congo red (CR)	98.76%	48 hours	D'Souza et al. (2017)
Laccase, lignin peroxidase, azoreductase and NADH-DCIP reductase	*Aspergillus ochraceus* NCIM-1146 (fungi) and *Providencia rettgeri* strain HSL1 (bacteria)	Textile effluent	92%	30 hours	Lade et al. (2016)

recalcitrant dyes and TIWW. In consortium treatment, the individual strain might be attacked at different position of dye compounds and metabolites formed by another existing bacterium culture for supplementary degradation of dyes.

For example, the bacterial consortium of *Bacilli, Betaproteobacteria* and *Gammaproteobacteria* was found to degrade and detoxify the Direct Blue 2 dye more rapidly than single culture (Cao et al., 2019). Recently, many workers have developed bacterial–yeast, fungal–yeast, bacterial–fungal and bacterial-algal consortia that are more effective in decolourizing and removing dyes and COD (Cao et al., 2019; Kishor et al., 2021a).

11.4.6 Plant Treatment

Plant treatment is an eco-friendly, green, solar energy-driven and cost-effective method to remediate wastewater pollutants (Chandanshive et al., 2016). A plant can degrade and detoxify the dyes, pesticides, toxic metals, landfill leachates, polycyclic aromatic hydrocarbons, chlorinated solvents and polychlorinated biphenyls (Kishor et al., 2018; Chandanshive et al., 2020). Phytotransformation, rhizofiltration, phytostabilization, phytovolatilization and phytoextraction steps are involved in the degradation of wastewater pollutants (Chandanshive et al., 2020; Kishor et al., 2021a). Several plants have the potential to utilize many persistent organic and inorganic pollutants as the sole source of energy (Kishor et al., 2018, 2021a; Haq et al., 2018).

For example, *Salvinia molesta* is able to remove Rubine GFL dye (Chandanshive et al., 2016). *Hyacinth* is also able to decolourize textile wastewater along with COD reduction (Safauldeen et al., 2019). Phytoremediation has major advantages such as cost-effectiveness, the use of solar energy, being green and eco-friendly, negligible nutrient requirements and less sludge production. Individual and mixture/ consortium of plants were also reported in the treatment of textile wastewater, as shown in Table 11.5.

TABLE 11.5

Different Plants Used in Phytoremediation/Phytotreatment of Textile Industry Wastewater/Dyes

Plant	Textile Wastewater/ Dye and Concentration	Decolourization and Treatment Efficiency (%)	Time (hour/ day)	References
Vetiveria zizanioides	Remazol red (RR) (100 mg/L) and textile wastewater (400 L)	Colour (93%), ADMI (74%), COD (74%), BOD (81%), TDS (66%) and TSS (47%)	48 and 72 hours	Chandanshive et al. (2020)
Bacopa monnieri (L)	Reactive 9 and direct 5 (40 mg/L)	100%	14 days	Shanmugam et al. (2020)
Plant–bacteria consortium (*Phragmites australis, Pseudomonas* sp. NT-38, *Rhodococcus* sp. NT-39, *and Acinetobacter* sp. TT-15)	Textile industry wastewater (TIWW)	Colour (86%), BOD (91%), COD (92%) and trace metals (87%)		Tara et al. (2019)
Hyacinth	Textile effluent	Colour (83%), and COD (89%)	7 days	Safauldeen et al. (2019)
Phragmites australis and Typha domingensis	Textile effluent	Colour (97%), BOD (92%) and COD (87%)	8 days	Tara et al. (2019)
Portulaca grandiflora	Textile wastewater	59%	30 days	Chandanshive et al. (2018)
Gaillardia grandiflora	Textile wastewater	73%	30 days	Chandanshive et al. (2018)
Azolla pinnata	Methylene blue (MB) (25 mg/L)	85%	24 hours	Al-Baldawi et al. (2018)
Co-plantation consortium (*Fimbristylis dichotoma* and *Ammannia baccifera)*	Textile effluent and methyl orange (MO) (50 mg/L)	95%	48 hours	Kadam et al. (2018)
Co-plantation (Consortium-TP) (*Typha angustifolia* and *Paspalum scrobiculatum*)	Congo red (CR) (100 mg/L) and textile industry effluent	Colour (94%), ADMI (76%), COD (70%), BOD (75%), TDS (57%) and TSS (47%)	48 and 96 hours	Chandanshive et al. (2017)
Scirpus grossus	Methylene blue (MB) (200–1,000 mg/L)	86%	72 days	Almaamary et al. (2017)
Typha angustifolia	Congo red (CR) (100 mg/L) and textile industry effluent	Colour (80%), ADMI (72%), COD (65%), BOD (68%), TDS (45%) and TSS (35%)	48 and 96 hours	Chandanshive et al. (2017)
Ipomoea aquatic	Brown 5R (B5R) (200 mg/L) and textile wastewater (510 L)	Colour (94%), ADMI (70%), COD (87%), BOD (76%) and TS (34%)	72 and 72 hours	Rane et al. (2014)
Salvinia molesta	Textile dye effluent and Rubine GFL (RGFL) (50 mg/L)	97%	72 hours	Chandanshive et al. (2016)
Physalis minima	Reactive black 8 (RB8) (30 mg/L)	76%	120 hours	Jha et al. (2015)
Alternanthera philoxeroides	Remazol red (RR) (70 mg/L)	100%	72 hours	Rane et al. (2015)

(Continued)

TABLE 11.5 (*Continued*)

Different Plants Used in Phytoremediation/Phytotreatment of Textile Industry Wastewater/Dyes

Plant	Textile Wastewater/ Dye and Concentration	Decolourization and Treatment Efficiency (%)	Time (hour/ day)	References
Blumea malcolmii Hook	Brilliant blue R (BBR) (40 mg/L)	98%	24 hours	Kagalkar et al. (2015)
Lemna minor	Triarylmethane (20 mg/L)	88%		Torbati (2015)
Plant–bacteria mixed culture (*Bacillus pumilus* and *Pogonatherum crinitum*)	Textile wastewater	BOD (78%), COD (70%) and TDS (13%)	12 days	Watharkar et al. (2015)

11.5 Key Issues, Challenges and Future Perspectives

TIs are facing many challenges from the public and government sector, such as increased cost of raw textile products, increasing demand for various types of textile fabrics, lack of advanced processing techniques, lack of specific dedicated industrial areas for the positioning of textile industries, lack of financial support from the government and waste treatment technologies in developing countries. The mitigation of these challenges requires large-scale financial supports from the government for the proper functioning of TIs, especially for small-scale industries. TIs should also use eco-friendly/natural colouring/auxiliary agents instead of synthetic agents as they may be helpful in the reduction of treatment cost for environmental safety. TIs should adopt recycling/reuse of treated wastewater to minimize the use of fresh groundwater for economic and environmental benefits.

11.6 Conclusions

TIWW is characterized by its dark colour, high pH, BOD, COD, TDS, TSS, chlorides, phosphates, total solids and different metals. TIWW has toxic recalcitrant pollutants, which persist in the environment for long duration, posing severe environmental and public health hazards. The treatment of TIWW is a key issue as there is no particular and economically feasible technique for the adequate degradation of TIWW. Many physicochemical and AOP methods are effective in the treatment of TIWW, but these require high operating costs and generate highly toxic secondary pollutants (sludge).

A sustainable (biological) approach is a green, emerging, cost-effective, eco-friendly and globally acceptable approach, but is less effective and takes more time. Therefore, more research is required from laboratory scale to pilot scale to minimize the environmental and public health hazards. Besides these, the use of hazardous, poorly degradable/non-degradable dyes and auxiliaries should also be avoided in TIs as it may be helpful in reducing the treatment cost for environmental clean-up. In fact, till date, there has been no meticulous and economically feasible approach/technique that is able to adequately degrade and decolourize the dyes and chemicals present in TIWW.

Acknowledgement

Mr. Roop Kishor gratefully acknowledges the financial support provided by the University Grants Commission (UGC), Government of India (GOI), New Delhi, India.

REFERENCES

Afshariani, F., and Roosta, A. 2019. Experimental study and mathematical modeling of biosorption of methylene blue from aqueous solution in a packed bed of microalgae *Scenedesmus*, *J. Clean. Prod.* 225: 133–142.

Al-Baldawi, I. A., Abdullah, S. R. S., Anuar, N., and Hasan, H. A. 2018. Phytotransformation of methylene blue from water using aquatic plant (*Azolla pinnata*). *Environ. Technol. Innov.* 11: 15–22.

Almaamary, E. A., Abdullah, S. R., Hasan, H., Rahim, R. A., and Idris, M. 2017. Treatment of methylene blue in wastewater using *Scirpus grossus*. *Malays J. Anal. Sci.* 21: 182–187.

Ameen, F., Dawoud, T.M., Alshehrei, F., Alsamhary, K., and Almansob, A. 2021. Decolorization of acid blue 29, disperse red 1 and Congo red by different indigenous fungal strains. *Chemosphere* 271: 129532.

Bharagava, R.N., Mani, S., Mulla, S.I., and Saratale, G.D. 2018. Degradation and decolorization potential of an ligninolytic enzyme producing *Aeromonas hydrophila* for crystal violet dye and its phytotoxicity evaluation. *Ecotoxicol. Environ. Saf.* 156: 166–175.

Cao, J., Sanganyado, E., Liu, W., Zhang, W., and Liu, Y. 2019. Decolorization and detoxification of Direct Blue 2B by indigenous bacterial consortium. *J. Environ. Manag.* 242: 229–237.

Chandanshive, V., Kadam, S., Rane, N., Jeon, B-H., Jadhav, J., and Govindwar, S. 2020. In situ textile wastewater treatment in high rate transpiration system furrows planted with aquatic macrophytes and floating phytobeds. *Chemosphere* 252: 126513.

Chandanshive, V.V., Kadam, S.K., Khandare, R.V., Kurade, M.B., Jeon, B-H., Jadhav, J.P., and Govindwar, S.P. 2018. In situ phytoremediation of dyes from textile wastewater using garden ornamental plants, effect on soil quality and plant growth. *Chemosphere* 210: 968–976.

Chandanshive, V.V., Rane, N.R., Gholave, A.R., Patil, S.M., Jeon, B.H. and Govindwar, S.P. 2016. Efficient decolorization and detoxification of textile industry effluent by Salvinia molesta in lagoon treatment. *Environ. Res.* 150: 88–96.

Chandanshive, V.V., Rane, N.R., Tamboli, A.S., Gholave, A.R., Khandare, R.V., and Govindwar, S.P. 2017. Co-plantation of aquatic macrophytes *Typha angustifolia* and *Paspalum scrobiculatum* for effective treatment of textile industry effluent. *J. Hazard. Mater.* 338: 47–56.

Chen, Y., Feng, L., Li, H., Wang, Y., Chen, G., and Zhang, Q. 2018. Biodegradation and detoxification of Direct Black G textile dye by a newly isolated thermophilic microflora. *Bioresour. Technol.* 250: 650–657.

D'Souza, E., Fulke, A.B., Mulani, N., Ram, A., Asodekar, M., Narkhede, N., and Gajbhiye, S.N. 2017. Decolorization of Congo red mediated by marine *Alcaligenes* sp. isolated from Indian West coast sediments. *Environ. Earth Sci.* 76 (20): 721.

Dauda, M.Y., and Erkurt, E.A. 2019. Investigation of Reactive Blue 19 Biodegradation and Byproducts Toxicity Assessment using Crude Laccase Extract from *Trametes versicolor*. *J. Hazard. Mater.* 393: 121555.

Devaraja, S., Bharath, M., Deepak, K., Suganya, B., Vishal, B.S., Swaminathan, D., and Meyyappan, N. 2017. Studies on the Effect of Red, Blue and White LED Lights on the Productivity of *Chlorella Vulgaris* to Treat Dye Industry Effluent. *Adv. Biotech Micro.* 6 (2).

Haq, I., Raj, A., and Markandeya 2018. Biodegradation of Azure-B dye by *Serratia liquefaciens* and its validation by phytotoxicity, genotoxicity and cytotoxicity studies. *Chemosphere* 196: 58–68.

Herrera-González, A.M., Caldera-Villalobos, M., and Peláez-Cid, A.A. 2019. Adsorption of textile dyes using an activated carbon and crosslinked polyvinyl phosphonic acid composite. *J. Environ. Manag.* 234: 237–244.

Holkar, C.R., Jadhav, A.J., Pinjari, D.V., Mahamuni, N.M., and Pandit. A.B. 2016. A critical review on textile wastewater treatments: possible approaches. *J. Environ. Manag.* 182: 351–366.

Iark, D., dos Reis Buzzo, A.J., Garcia, J.A.A., Côrrea, V.G., Helm, C.V., Corrêa, R.C.G., Peralta, R.A., Moreira, R.D.F.P.M., Bracht, A., and Peralta, R.M. 2019. Enzymatic degradation and detoxification of azo dye Congo red by a new laccase from *Oudemansiella canarii*. *Bioresour. Technol.* 289: 121655.

Jha, P., Modi, N., Jobby, R., and Desai, N. 2015. Differential expression of antioxidant enzymes during. Degradation of azo dye reactive black 8 in hairy roots of *Physalis minima* L. *Int. J. Phytoremed.* 17: 305–312.

Kadam, S.K., Watharkar, A.D., Chandanshive, V.V., Khandare, R.V., Jeon, B.H., Jadhav, J.P., and Govindwar, S.P. 2018. Co-planted floating phyto-bed along with microbial fuel cell for enhanced textile effluent treatment. *J. Clean. Prod.* 203: 788–798.

Kagalkar, A.N., Khandare, R.V., and Govindwar, S.P. 2015. Textile dye degradation potential of plant laccase significantly enhances upon augmentation with redox mediators. *RSC Adv.* 5(98): 80505–80517.

Khan, R., and Fulekar, M.H. 2017. Mineralization of a sulfonated textile dye Reactive Red 31 from simulated wastewater using pellets of *Aspergillus bombycis. Bioresour. Bioprocess.* 4(1): 1–11.

Khan, S., and Malik, A. 2017. Toxicity evaluation of textile effluents and role of native soil bacterium in biodegradation of a textile dye. *Environ. Sci. Pollut. Res.* 25: 4446–4458.

Kishor, R., Bharagava, R.N., Ferreira, L.F.R., Bilal, M., and Purchase, D. 2021d. Molecular techniques used to identify perfluorooctanoic acid degrading microbes and their application in a wastewater treatment reactor/plant. In *Wastewater Treatment Reactors*. 253–271.

Kishor, R., Bharagava, R.N., and Saxena, G. 2018. Industrial wastewaters: the major sources of dye contamination in the environment, ecotoxicological effects, and bioremediation approaches. In Bharagava, R.N. (ed) *Recent Advances in Environmental Management* (pp. 1–25). CRC Press/Taylor & Francis Group, Boca Raton.

Kishor, R., Purchase, D., Ferreira, L.F., Mulla, S.I., Bilal, M., and Bharagava, R.N. 2020. Environmental and health hazards of textile industry wastewater pollutants and its treatment approaches. In Hussain, C.M. (ed) *Handbook of Environmental Materials Management*. Springer Nature, Switzerland. https://doi.org/10.1007/978-3-319-58538-3_230-1.

Kishor, R., Purchase, D., Saratale, G.D., Ferreira, L.F.R., Bilal, M., Iqbal, H.M., and Bharagava, R.N. 2021b. Environment friendly degradation and detoxification of Congo red dye and textile industry wastewater by a newly isolated *Bacillus cohnni* (RKS9). *Environ. Technol. Innov.* 22: 101425.

Kishor, R., Purchase, D., Saratale, G.D., Saratale, R.G., Ferreira, L.F.R., Bilal, M., Chandra, R., and Bharagava, R.N. 2021a. Ecotoxicological and health concerns of persistent coloring pollutants of textile industry wastewater and treatment approaches for environmental safety. *J. Environ. Chem. Eng.* 105012.

Kishor, R., Saratale, G.D., Saratale, R.G., Ferreira, L.F.R., Bilal, M., Iqbal, H.M., and Bharagava, R.N., 2021c. Efficient degradation and detoxification of methylene blue dye by a newly isolated ligninolytic enzyme producing bacterium *Bacillus albus* MW407057. *Colloid. Surface.* 111947.

Lade, H., Kadam, A., Paul, D., and Govindwar, S. 2016. Exploring the potential of fungal-bacterial consortium for low-cost biodegradation and detoxification of textile effluent. *Arch. Environ. Prot.* 42(4): 12–21.

Lai, C-Y., Wu, C-H., Meng, C-T., and Lin, C-W. 2017. Decolorization of azo dye and generation of electricity by microbial fuel cell with laccase-producing white-rot fungus on cathode. *Appl. Energy* 188: 392–398.

Leo, V.V., Passari, A.K., Muniraj, I.K., Uthandi, S., Hashem, A., Abd_Allah, E.F., Alqarawi, A. A., and Singh, B. P. 2019. Elevated levels of laccase synthesis by *Pleurotus pulmonarius* BPSM10 and its potential as a dye decolorizing agent. *Saudi J. Biol. Sci.* 26(3): 464–468.

Liu, W., Liu, C., Liu, L., You, Y., Jiang, J., Zhou, Z., and Dong, Z. 2017. Simultaneous decolorization of sulfonated azo dyes and reduction of hexavalent chromium under high salt condition by a newly isolated salt tolerant strain *Bacillus circulans* BWL1061. *Ecotoxicol. Environ. Saf.* 141: 9–16.

Louati, I., Elloumi-Mseddi, J., Cheikhrouhou, W., Hadrich, B., Nasri, M., Aifa, S., Woodward, S., and Mechichi, T. 2020. Simultaneous cleanup of Reactive Black 5 and cadmium by a desert soil bacterium. *Ecotoxicol. Environ. Saf.* 190: 110103.

Navada, K.K., and Kulal, A. 2019. Enhanced production of laccase from gamma irradiated endophytic fungus: a study on biotransformation kinetics of aniline blue and textile effluent decolourisation. *J. Environ. Chem. Eng.* 103550.

Paz, A., Carballo, J., Perez, M. J., and Domínguez, J. M. 2017. Biological treatment of model dyes and textile wastewaters. *Chemosphere* 181: 168–177.

Rane, N. R., Chandanshive, V. V., Khandare, R. V., Gholave, A. R., Yadav, S. R., and Govindwar, S. P. 2014. Green remediation of textile dyes containing wastewater by *Ipomoea hederifolia* L. *RSC Adv.* 4(69): 36623–36632.

Rane, N.R., Patil, S.M., Chandanshive, V.V., Kadam, S.K., Khandare, R.V., Jadhav, J.P., and Govindwar, S. P. 2015. *Alternanthera* sp. rooted soil bed and *Ipomoea aquatica* rhizofiltration coupled phytoreactors for efficient treatment of textile wastewater. *Water Res.* 96: 1–11.

Revathi, S., Kumar, S.M., Santhanam, P., Kumar, S.D., Son, N., Kim, M.K. 2017. Bioremoval of the indigo blue dye by immobilized microalga *Chlorella vulgaris* (PSBDU06). *J. Sci. Ind. Res.* 76(1): 50–56.

Robledo-Padilla, F., Aquines, O., Silva-Núñez, A., Alemán-Nava, G.S., Castillo-Zacarías, C., Ramirez-Mendoza, R.A., Zavala-Yoe, R., Iqbal, H., and Parra-Saldívar, R. 2020. Evaluation and predictive modeling of removal condition for bioadsorption of indigo blue dye by *Spirulina platensis*. *Microorganisms* 8(1): 82.

Rosu, C.M., Avadanei, M., Gherghel, D., Mihasan, M., Mihai, C., Trifan, A., Miron, A., and Vochita, G. 2018. Biodegradation and detoxification efficiency of azo-dye reactive orange 16 by *Pichia kudriavzevii* CR-Y103. *Water. Air. Soil. Pollut.* 229: 1–8.

Rybczyńska-Tkaczyk, K., Korniłłowicz-Kowalska, T., Szychowski, K. A., and Gmiński, J. 2020. Biotransformation and toxicity effect of monoanthraquinone dyes during *Bjerkandera adusta* CCBAS 930 cultures. *Ecotoxicol. Environ. Saf.* 191: 110203.

Safauldeen, S.H., Abu Hasan, H., Abdullah, S.R. 2019. Phytoremediation efficiency of water hyacinth for batik textile effluent treatment. *J. Ecol. Eng.* 20(9): 177–187.

Saratale, R.G., Saratale, G.D., Govindwar, S.P., and Kim, D.S. 2015. Exploiting the efficacy of *Lysinibacillus* sp. RGS for decolorization and detoxification of industrial dyes, textile effluent and bioreactor studies. *J. Environ. Sci. Health* 50(2): 176–192.

Saxena, G., Kishor, R., and Bharagava, R.N. 2020. Application of microbial enzymes in degradation and detoxification of organic and inorganic pollutants. In Bharagava, R.N. and Saxena, G. (eds.), *Bioremediation of Industrial Waste for Environmental Safety* (pp. 41–51). Springer, Singapore.

Sen, S.K., Patra, P., Das, C.R., Raut, S., and Raut, S., 2019. Pilot-scale evaluation of bio-decolorization and biodegradation of reactive textile wastewater: an impact on its use in irrigation of wheat crop. *Water Resour. Ind.* 21: 100106.

Sen, S.K., Raut, S., Bandyopadhyay, P., and Raut, S., 2016. Fungal decolouration and degradation of azo dyes: a review. *Fungal Biol. Rev.* 30 (3): 112–133.

Shanmugam, L., Ahire, M., and Nikam, T. 2020. *Bacopa monnieri* (L.) Pennell, a potential plant species for degradation of textile azo dyes. *Environ. Sci. Pollut. Res.* 27: 9349–9363.

Sinha, S., Singh, R., Chaurasia, A.K., and Nigam, S. 2016. Self-sustainable *Chlorella pyrenoidosa* strain NCIM 2738 based photobioreactor for removal of Direct Red-31 dye along with other industrial pollutants to improve the water-quality. *J. Hazard. Mater.* 306: 386–394.

Song, L., Shao, Y., Ning, S., and Tan, L. 2017. Performance of a newly isolated salt-tolerant yeast strain *Pichia occidentalis* G1 for degrading and detoxifying azo dyes. *Bioresour. Technol.* 233: 21–29.

Sun, Y., Cheng, S., Lin, Z., Yang, J., Li, C., and Gu, R. 2020. Combination of plasma oxidation process with microbial fuel cell for mineralizing methylene blue with high energy efficiency. *J. Hazard. Mater.* 384: 121307.

Tara, N., Arslan, M., Hussain, Z., Iqbal, M., Khan, Q. M., and Afzal, M. 2019. On-site performance of floating treatment wetland macrocosms augmented with dye-degrading bacteria for the remediation of textile industry wastewater. *J. Clean. Prod.* 217: 541–548.

Tian, F., Guo, G., Zhang, C., Yang, F., Hu, Z., Liu, C., and Wang, S. 2018. Isolation, cloning and characterization of an azoreductase and the effect of salinity on its expression in a halophilic bacterium. *Int. J. Biol. Macromol.* 123:1062–1069.

Ting, A.S.Y., Lee, M.V.J., Chow, Y.Y., and Cheong, S.L. 2016. Novel exploration of endophytic *Diaporthe* sp. for the biosorption and biodegradation of triphenylmethane dyes. *Water Air Soil Pollut.* 227(4): 109.

Torbati, S. 2015. Feasibility and assessment of the phytoremediation potential of duckweed for triarylmethane dye degradation with the emphasis on some physiological responses and effect of operational parameters. *Turk. J. Biol.* 39: 438–446.

Wang, T-N., Lu, L., Wang, J-Y., Xu, T-F., Li, J., and Zhao, M. 2015. Enhanced expression of an industry applicable Cot A laccase from *Bacillus subtilis* in Pichia pastoris by non-repressing carbon sources together with pH adjustment: recombinant enzyme characterization and dye decolorization. *Proc. Biochem.* 50: 97–103.

Wanyonyi, W.C., Onyari, J.M., Shiundu, P.M., and Mulaa, F.J. 2019. Effective biotransformation of reactive black 5 dye using crude protease from *Bacillus cereus* strain KM201428. *Energy Proc.* 157: 815–824.

Watharkar, A.D., Khandare, R.V., Waghmare, P.R., Jagadale, A.D., Govindwar, S.P., and Jadhav, J.P. 2015. Treatment of textile effluent in a developed phytoreactor with immobilized bacterial augmentation and subsequent toxicity studies on *Etheostoma olmstedi* fish. *J. Hazard. Mater.* 283: 698–704.

Zhuang, M., Sanganyado, E., Zhang, X., Xu, L., Zhu, J., Liu, W., and Song, H. 2020. Azo dye degrading bacteria tolerant to extreme conditions inhabit nearshore ecosystems: optimization and degradation pathways. *J. Environ. Manag.* 261: 110222.

Zhuo, R., Zhang, J., Yu, H., Ma, F., and Zhang, X. 2019. The roles of *Pleurotus ostreatus* HAUCC 162 laccase isoenzymes in decolorization of synthetic dyes and the transformation pathways. *Chemosphere* 234: 733–745.

12

Ecological and Health Implications of Heavy Metals Contamination in the Environment and Their Bioremediation Approaches

Sushila Saini
JVMGRR College

Geeta Dhania
M.D. University

CONTENTS

DOI: 10.1201/9781003181224-12

12.1 Introduction

The ever increasing population, industrialization and speedy urbanization are responsible for the release of a large number of pollutants in the environment. The disposal of pollutants from various industries, mining, vehicular pollution, lead–acid batteries, domestic waste and waste from agricultural practices has resulted in the contamination of soil, air and water. These wastes contain organic and inorganic substances, radionuclides and heavy metals. Heavy metals are elements with high atomic weight and relatively high density in comparison with water. Heavy metals released from natural and man-made sources are considered one of the major areas of interest by scientists and engineers as heavy metals are responsible for the degradation of environment and human health.

The heavy metals chiefly considered hazardous are lead (Pb), cadmium (Cd), mercury (Hg), arsenic (As), chromium (Cr), nickel (Ni) and aluminium (Al). Metals such as Cu, Zn, Fe and Se play an important role in human metabolism at low concentration, but at higher concentration, cause severe damage to kidney and liver, anaemia, intestine and stomach pain, reduced reproductive fitness, development of carcinoma and finally death. Heavy metals persist for a long period of time in the environment and inside organisms, which enhances their bioaccumulation and biomagnification. Mostly, humans, animals and fishes are badly affected from these metals as they are top consumers in food chain on land and in aquatic ecosystem, respectively (Sumiahadi and Acar 2018).

Heavy metals-contaminated sites are difficult to remediate. The conventional methods such as excavation, landfilling, thermal treatment and electro-reclamation are not appropriate due to high expenditure and low efficiency, and they result in residuals having toxic effects (secondary pollution). Bioremediation that utilizes various microbes, plant species or enzymes is a green, splendid alternative technique to transform toxic heavy metals into less toxic state for cleaning up of contaminated sites. It is the most cost-effective environmental management tool for the revival of environment (Shishir et al. 2019). This chapter provides a detailed overview of the agents responsible for the release of heavy metals into the environment, health impacts of exposures to heavy metals and the use of bioremediation techniques as a solution to emerging contamination problems.

12.2 Sources and Ecotoxicological Effects of Different Heavy Metals

There are a number of anthropogenic sources of toxic metals contamination in the environment, and after entering the environment (air, water and soil), metals may enter the human/animal body through different routes causing various diseases (Figure 12.1).

12.2.1 Nickel

12.2.1.1 Sources of Nickel Exposure

Nickel compounds are used in the production of nickel–cadmium batteries, in electroplating, coins, inexpensive jewellery, electronic equipment and medical prostheses. Nickel contamination in the

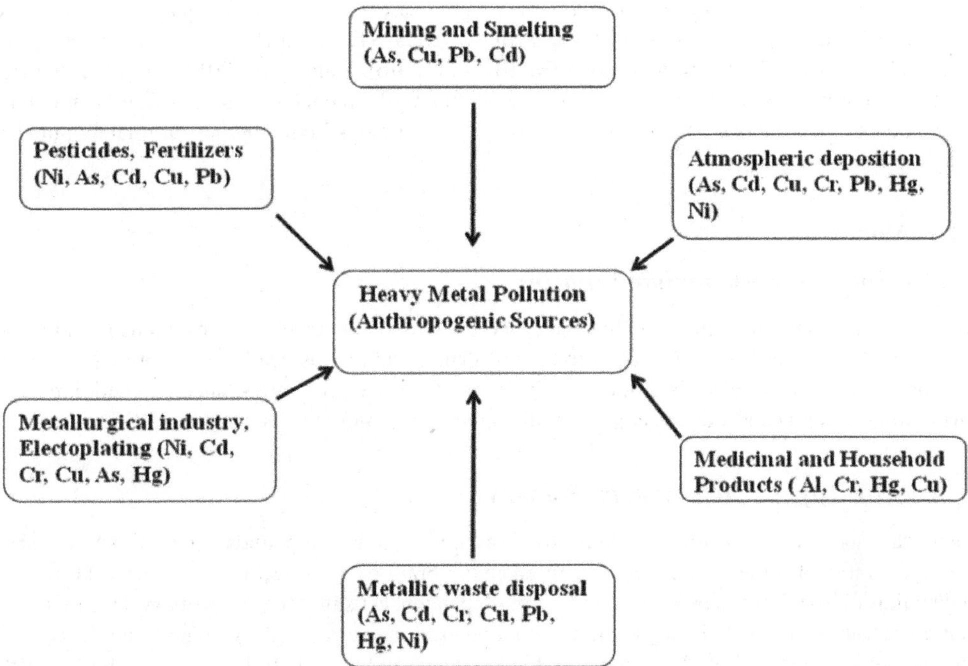

FIGURE 12.1 Sources of heavy metal pollution in the environment.

environment occurs during the production, processing and recycling of nickel-containing products, fossil fuel burning, application of fertilizers and waste disposal of nickel products (Rathor et al. 2014).

12.2.1.2 Toxicity of Nickel in Humans

Nickel has been recognized as a contact irritant, and it produces allergy in the form of contact dermatitis known as 'nickel itch'. Allergy may occur from various products containing nickel such as coins, jewellery (necklaces, earrings, bracelets and wristwatches), buttons, mobile phones, implants, prostheses and medications. Ni and Ni compounds induce ROS, mitochondria, endoplasmic reticulum and Fas (transmembrane protein)-mediated and c-Myc (transcription factor)-mediated apoptosis (Guo et al. 2016). Inhalation of nickel by nickel refinery workers has shown a high risk of lung cancer, cancer of nasal cavity, kidney cancer and prostate cancer. Ni^{2+} ions are responsible for mutagenesis, apoptosis, chromosome aberrations, formation of Z-DNA, and alteration of the expression of micro-RNA resulting in carcinogenesis. Of all the compounds of nickel, the most toxic is nickel carbonyl. Its acute toxic effects include vertigo, sleeplessness, nausea, vomiting, chest pain, abdominal pain, sweating, blurring, weakness and degeneration of parenchymal tissues affecting liver, kidneys, adrenal glands and spleen. Acute inhalation of nickel carbonyl leads to death of individuals due to cerebral oedema and punctate cerebral haemorrhages (Kumar and Trivedi 2016).

12.2.1.3 Toxicity of Nickel in Animals

Administration of nickel and nickel nanoparticles to male Wistar rats resulted in negative effects on testicular structure resulting in abnormalities of spermatozoa along with reduced mobility and count. Testicular toxicity of nickel compounds may be due to enhanced production of reactive oxygen species causing oxidative damage to macromolecules as well as damaging DNA (Casalegno et al. 2015). Exposure of rats and mice to nickel chloride by feed resulted in suppression of activity of natural killer cells and affected the T cell system (Guo et al. 2016).

Administration of nickel to *Labeo rohita* showed toxic effects with evident changes in biochemical parameters such as amplified levels of liver enzymes (ALT, AST and ALP) along with decreased body weight and change in plasma protein levels (Malik et al. 2016). Tang et al. (2014) observed that treatment of avian broilers with dietary $NiCl_2$ in excess of 300 mg/kg reduced the broiler weight and thymus growth and caused oxidative damage and histopathological lesions in the thymus, ultimately impairing the thymic function.

12.2.2 Aluminium

12.2.2.1 Sources of Aluminium Exposure

In recent years, the use of aluminium has increased many folds in our lives. Cans, utensils, aluminium foil, yellow cheese, ready-to-make cake mixes, soft drinks, salt, herbs, tea leaves, tooth paste, housing materials, components of electrical devices, airplanes, decorative cosmetics, antacids, antidiarrhoeals, aspirin, infant formula and vaccine adjuvants all contain aluminium (Mahor and Ali 2015).

12.2.2.2 Toxicity of Aluminium in Humans

Various data show that exposure of humans to aluminium and its compounds can be linked to various neurodegenerative disorders. Aluminium can enter the brain by passing the blood–brain barrier, and being similar to iron, it replaces iron from the iron-dependent cells involved in memory. Due to its accumulation, it causes microtubule depletion, amyloid deposition, neurofibrillary tangles and degeneration of neurons resulting in multiregion atrophy in hippocampus and the cerebral cortex, which is characteristic of Alzheimer's pathology (Bondy 2016). Vaccine adjuvants and infant formula contain aluminium, which can cause chronic brain inflammation. Strong correlation exists between the amount of aluminium given to infants by aluminium-adjuvanted vaccines and autism rates (Miller 2016). Aluminium can be transferred by placenta and milk to foetus. Overconsumption of antacids during pregnancy may cause toxicity to foetus. Fanni et al. (2014) observed that it can cross placenta leading to accumulation in foetal tissue, which results in delayed ossification, malformations, mental retardation and in utero death.

12.2.2.3 Toxicity of Aluminium in Animals

Treatment of albino rats with aluminium resulted in the suppression of spermatogenesis and absence of spermatocytes. Administration of aluminium chloride to rats leads to lower deposition of calcium, phosphorus and magnesium in bone, resulting in significantly lower bone mineral density. Besides affecting the bone, administration of aluminium chloride and aluminium hydroxide to rats caused an increase in the levels of blood urea and creatinine, which are markers of renal dysfunction. The renal failure occurs due to shrunken glomeruli, intraglomerular congestion, mesangial hyperplasia and obliteration of the filtration slits (Narayanan 2014).

Mass vaccination of sheep provoked 'sheep adjuvant syndrome' that shows similarity with human neurological disorders associated with Al adjuvants. This disorder shows severe neurobehavioral complications ultimately resulting in coma and death (Shaw et al. 2014). Aluminium administration in chick embryos causes defects in bone mineralization leading to malformations of tibias and femurs (Fanni et al. 2014).

12.2.3 Lead

12.2.3.1 Sources of Lead Exposure

Lead is one of the most potent toxic heavy metals due to its persistent nature and bioaccumulation. Lead exposure occurs by various sources such as leaded gasoline, smelting, lead-based paints, pottery glazing, water pipes, lead-containing pipes, lead batteries, tank linings, printing of books, toys, ammunition and fishing sinkers. Lead gets entry into the body by ingestion, inhalation and skin contact (Wani et al. 2015).

12.2.3.2 Toxicity of Lead in Humans

Nervous system is the most affected part due to lead exposure. Children are more vulnerable for lead poisoning as in children lead is absorbed more quickly than in adults. Lead toxicity causes degeneration of some part of brain (encephalopathy) that is responsible for reduced concentration, loss of memory, hallucinations, irritability, muscular shaking, depression, brain damage and even death (Mason et al. 2014).

Reproductive system is severely affected by lead toxicity in both sexes. In males, it is responsible for suppressed rates of serum testosterone, reduced plasma luteinizing hormone, testicular tissue alteration, increased prostate weight, reduced libido, chromosomal damage, reduced sperm counts, deteriorated sperm quality and infertility. In females, toxic levels of lead cause unbalanced oestrus cycles, miscarriages, ovarian follicular cysts, premature membrane rupture and lower birth weight. In persons with lead poisoning, kidney biopsy reveals loss of PCT (proximal convoluted tubules), interstitial fibrosis, excretion of urate (suggesting gout), hyperuricaemia, hypertension and renal breakdown. Lead poisoning causes anaemia by damaging the membrane of RBC, making them fragile (Wani et al. 2015).

12.2.3.3 Toxicity of Lead in Animals

In rats, administration of lead acetate resulted in a decrease in mean body weight and the effect increased with increasing dose. Levels of AST (aspartate transaminase) and ALT (alanine transaminase) rose due to jaundice, hepatitis and liver cirrhosis. A significant reduction in the total erythrocyte count (TEC), decreased haemoglobin (Hb) and packed cell volume result in hypochromic microcytic anaemia (Alwaleedi 2015). High levels of lead were found in blood and milk of cattle that were reared near industrial areas involving manufacturing of batteries and iron pipes. In cattle, clinical signs of lead poisoning include tongue paralysis, difficult breathing, blindness, excessive salivation, teeth grinding, nasal discharge, muscle spasms, head pressing and incoordination (Barbosa et al. 2014).

12.2.4 Cadmium

12.2.4.1 Sources of Cadmium Exposure

Cadmium mainly exists as cadmium sulphide (CdS or greenockite) in zinc deposits. Cadmium compounds find their use in nickel–cadmium batteries, electroplating, nuclear reactor, as an anticorrosive agent, paint pigments, heat-resistant plastics, phosphate fertilizers, etc. Cadmium intoxication occurs by contaminated water, food and cigarette smoking (Govind et al. 2014).

12.2.4.2 Toxicity of Cadmium in Humans

Kidney is the primary organ affected by chronic cadmium exposure, resulting in renal tubular dysfunction. Prolonged exposure to cadmium damages the reabsorptive function of renal tubules, resulting in increased excretion of low molecular weight proteins, particularly α_2-, β_2- and γ-globulins, ultimately resulting in renal failure (Alfvén et al. 2002). Individuals residing in contaminated areas show more cases of urinary stones and hypertension than control groups. By interfering with calcium metabolism, it causes reduced calcium levels in bones, leading to osteomalacia or osteoporosis with multiple fractures (Swaddiwudhipong et al. 2015).

In a study on 200 pregnant women in Nigeria showed that cadmium increases rates of premature births with lower birth weights, head circumference and lengths than newborns of unexposed women. Pregnant women that gave birth to babies with lower birth weight had higher maternal blood cadmium, high cadmium/zinc ratio and lower zinc concentration (Emokpae et al. 2016). In children, cadmium exposure elevated the development of inflammatory disease of ear called otitis media. Song et al. (2015) by investigating the effect of cadmium on epithelial cells of human middle ear found that incubation with 20 μM Cd^{2+} for 24 hours induces apoptosis and necrosis, resulting in reduced viability of these cells. Cd^{2+} increased ROS production, expression of inflammatory cytokine and mucin gene in middle ear epithelial cells, causing development of otitis media.

12.2.4.3 Toxicity of Cadmium in Animals

In animals, cadmium intoxication is associated with kidney and liver damage, enlarged joints, osteoporosis, neurotoxicity, degeneration of testes, reduced growth and increased mortality. Tribowo et al. (2014) by an experimental study in rats observed that Cd exposure leads to oxidative damage in rat ovarian cells. Cd induces oxidative damage by enhancing peroxidation of membrane lipids due to inhibition of antioxidant enzymes. Rats exposed to cadmium in drinking water revealed a significant reduction in the levels of cholesterol, triglycerides and phospholipids in the plasma and erythrocyte, respectively. In cattle, a high Cd exposure resulted in dehydration, rough and scaly skin, mouth lesions, shrunken scaly scrotum, renal function impairment, hypertension and enlarged joints (Lane et al. 2015).

12.2.5 Chromium

12.2.5.1 Sources of Chromium

In the environment, chromium exists in two valence forms, i.e. trivalent chromium (Cr(III)) and hexavalent chromium (Cr(VI)). Chromium is released into the environment by various activities such as ferrochrome production, magnetic tapes production, cement production, preparation of metal alloys by stainless steel electroplating, formation of food preservatives, tanning of leather, pigment production, automobile brakes lining and anticorrosive agents in cooking goods. Chromates are produced during smelting, mining, roasting and extraction (ATSDR 1998).

12.2.5.2 Toxicity of Chromium in Human

Trivalent chromium is an essential micronutrient, whereas hexavalent chromium is toxic to both humans and animals. Toxicity of chromium is found in people who live in areas close to waste disposal sites and chromium manufacturing industries. Chromium enters through respiratory tract in workers who work in leather, steel and textile industries and causes shortness of breath, coughing and wheezing. Allergic reactions such as skin rash occur in people using products containing chromium from leather and textile industries. Acute chromium toxicity leads to respiratory tract and skin inflammation, upset of gastrointestinal tract, coma, renal failure, cardiovascular diseases, teratogenicity, carcinogenicity and haematological effects (Douglas 2018).

Chromosomal aberrations, sister chromatid exchange, DNA damage, gene mutation, cell transformation and dominant lethal mutations to animal and human cells can be induced by many chromate salts. In China, chromium poisoning in female workers was observed, which is found to lead to increased risk of abortion and miscarriage as compared to workers not exposed (Yang et al. 2013). Cr(VI) can easily enter the cells by non-specific anion channel of cytomembranes. Inside the cell, various ROS (reactive oxygen species) formed, resulting in the formation of DNA–Cr–DNA crosslinks, Cr–DNA adducts and mutations. These then cause the activation of DNA-dependant protein kinases and P^{53} gene, ultimately leading to apoptosis. When hexavalent chromium compounds are reduced to pentavalent form, they combine with DNA and cellular processes are interrupted (Macfie et al. 2010).

12.2.5.3 Toxicity of Chromium in Animals

Aquatic ecosystem is contaminated with chromium through effluents discharge from various industries such as textile, steel, pharmaceutical and tanneries. Chromium affects fish health in many ways, i.e. decrease in haematological parameters, severe anaemia, production of ROS, loss of immune system, growth inhibition, histological and morphological alterations and decline in the levels of protein, lipid and liver glycogen (Praveena et al. 2013).

Chromium in the form of $Na_2Cr_2O_7$ induced adverse effects on the rat kidney due to the significant increase in MDA (malondialdehyde) in kidney. A reduction in antioxidants such as SOD (superoxide dismutase) and GSH (glutathione) along with an increase in MDA in the kidney of rats exposed to chromium administration was reported by Balakrishnan et al. (2013).

12.2.6 Arsenic

12.2.6.1 Sources of Arsenic

Arsenic (As) is a metalloid that exists in the form of inorganic and organic compounds. Inorganic As is more harmful than the organic form. Arsenic exists in three oxidation states, i.e. As(III) – trivalent arsenate, As(V) – pentavalent arsenate and elemental forms. As(III) is ten times more toxic than As(V), while elemental form is non-toxic in nature. Arsenic is found in soil, sediment and bedrock, and it dissolves in the water of aquifers, making the groundwater contaminated. In the food chain, it also enters through the use of pesticides and herbicides and results in bioaccumulation. Coal burning from thermal power plants and emissions from various industries cause air arsenic poisoning. Mining and ore smelting for industrial purposes are also related to As poisoning in humans (Khalida et al. 2018).

12.2.6.2 Toxicity of Arsenic in Humans

The effects of arsenic on health depend upon the level of arsenic, volume intake and health status of individuals. Humans are exposed to arsenic by ingestion, skin absorption and inhalation. It has been found that 80%–90% of As(III) and As(V) can be absorbed in humans and animals from gastrointestinal tracts. Arsenic toxicity is increasing at an alarming rate in Asian countries such as India, Pakistan and Bangladesh. In Bangladesh, a close association between arsenic and stillbirth, neonatal death and spontaneous abortion has been found. In India, West Bengal is highly affected state with 50% area exposed to groundwater arsenic contamination. Inorganic arsenic has been found in rice (main staple diet), both in raw (93.8%) and in cooked form (88.1%) (Basu et al. 2015). Due to exposure, West Bengal population suffers from liver diseases (non-cirrhotic portal fibrosis), lung diseases (chronic obstructive pulmonary disease and bronchitis), anaemia, hypertension, oedema, weakness and neuropathy.

Arsenic inhibits the process of DNA repair and produces ROS by metabolic process in liver and spleen. Accumulation of free radicals disrupts gene expression and results in cell death. Arsenic increases the risk of carcinogenesis via binding with DNA binding protein, which inhibits the process of DNA repair mechanism (Chen et al. 2019).

12.2.6.3 Toxicity of Arsenic in Animals

Animals have been exposed to arsenic toxicity by veterinary drugs and feed additives. Arsenic toxicity in animals causes damage to digestive and nervous systems. It mainly accumulates in liver, kidneys and skin, causing cancer of bladder, skin, liver and pancreatic cells. Arsenic has endocrine disrupting properties, which suppress the transcription of 17β-oestradiol-inducible vitellogenin II gene when studied on chicken embryos supplemented with $NaAsO_2$ (Rahman et al. 2016). Bera et al. (2010) found high arsenic concentrations in the hair and urine of cattle in endemic zones as high as 0.461–0.984 ppm. Arsenic leads to respiratory distress, severe icterus, ataxia, dehydration, haemolysis and sudden death in cattle. Urinary bladder cancer has been caused by methylated metabolite of As by the formation of ROS in rats that are exposed to arsenic. Arsenic toxicity causes disruption in hepatic function by cross-linking with enzymes and DNA damage by oxidative stress in case of rats (Patlolla et al. 2012). Keshavarzi et al. (2015) observed hyperkeratinization and pustule formation in sheep consuming arsenic-contaminated water. There is a reduction in body weight, heart rate and body temperature in sheep affected with arsenic.

12.2.7 Mercury

12.2.7.1 Sources of Mercury

In the environment, it exists in three forms, i.e. organic, inorganic and elemental mercury (Hg) (Park and Zheng 2012). It occurs in the Earth's crust and is released into the environment by weathering of rocks and volcanic activities. Various industries such as thermal power plants, cement industries and gold mining release mercury into the environment. Most of the people are exposed to high amount of mercury by fungicides, incandescent lights, hair dyes, thermometers, petroleum products, vaccination, consumption of saltwater fish, silver dental amalgam, barometers, fossil fuel emission, batteries and incineration of medical wastes.

12.2.7.2 Toxicity of Mercury in Humans

Methylmercury (MeHg) is an organic form of mercury and is neurotoxic in nature. Methylmercury accumulates in food chain and reaches higher concentration by biomagnification. The International Agency for Research on Cancer (IARC) has classified methylmercury as 'possibly carcinogenic to humans' (Group 2B). Methylmercury's toxicity was highlighted in the 1950s in Minamata (Japan) when wastes from the chemical factory were discharged into the local bay (Yokoyama 2018). Mercury can cause mental retardation, urological defects, hearing loss, developmental defects, blindness, dysarthria and even death. Mercury causes fetotoxicity by low birth weight, spontaneous abortion, miscarriage and stillbirth. Mercury can easily cross the placental barrier and inhibit the development of foetal brain, resulting in psychomotor retardation and cerebral palsy. Children are more sensitive to MeHg, and exposure during pregnancy period can lead to delay in development, low IQ (intelligence quotient) and ADHD (attention deficit hyperactivity disorder) (WHO 2011).

The main route of exposure to elemental mercury is by inhalation of the vapours. About 80% of inhaled vapours are absorbed by the lung tissues. Its vapour can easily penetrate the blood–brain barrier, and it is a well-documented neurotoxicant (Park and Zheng 2012). Women exposed to elemental mercury vapours have a higher possibility of having abortion as compared to women who are not exposed. Otebhi and Osadolor (2016) reported a 9.5% increase in miscarriage in exposed women. It has been found that umbilical cord blood contains a high amount of mercury that results in an abnormal amino acid transfer, abnormal hormone secretion, high haematocrit, abnormal enzymes activity and high plasma albumin levels of the placenta.

An increased level of mercury in body leads to various types of autoimmune diseases such as autism, arthritis, eczema, multiple sclerosis, systemic lupus erythematosus, allergy, rheumatoid arthritis, ADHD, psoriasis, epilepsy and scleroderma (Amadi et al. 2017). Various reports have been available on the effects of mercury on kidney, such as tubular dysfunction, nephritic syndrome, glomerular disease, syncretistic nephrotic syndrome, membranous glomerulonephritis and subacute-onset nephritic syndrome (Miller et al. 2013).

12.2.7.3 Toxicity of Mercury in Animals

Fish are an important component of human diet in many parts of the world and are a rich source of protein and omega-3 fatty acids. Mercury present in fish is the major threat to the life of humans and other animals. Mercury causes loss of appetite and weight, muscular incoordination, unstable gait and lameness in sheep, pig, chicken, cattle and turkey. Organic mercury compounds lead to embryotoxicity and teratogenicity in fish, birds and mammals. Exposure to mercury suppresses the immune function in marine mammals. Heinz et al. (2009) found out that American kestrel, snowy egret and tricoloured heron are extremely sensitive to methylmercury, whereas Canada goose and laughing gull are less sensitive.

12.3 Bioremediation Approaches for Heavy Metals-Contaminated Environments

Heavy metals are difficult to remove from polluted sites as they cannot be degraded, thus persisting in the ecosystem. Remediation of heavy metals from the environment using microorganisms or plants is called bioremediation, which offers high specificity for the removal of heavy metals. These technologies have recently become very popular and are a promising option for the treatment of soils contaminated by heavy metals. The use of bacteria, fungi and algae for heavy metal remediation is called microremediation, while phytoremediation refers to using plants for remediation, and the interaction of both plants and microbes for enhancing remediation is known as rhizoremediation.

12.3.1 Microbial Bioremediation

Remediation by employing microbes can be carried out in situ or ex situ. In situ bioremediation is done at the site of contamination either by supplementing contaminated soils with nutrients to stimulate the activity of native microorganisms, or by adding new microorganisms in the existing pool to augment the biodegradative capability of indigenous microbial population. Ex situ bioremediation is done by transferring the polluted material to another area for management. Microbes adopt various ways to interact with heavy metals, such as by changing their oxidation states, by immobilizing them, by volatilizing or by increasing the solubility of the toxic metal, thus allowing them to flush away easily from the site. The interaction of microbes with metals is a complicated process, and it depends on various factors such as metal type, temperature, pH, moisture content, soil structure and nutrients (Gupta et al. 2016).

12.3.1.1 Microremediation Mechanism of Heavy Metals-Contaminated Soil

12.3.1.1.1 Biosorption

Biosorption is the passive uptake and adherence of metal ions to microbial cell wall and extracellular substances independent of cell metabolic activity. The cell wall of microbes consist of polysaccharides, lipids and proteins that have various negatively charged sites such as hydroxyl, amino, alcohol, carboxyl, ester, thioether, thiol and phosphate, for which heavy metals show affinities. As the microbes, i.e. bacteria, fungi and algae, vary in their cell wall properties, this causes a major difference in the type and amount of metal binding to them. Biosorption process mostly remediates the soil faster than other bioremediation approaches. For example, 60% of Cu^{2+}ions can be adsorbed by *Bacillus* species in the first 5 minutes and adsorption equilibrium was achieved within 10 minutes (He and Tebo 1998). Biosorption is a surface phenomenon that is independent of cell metabolism as it involves the use of dead or alive biomass, making it a less expensive process, and after completion, the biomass can be regenerated to be used in other processes. However, the selectivity of biosorption is generally low as binding takes place mainly by physiochemical interaction, but by the alteration of biomass, selectivity to metal ions can be increased (Luka et al. 2018).

12.3.1.1.2 Bioaccumulation

It is cell metabolism-dependent process that involves metal uptake into the cells by crossing the cell membrane followed by complex formation by various compounds in the cell cytoplasm. It is also called active uptake as it is an energy-dependent process and occurs exclusively in living cells. As compared to biosorption, bioaccumulation is an expensive process because it involves living cells and their reuse is limited. Bioaccumulation includes intracellular sequestration, localization, binding with proteins and peptides (metallothioneins and phytochelatins) and formation of complexes. In *E. coli*, the range of accumulation of Cd is regulated by the expression of different peptides and proteins (Mejare and Bulow 2001). In case of biosorption, as the metals are bounded only to cell surface, under certain environmental conditions the process is reversible (surface-bound toxic metal return to the environment). However, in bioaccumulation as the toxic metals are sequestered inside microorganisms their return to environment is not possible. So some scientists prefer bioaccumulation to biosorption.

12.3.1.1.3 Bioleaching

Bioleaching involves production of organic acids by microbial population, leading to dissolution of heavy metals from polluted sites, and it is one of the methods that show high prospective for the removal of heavy metals. It is also called biooxidation or biomining. Microbes lead to mobilization of heavy metals from solid matrix into liquid phase for their extraction and reuse when water is filtered through it. Bioleaching depends on temperature, concentration, pH, oxygen and carbon dioxide content, nutrient availability, mineral composition, mineral type and the presence of inhibitors. Bioleaching process can be enhanced by providing nutrients and energy to microbes. In the presence of glucose and other nutrients, the leaching rate of Cd increases from 9% to 36%. Various studies have shown that microbes can efficiently remove heavy metals and thus reduce their bioavailability. For example, as compared to chemical leaching, the rate of metal solubilization was observed to be 88.5% Zn, 79.9% Cu, 50.1% Pb and 33.2% Cr after 12 days of bioleaching with iron-oxidizing microorganisms (Wen et al. 2013). Chen and Lin (2004) obtained 97%–99% of Cu, 96%–98% of Zn, 62%–68% of Mn, 73%–87% of Ni and 31%–50% of Pb from sediments after 8 days of leaching.

12.3.1.1.4 Biotransformation

Microbes can lead to a change in the oxidation state of metal ions, rendering them harmless, or precipitate them. Different bacteria utilize metals as electron acceptor or donor for energy generation. Aerobic bacteria use oxygen as electron acceptor, whereas in anaerobic bacteria, metals in oxidized form serve as a terminal acceptor of electrons. Microbes reduce the state of metals and change their solubility similar to *Geobacter* species, which reduce uranium (U^{6+}) from soluble to insoluble state (U^{4+}) (Lovley et al. 1991). Similarly, mercury-resistant fungi such as *Hymenoscyphus ericae, Verticillium terrestre* and *Neocosmospora vasinfecta* have the ability to convert Hg(II) into a harmless state (Kelly et al. 2006).

12.3.1.1.5 Biofilms

A biofilm is a group of microorganisms that are attached to a biological or inert surface by secretion of extracellular polymeric substances (EPS), which are composed of water, proteins, carbohydrates and extracellular DNA. Biofilm-forming microbes are more tolerant to harsh environment such as mechanical and chemical stress than free-living microbes, even at the concentration which is lethal. Thus, bacterial biofilms existing near heavily contaminated sites show better adaptation, survival and tolerance to the harsh environment. Microbial population within biofilms shows variable gene expression, which may be responsible for the degradation of various pollutants by various metabolic pathways. Biofilms made from conglomeration of *Bacillus subtilis* and *B. cereus* have the capability to take away 98% of Cr(III) from contaminated sites (Das et al. 2012). Grujić et al. (2017) observed that in case of *Rhodotorula mucilaginosa*, the metal removal efficiency of biofilm cells was 91.75%–95.39% as compared to 4.79%–10.25% removal efficiency of planktonic cells.

12.3.1.2 Remediation of Heavy Metals by Bacteria

Different bacterial species are important biosorbents because they are small in size, are ubiquitous and have the adaptability to grow and develop in a broad range of environmental conditions. Bacteria possess specific genes for heavy metal resistance in their plasmids or on chromosome. Due to their high surface-to-volume ratio and active chemisorption sites at cell wall, bacteria have high biosorption ability. Mixed culture of bacteria are more stable, efficient and metabolically more capable of biosorption of heavy metals under field conditions. Besides biosorption, bacterial cells have the ability to accumulate metals intracellularly by binding of metal ions to various compounds in the cytoplasm. *Rhizobium leguminosarum* cells are capable of sequestering Cd^{+} ions by glutathione (Lima et al. 2006). *Pseudomonas putida* is able to sequester Cu, Cd and Zn ions with the help of proteins rich in cysteine amino acid (Higham et al. 1986).

12.3.1.3 Remediation of Heavy Metals by Fungi

Fungi have extensively been used for biosorption and bioaccumulation of heavy metals at low cost and in an eco-friendly way. Fungi are even more tolerant to heavy metals than bacteria as they can alter themselves to survive under extreme pH, temperature, nutrient availability and high metal concentrations. Fungal cell wall consists of chitin, nitrogen-containing polysaccharides, proteins and lipids that comprise metal-binding sites such as –OH, –COOH, –RCOO, $–NH_2$ and –SH, leading to elimination of inorganic metal ions (Igiri et al. 2018). Members of ascomycetes and basidiomycetes are the most efficiently reported fungi for heavy metal removal. The efficient biosorption of Cd^{2+} and Cr^{2+} from contaminated sites by *Aspergillus* and *Rhizopus* has been reported by Zafar et al. (2007). Root exudates of fungi release certain organic compounds that mobilize heavy metals by forming metal complexes, thus reducing their bioavailability.

12.3.1.4 Remediation of Heavy Metals by Algal Cells

The use of a range of algae and cyanobacteria for the removal of heavy metals is known as phycoremediation. Algae are autotrophic and capable of producing enormous biomass that has immense capability to adsorb heavy metals. Brown algae are more efficient as a biosorbent than red and green algae as they have a high amount of carboxyl and sulphate active groups that show a high sorption efficiency. Algal walls have numerous active sites such as hydroxyl, carboxyl, sulphate and phosphate groups that act as metal-binding sites. Biomass of *Chlorella vulgaris* has a high efficiency to adsorb Cd (95.5%), Cu (97.7%) and lead (99.7%) from mixed solution of these ions (Goher et al. 2016). In the last 10 years, various algae have been used by scientists for the removal of heavy metals from contaminated sites.

12.3.2 Phytoremediation

It is a technology used to clean up the heavy metals and organic pollutants using plants and microorganisms (rhizospheric). It is a low-cost, eco-friendly and efficient technology for the re-establishment of contaminated environment from heavy metals. This method is found to be effective in areas with low contamination as compared to intensely contaminated areas. Plants have developed complex tolerance strategies to modulate the internal concentration of trace heavy metals, which include detoxification processes, accumulation, complexation with chelators and compartmentalization of toxic metals in the vacuoles.

12.3.2.1 Mechanism of Phytoremediation

Various mechanisms for the removal of heavy metals from contaminated sites are phytoaccumulation/phytoextraction, phytodegradation, phytovolatilization, phytostabilization, phytostimulation and phytofiltration, as shown in Figures 12.2 and 12.3.

12.3.2.1.1 Phytoextraction/Phytoaccumulation

It involves the process of uptake of metal pollutants along with water and nutrients through roots by hyperaccumulator plants. The absorbed components are accumulated into various parts of plants such as shoots and leaves without any phototoxic effects on plants. Various experiments are conducted by

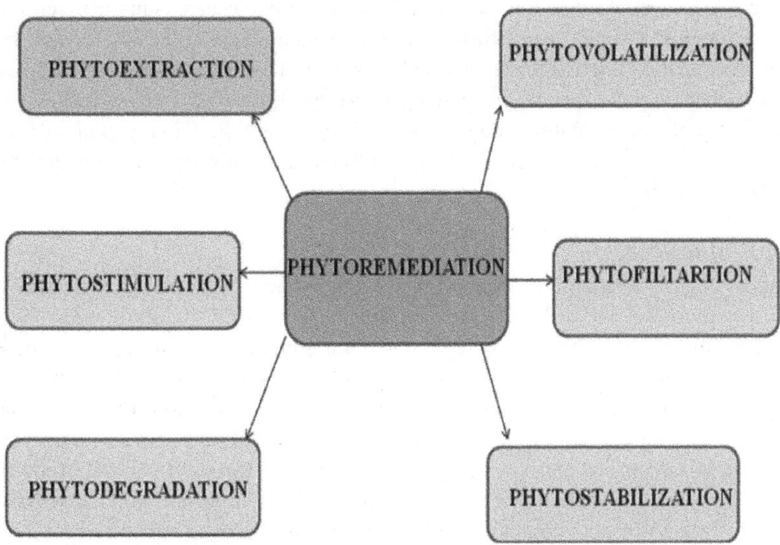

FIGURE 12.2 Phytoremediation techniques.

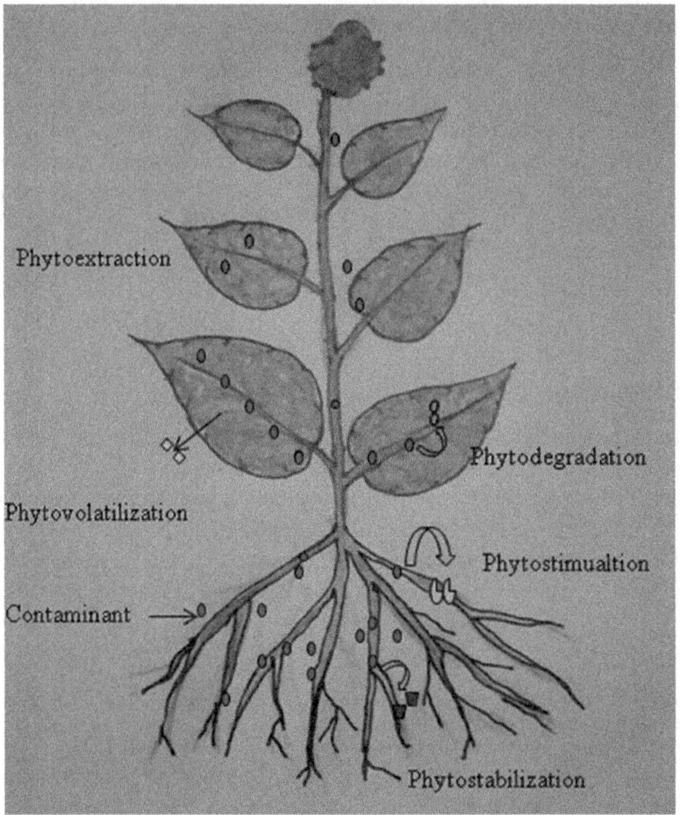

FIGURE 12.3 Different mechanisms of phytoremediation.

researchers to find out the potential plants (aquatic or terrestrial) with high accumulation efficiency. The study by Singh and Sinha (2005) on *Brassica juncea* found that it has good potential to accumulate heavy metals such as copper, lead and nickel from the contaminated sites. Aquatic macrophytes such as *Eichhornia crassipes* and *Centella asiatica* were found to accumulate up to 99% copper from polluted sites (Mokhtar et al. 2011). These aquatic macrophytes have the capacity to accumulate heavy metals 1,00,000 times higher than the quantity of concerned water. *Amaranthus spinosus* has good potential for phytoremediation of soils contaminated by low levels of Cd and Pb (Huang et al. 2019). Addition of oxalic acid (2.5 mmol/kg) was suitable for facilitating phytoremediation of soil arsenic by hyperaccumulator *Pteris vittata* L. (Liang et al. 2019).

12.3.2.1.2 Phytostimulation

Phytostimulation occurs where plant interaction in rhizosphere with microorganisms helps in the removal of heavy metals from soils. It has been found that fungal endophytes produce phytohormones, ACC deaminase, siderophores and antimicrobial compounds that help in plant growth. The ACC deaminase helps in reducing ethylene production by consumption of its precursor, and siderophores help in the accumulation of iron by plants. Zhu et al. (2015) found that soil-borne *Mucor* sp. (Fungi Z4 and Z8) alleviates Cd and Pb stress in Guizhou oilseed rape and accelerates the accumulation of heavy metals. Mustard plants when grown along with *Mucor* sp. MHR-7 reduce the uptake and toxicity of heavy metals by plants as well as produce ACC deaminase and indole acetic acid and help in the solubilization of phosphates. Simply by reducing the concentration of ethylene, it helps in the elongation of roots, as ethylene inhibits cell division and synthesis of DNA when present at high concentration.

12.3.2.1.3 Phytostabilization

It is a process where plants help in the immobilization of pollutants from contaminated sites. Plant roots through root hairs absorb, adsorb, accumulate, precipitate and stabilize the pollutant in rhizosphere. It helps in preventing its movement into food chain and reduces its availability. Microbial flora associated with phytostabilization help in plant growth and increase its tolerance towards metals (Tripathi et al. 2005). This method has the limitation that the contaminant still remains in the soil.

12.3.2.1.4 Phytovolatilization

This process deals with the removal of contaminants by plants and their release into atmosphere in less toxic form. Toxic heavy metals such as mercury and selenium can be converted into less toxic volatile forms as mercuric oxide and dimethyl selenide and released into atmosphere. LeDuc and Terry (2005) used the aquatic plant *Typha latifolia* L. to remove selenium from contaminated soil. Tobacco plant (*Nicotiana tabacum* L.) and *Arabidopsis thaliana* were genetically modified by bacterial organomercurial lyase and mercuric reductase to absorb mercury from soil and release volatile mercury into the atmosphere (Rahman et al. 2008). Indian mustard and muskgrass absorb selenium and mercury and convert them into gaseous form (Ghosh and Singh 2005).

12.3.2.1.5 Phytodegradation

In phytodegradation, plants uptake the contaminants and metabolize or degrade them into less toxic form by producing various enzymes (Ahmadpour et al. 2012). It is also called phytotransformation. As heavy metals are non-biodegradable, this method is used to remove organic contaminants such as pesticide, herbicide, benzene, toluene and chlorinated solvents.

12.3.2.1.6 Phytofiltration

This technique is used to clean up the contaminated environment using roots (rhizofiltration), seedlings (blastofiltration) and excised plant shoots (caulofiltration). The trees with long roots are used, which absorb a large amount of underground water along with contaminants. *Populus*, *Eucalyptus* and *Salix* species are high transpiring trees that are found to be good at phytofiltration as they reduce the entry of contaminants into water table (Muthusaravanan et al. 2018).

12.3.2.2 Enrichment of Phytoremediation Technique

Various chemical and biological methods are used to increase the phytoremediation process.

1. Certain chemicals such as EDTA, humic acids and plant nutrients (fertilizers) are used to increase the metals uptake by plants.
2. The use of genetically modified plants can accumulate a large amount of heavy metals.
3. Microbial consortium is used to enhance the metals uptake.
4. *Jatropha, Ricinus* and *Populus* plants are used for phytoremediation, and after remediation, they can be used as a fuel.
5. Earthworms help in improving the soil physical and chemical properties, which ultimately results in increased the plant biomass. Phytoremediation efficiency of switchgrass has been increased by the use of vermicomposting in heavy metals-contaminated soil (Shrestha et al. 2019).
6. Crop rotation, co-cropping and tillage help increase phytoremediation.

12.3.3 Rhizomediation (Microbes and Plants)

Symbiosis between microbes and plants plays an important role in the remediation of heavy metals-contaminated sites. Plants by secreting exudates such as carbohydrates, flavonoids and amino acids can enhance microbial activities many folds. Rhizospheric microbes boost plant growth by siderophore production, nitrogen fixation, solubilization of potassium and phosphates and increased phytohormone production (indole acetic acid). Rhizosphere microbes through the regulation of the level of ethylene by ACC deaminase result in the generation of larger roots for healthy growth of seedlings. Remediation of heavy metal cadmium by using symbiotic plant–microbe relationship was done by Wu et al. (2006). They used *Pseudomonas putida* 06909, a rhizobacterium isolated from citrus root having metal-binding peptide (EC20) expression. Inoculation of engineered bacterium to sunflower roots increases the cadmium accumulation by up to 40% and decrease the phytotoxicity of cadmium. Rhizomediation is a cost-effective, convenient tool with little or no side effects, which make it a clean and green future technology. Heavy metals such as lead, cadmium, tin and copper are converted from one oxidation state to another by various mechanisms such as chelating, siderophore production, biofilm production and production of acids (Oberai and Khanna 2018). Rhizomediation is affected by various chemical, physical and biological factors such as condition of soil, pH, temperature, type of the pollutant, organic matter and indigenous microflora.

12.4 Challenges and Future Prospects

The contamination of environment by toxic heavy metals is increasing at an alarming rate due to various anthropogenic reasons. Heavy metals are non-biodegradable in nature, so their persistence for long time results in ecotoxicological and health hazards. Various mechanisms are used to remove these pollutants from the environment, of which bioremediation and phytoremediation have been found to be effective. Selection of microbial flora in the polluted sites having resistance gene is used to improve the bioremediation efficiency. Microbial consortium has more capabilities to transform and remove heavy metals, and there is need to carry out such studies for harnessing the potential of different consortia. Phytoremediation has various advantages such as remediation of habitat with reduced costs and in situ clean-up rather than transporting the pollutant to another site. Hyperaccumulator plants are used for the effective phytoremediation process. Phytoextraction of heavy metals in aerial parts of plant and disposal approaches such as burning (energy source) and ashing (use in brick formation) are not applicable for volatile metals; therefore, research studies are needed to develop methods for extraction of metals from phytoremediator plants. The use of genetically modified and transgenic plants and microbes in future may improve the applicability of these processes. Transgenic technology, which is effective for bioremediation process, needs to be accepted by public through environmentalists and researchers by changing their view. More studies are required to understand the mechanisms underlying in these processes to enhance their effectiveness and to predict their possible side effects.

12.5 Conclusions

High levels of heavy metals in humans affect kidneys, haematopoietic system, gastrointestinal tract (GIT) and reproductive system and cause acute or chronic damage to the nervous system. They enhance the ROS production and compete with essential elements and make them unavailable to body, resulting in illness. Their toxicity is responsible for the development of cancer of liver, skin, lung, pancreas and prostate glands. Heavy metals released from anthropogenic activities can highly contaminate the aquatic ecosystems, thus adversely affecting the ecological balance and biodiversity of aquatic flora and fauna. The heavy metals such as As, Cd, Cr, Pb, Mn, Hg, Ni and Zn cause severe toxicity to animals and may alter the physiological and biochemical functions. As heavy metals pose threat to health and biodiversity of humans and animals, respectively, there is need to develop more efficient methods for their removal from the ecosystem. Bioremediation involving the use of microbes and plants are new clean-up technologies compared to conventional concepts for the removal of metals and stabilization of contaminated environments. Microbes and plants have inherent capacity that permits them to stay alive under heavy metal stress and effectively remediate polluted environments. Exploring more efficient microbes and plant species that can effectively remove heavy metals is required. More research needs to be focused on the mechanism underlying these processes and the suitable conditions required by microbes for the maximal removal of heavy metals.

REFERENCES

Agency for Toxic Substances and Disease Registry (ATSDR). 1998. *Toxicological Profile for Chromium*. U.S. Public Health Service, U.S. Department of Health and Human Services, Atlanta, GA.

Ahmadpour, P., Ahmadpour, F., Mahmud, T.M.M., Abdu, A., Soleimani, M., and Tayefeh, F.H. 2012. Phytoremediation of heavy metals: A green technology. *Afr J Biotechnol* 11(76): 14036–14043.

Alfvén, T., Järup, L., and Elinder, C.G. 2002. Cadmium and lead in blood in relation to low bone mineral density and tubular proteinuria. *Environ Health Perspect* 110(7): 699–702

Alwaleedi, S. A. 2015. Haemato-biochemical changes induced by lead intoxication in male and female albino mice. *Int J Recent Sci Res* 6(5): 3999–4004.

Amadi, C.N., Igweze, Z.N., and Orisakwe, O.E. 2017. Heavy metals in miscarriages and stillbirths in developing nations. *Middle East Fertil Soc J* 22: 91–100.

Balakrishnan, R., Satish Kumar, C.S., Rani, M.U., Srikanth, M.K. Boobalan, G., and Reddy, A.G. 2013. An evaluation of the protective role of α-tocopherol on free radical induced hepatotoxicity and nephrotoxicity due to chromium in rats. *Indian J Pharmacol* 45(5): 490–495.

Barbosa, D., Bomjardim, H.A., Campos, K.F., Duarte, M.D., Júnior, P.S.B., Gava, A., Salvarani, F.M., and Oliveira, C. M. C. 2014. Lead poisoning in cattle and chickens in the state of Pará, Brazil. *Pesqui Vet Brasil* 34(11): 1077–1080.

Basu, A., Sen, P., and Jha, A. 2015. Environmental arsenic toxicity in West Bengal, India: A brief policy review. *Indian J Public Health* 59(4): 295–298.

Bera, A.K., Rana, T, Das, S., Bhattacharya, D., Bandyopadhyay, S., and Pan, D. 2010. Ground water arsenic contamination in West Bengal, India: A risk of sub-clinical toxicity in cattle as evident by correlation between arsenic exposure, excretion and deposition. *Toxicol Ind Health* 26: 709–716.

Bondy, S.C. 2016. Low levels of aluminum can lead to behavioral and morphological changes associated with Alzheimer's disease and age-related neurodegeneration. *Neurotoxicol* 52: 222–229.

Casalegno, C., Schifanella, O., Zennaro, E., Marroncelli, S., and Briant, R. 2015. Collate literature data on toxicity of chromium (Cr) and nickel (Ni) in experimental animals and humans. *Supporting Publications* EN-478: 1–287.

Chen S.Y., and Lin J.G. 2004. Bioleaching of heavy metals from contaminated sediment by indigenous sulfur-oxidizing bacteria in an air-lift bioreactor: Effects of sulfur concentration. *Water Res* 38(14–15): 3205–3214.

Chen, Q.Y., DesMarais, T., and Costa, M. 2019. Metals and mechanisms of carcinogenesis. *Annu Rev Pharmacol Toxicol* 59: 537–554.

Children's health and the environment WHO training package for the health sector 2011. World Health Organization. www.who.int/ceh.

Das N., Basak L.V.G., Salam J.A., and Abigail M.E.A. 2012. Application of biofilms on remediation of pollutants – an overview. *J Microbiol Biotechnol Res* 2(5): 783–790.

Douglas, D. 2018. *Firefighter Chemical Review – ARP1701*. 1–106

Emokpae, M.A., Agbonlahor, O.J., and Evbuomwan, A.E. 2016. The relationship between maternal blood cadmium, zinc levels and birth weight of babies in non-occupationally exposed pregnant women in Benin city, Nigeria. *J Med Biomed Res* 15(1): 55–61.

Fanni, D., Ambu, R., Gerosa, C., Nemolato, S., Iacovidou, N., Eyken, P.V., Fanos, V., Zaffanello, M., and Faa, G. 2014. Aluminum exposure and toxicity in neonates: A practical guide to halt aluminum overload in the prenatal and prenatal periods. *World J Pediatr* 10(2): 101–107.

Ghosh, M., and Singh, S.P. 2005. A review on phytoremediation of heavy metals and utilization of it's by products. *Asian J Energy Env* 6: 18.

Goher, M.E., El-Monem, A.M.A., Abdel-Satar, A.M., Ali, M.H., Hussian, A.E.M., and Napi´orkowska-Krzebietke, A. 2016. Biosorption of some toxic metals from aqueous solution using nonliving algal cells of *Chlorella vulgaris*. *J Elem* 21(3): 703–714.

Govind P, Madhuri S., and Shrivastav, A.B. 2014. *Fish Cancer by Environmental Pollutants*, 1st ed., Narendra Publishing House, Delhi, India.

Gruji'c, S., Vasi'c, S., Radojevi'c I., ˇComi'c, L., and Ostoji'c, A. 2017. Comparison of the *Rhodotorula mucilaginosa* biofilm and planktonic culture on heavy metal susceptibility and removal potential. *Water Air Soil Pollut* 228(2): 73.

Guo, H., Chen, L., Cui, H., Peng, X., Fang, J., Zuo, Z., Deng, J., Wang, X., and Wu, B. 2016. Research advances on pathways of nickel-induced apoptosis. *Int J Mol Sci* 17(10): 1–18.

Gupta, A., Joia, J., Sood, A., Sood, R., Sidhu, C., and Kaur, G. 2016. Microbes as potential tool for remediation of heavy metals: A review. *J Microb Biochem Technol* 8: 364–372.

He, L.M., and Tebo, B.M. 1998. Surface charge properties of and Cu (II) adsorption by spores of the marine *Bacillus* sp strain SG-1. *Appl Environ Microbiol* 64: 1123–1129.

Heinz, G.H., Hoffman, D.J., Klimstra, J.D., Stebbins, K.R., Kondrad, S.L., and Erwin, C. A. 2009. Species differences in the sensitivity of avian embryos to methyl mercury. *Arch Environ Contam Toxicol* 56(1): 129–138.

Higham, D.P., Sadler, P.J., and Scawen, M.D. 1986. Cadmium-binding proteins in *Pseudomonas putida*: Pseudothioneins. *Environ Health Perspect* 65: 5–11.

Huang, Y., Xi, Y., Gan, L., Johnson, D., Wu, Y., Ren, D., and Liu, H. 2019. Effects of lead and cadmium on photosynthesis in *Amaranthus spinosus* and assessment of phytoremediation potential. *Int J Phytoremediat* 21(10): 1041–1049.

Igiri, B. E., Okoduwa, S.I.R., Idoko, G.O., Akabuogu, E.P., Adeyi, A.O., and Ejiogu, I. K. 2018. Toxicity and bioremediation of heavy metals contaminated ecosystem from tannery wastewater: A review. *J Toxicol* Article ID 2568038, 16 p.

Kelly, D.J.A., Budd, K., and Lefebvre, D.D. 2006. The biotransformation of mercury in pH-stat cultures of microfungi. *Can J Bot* 84: 254–260.

Keshavarzi, B., Afsaneh, S., Zahra, A., Farid, M., Alireza, R.S., and Mehrdad P. 2015. Chronic arsenic toxicity in sheep of Kurdistan Province, Western Iran. *Arch Environ Contam Toxicol* doi: 10.1007/s00244-015-0157-4.

Khalida, B., Poorvi, V., Priya, M., and Bansuri, G. 2018. Biodegradation of heavy metals by microorganism and their role in controlling environmental pollution. *IJSRR* 7(1): 275–284.

Kumar, S., and Trivedi, A.V. 2016. A review on role of nickel in the biological system. *Int J Curr Microbiol App Sci* 5(3): 719–727.

Lane, E.A., Canty, M.J., and More, S.J. 2015. Cadmium exposure and consequence for the health and productivity of farmed ruminants. *Res Vet Sci* 101: 132–13.

LeDuc, D.L., and Terry, N. 2005. Phytoremediation of toxic trace elements in soil and water. *J Ind Microbiol Biotechnol* 32: 514–520.

Liang, Y., Wang, X., Guo, Z., Xiao, X., Peng, C., Yang, J., Zhou C., and Zeng, P. 2019. Chelator-assisted phytoextraction of arsenic, cadmium and lead by *Pteris vittata* L. and soil microbial community structure response. *Int J Phytoremediat* 21(10): 1032–1040.

Lima, A.I.G., Corticeiro, S.C., and de Almeida Paula Figueira, E.M. 2006. Glutathione-mediated cadmium sequestration in *Rhizobium leguminosarum*. *Enzyme Microb Techno* 39(4): 763–769.

Lovley, D.R., Philips, E.J, Gorby, Y.A., and Landa, E.R. 1991. Microbial reduction of uranium. *Nature* 350: 413–416.

Luka, Y., Highina, B. K., and Zubairu, A. 2018. Bioremediation: A solution to environmental pollution: A Review. *Am J Eng Res* 7(2): 101–109.

Macfie, A., Hagan, E., and Zhitkovich, A. 2010. Mechanism of DNA-Protein crosslinking by chromium. *Chem Res Toxicol.* 23(2): 341–347.

Mahor, G., and Ali, S.A. 2015. An update on the role of medicinal plants in amelioration of aluminium toxicity. *Biosci Biotechnol Res* 8(2): 175–188.

Malik, H., Sajjad, S., Akhtar, S., and Bilal, S. 2016. Effect of nickel toxicity on growth parameters and hepatic enzymes in major carp. *Indian J Anim Res* 50(3): 370–373.

Mason, L.H., Harp, J.P., and Han, D.Y. 2014. Pb neurotoxicity: Neuropsychological effects of lead toxicity. *Biomed Res Int* Article ID 840547, 8 p. doi: 10.1155/2014/840547.

Mejare, M., and Bulow, L. 2001. Metal binding proteins and peptides in bioremediation and phytoremediation of heavy metals. *Trends Biotechnol* 19: 67–73.

Miller, N.Z. 2016. Aluminum in childhood vaccines is unsafe. *JPANDS* 21(4): 109–117.

Miller, S., Pallan, S., Gangji, A.S., Lukic, D., and Clase, C.M. 2013. Mercury-associated nephrotic syndrome: A case report and systematic review of the literature. *Am J Kidney Dis* 62(1): 135–138.

Mokhtar H., Morad, N., and Fizri, F.F.A. 2011. Phytoaccumulation of copper from aqueous solutions using *Eichhornia crassipes* and *Centella asiatica*. *Int J Environ Sci Dev* 2:205–210.

Muthusaravanan, S., Sivarajasekar, N., Vivek, J.S., Paramasivan, T., Naushad, M., Prakashmaran, J., Gayathri, V., and Al-Duaij, O.K. 2018. Phytoremediation of heavy metals: Mechanisms, methods and enhancements. *Environ Chem Lett* 16: 1339–1359.

Narayanan, S. 2014. Comparative study on effect of aluminium chloride and aluminium hydroxide on serum biochemical parameters in *Wistar albino* rats. *Int J Pharm Biol Sci* 5(1): 253–258.

Oberai, M., and Khanna, V. 2018. Rhizomediation – Plant microbe interactions in the removal of pollutants. *Int J Curr Microb Appl Sci* 7(1): 2280–2287.

Otebhi, O.E., and Osadolor, H.B. 2016. Select toxic metals of pregnant women with history of pregnancy complications in Benincity, South-South Nigeria. *J Appl Sci Environ Manage* 20(1): 5–10.

Park, J-D., and Zheng, W. 2012. Human exposure and health effects of inorganic and elemental mercury. *J Prev Med Public Health* 45(6): 344–352.

Patlolla, A.K., Todorov, T.I., Tchounwou, P.B., van der Voet, G., and Centeno, J.A. 2012. Arsenic-induced biochemical and genotoxic effects and distribution in tissues of Sprague-Dawley rats. *Microchemical Journal* 105: 101–107.

Praveena, M., Sandeep, V., Kavitha, N., and Jayantha Rao, K. 2013. Impact of tannery effluent, chromium on hematological parameters in a fresh water fish, *Labeo rohita* (Hamilton). *Res J Animal Veterinary and Fishery Sci* 1(6): 1–5.

Rahman, A., Kumarathasan, P., and Gomes, J. 2016. Infant and mother related outcomes from exposure to metals with endocrine disrupting properties during pregnancy. *Sci Total Environ* 569–570: 1022–1031.

Rahman, R. A., Abou-shanab, R., and Moawad, H. 2008. Mercury detoxification using genetic engineered *Nicotiana tabacum*. *Global Nest J* 10(3): 432–438.

Rathor, G., Chopra, N., and Adhikari, T. 2014. Nickel as a pollutant and its management. *Int Res J Environ* 3(10): 94–98.

Shaw, C A., Seneff, S., Kette, S.D., Tomljenovic, L., OllerJr, J.W., and Davidson, R. M. 2014. Aluminum-induced entropy in biological systems: Implications for neurological disease. *J Toxicol* Article ID 491316, 27 p. doi: 10.1155/2014/491316e.

Shishir, T.A., Mahbu, N., and Kamal, N.E, 2019. Review on bioremediation: A tool to resurrect the polluted rivers. *Pollution* 5(3): 555–568.

Shrestha, P., Bellitürk, K., and Görres, J.H. 2019. Phytoremediation of heavy metal-contaminated soil by switchgrass: A comparative study utilizing different composts and coir fiber on pollution remediation, plant productivity, and nutrient leaching. *Int J Environ Res Public Health* 16(7): 1261.

Singh, S., and Sinha, S. 2005. Accumulation of metals and its effects in *Brassica juncea* (L.) Czern. (cv. Rohini) grown on various amendments of tannery waste. *Ecotoxicol Environ Saf* 62: 118–127.

Song, J.J., Kim, J.Y., Jang, A.S., Kim, S.H., Rah, Y.C., Park, M., and Park, M.K. 2015. Effect of cadmium on human middle ear epithelial Cells. *J Int Adv Otol* 11(3): 183–187.

Sumiahadi, A., and Acar, R. 2018. A review of phytoremediation technology: Heavy metals uptake by plants. *IOP Conf Ser Earth Environ Sci* 142: 012023.

Swaddiwudhipong, W., Nguntra, P., Kaewnate, Y., Mahasakpan, P., Limpatanachote, P., Aunjai, T., Jeekeeree, W., Punta, B., Funkhiew, T., and Phopueng, I. 2015. Human health effects from cadmium exposure: Comparison between persons living in cadmium-contaminated and non-contaminated areas in northwestern Thailand. *Southeast Asian J Trop Med Public Health* 46(1): 133–142.

Tang, K., Li, J., Yin, S., Guo, H., Deng, J., and Cui, H. 2014. Effects of nickel chloride on histopathological lesions and oxidative damage in the thymus. *Health* 6: 2875–2882.

Tribowo, J.A., Arizal, M.H., Nashrullah, M., Aditama, A.R., and Utama, D.G., 2014. Oxidative stress of cadmium-induced ovarian rat toxicity. *Int J Chem Eng* 5(3): 254–258.

Tripathi, M., Munot, H.P., Shouche, Y., Meyer, J.M., and Goel, R. 2005. Isolation and functional characterization of siderophore-producing lead-and cadmium resistant *Pseudomonas putida* KNP9. *Curr Microbiol* 50: 233–237

Wani, A.L., Ara, A., and Usmani, J.A. 2015. Lead toxicity: A review. *Interdiscip Toxico* 8(2): 55–64.

Wen, Y.-M., Cheng, Y., Tang, C., and Zu-Liang, C. 2013. Bioleaching of heavy metals from sewage sludge using indigenous iron-oxidizing microorganisms. *J Soil Sediment* 13(1): 166–175.

Wu, C.H., Wood, T.K., Mulchandani, A., and Chen, W. 2006. Engineering plant microbe symbiosis for rhizomediation of heavy metals. *Appl Environ Microbiol* 72: 1129–1134.

Yang, Y., Liu, H., Xiang, X., and Liu, F. 2013. Outline of occupational chromium poisoning in China. *Bull Environ Contam Toxicol* 90: 742–749.

Yokoyama, H. 2018. Lecture on methylmercury poisoning in minamata (MPM). *Mercury Pollut Minamata* 5–51.

Zafar, S., Aqil, F., and Ahmad, I. 2007. Metal tolerance and biosorption potential of filamentous fungi isolated from metal contaminated agricultural soil. *Bioresour Technol* 98: 2557–2561.

Zhu, S.-c., Jian-xin, T., Xiao-xi, Z., Ben-jie, W., Shao-di, Y., and Bin, H., 2015. Isolation of *Mucor circinelloides* Z4 and *Mucor racemosus* Z8 from heavy metal-contaminated soil and their potential in promoting phytoextraction with Guizhou oilseed rap. *J Cent South Univ* 22(1): 88–94.

13

Biogenic and Non-Biogenic Waste for the Synthesis of Nanoparticles and Their Applications

Abhishek Kumar Bhardwaj and Ram Naraian
Veer Bahadur Singh Purvanchal University

Shanthy Sundaram and Rupali Kaur
University of Allahabad

CONTENTS

13.1 Introduction

The exponential development and industrialization of the world, basically cities, generate various kinds of municipal and industrial wastes having a tremendous amount of construction and demolition debris, plastic waste, e-waste, biomedical waste and other industrial hazardous and non-hazardous wastes. On the contrary, villages generate bio-wastes containing crop residue, food waste and other agro-wastes. For waste management, the world is looking forward to developing and deploying various technologies to reduce, reuse and recycle materials, generate energy and extract valuable resources (Bhardwaj and Naraian, 2020). The international market for recycling and extraction of valuable materials is growing steadily.

The modern scientific community of nanotechnology is focusing on the sustainable environment; therefore, they are developing innovative and groundbreaking methods for the synthesis of nanomaterials (Bhardwaj et al., 2021). However, several green methods of NPs synthesis were suggested by

DOI: 10.1201/9781003181224-13

researchers using enzymes, microorganisms, fungi, plants, fruits and fruit peels due to the naturally occurring potential of biological extracts (Bhardwaj et al., 2018). Similarly, it is reported that the synthesis of NPs using less valuable waste including natural extract of peels and pulp waste from food industries, and other agro-industrial material that produced in huge amount of functional solid waste can be environment-friendly, cost-effective, and scalable nanomaterial synthesis process (Bhardwaj et al., 2017b). At the same time, growing human population and industrialization are the major factors for the increase in solid wastes in our planet earth. The improper management of the huge amount of solid wastes produced lead to public health- and environment-related problems. Various metals and heavy metals are released into environment by several industrial practices, including electroplating, painting, mining, tanning, smelting, dyeing, printing and battery and fertilizer manufacturing (Souda and Sreejith, 2015; Neves et al., 2020).

Thus, there is an urgent need for ideas and research to recycle and reutilize the wastes in the developed and developing countries, which would protect us from an unsustainable 'tsunami of waste in future'. The nanoparticles synthesis from the waste materials can be a big thought in future to utilize any type of waste as useful products.

Nanotechnology has emerged as a revolutionary technology with versatile applications in medical, clinical, storage and environmental fields (Bhardwaj and Naraian 2021). Nanoparticles are defined as particles in the nanoscale range (typically between 1 and 100 nm) with unique shape, size and orientation, which exhibit properties that are different from bulk solids (Bhardwaj et al., 2019). Hence, they show various physical and optical properties such as high surface-to-volume ratio, surface plasmon resonance (SPR) and surface-enhances Raman scattering (SERS) (Shukla et al., 2018). These unique properties of nanomaterials enable a wide range of applications in the field of sensing, bio-imaging, medicine, cosmetics, water purification and preservation of food and beverages (Bhardwaj et al., 2017a). This chapter covers the recent highlights of the synthesis of metal NPs, non-metals NPs, carbon nanostructure and other NPs using various available waste of biological, non-biological origin of solid waste or liquid waste. This synthesis approach resembles, recovery of valuable materials from waste and on the other hand, minimizes hazardous wastes in the process of nanomaterial synthesis.

13.2 Types of Waste Materials that Have the Potency to Be Utilized in Metal NPs Synthesis

From the last decade, scientist have been showing interest in the synthesis of nanomaterials such as metal NPs, polymeric NPs, nanocomposites and carbon nanostructure tubes/sheets/fibres using waste materials of biological and non-biological origin. Each and every waste is made up of unique set of elements and molecules where some elements are maximum in quantity than others. The recycling and recovery of these waste materials valuable nanomaterials into (carbon, copper, gold, iron, lead, platinum, silicon, titanium and zinc) is possible due to the processes being developed by the efforts of researchers/scientists. The complete process of synthesis of nanomaterials from biogenic and non-biogenic sources is shown with the help of a graphical scheme in Figure 13.1.

13.2.1 Wastes of Biological Origin

Biomass produced from the agro-waste and from restaurant and kitchen waste contains several primary and secondary metabolites such as phenolic acids, terpenoids, alkaloids and flavonoids (Kuppusamy et al., 2016; Bhardwaj et al., 2018). These green wastes having primary and secondary metabolites: primary metabolites play a significant role in the usual development, growth and reproduction of plants. In addition to secondary metabolites are compounds produced only by some selected plant groups which produce generally against defence from predators or during resist the change in environment (Garay et al., 2007). The members of the flavonoid family (flavonols, flavones, isoflavonoids, chalcones and flavanones) have the ability to reduce metal ions into NPs and form a chelate.

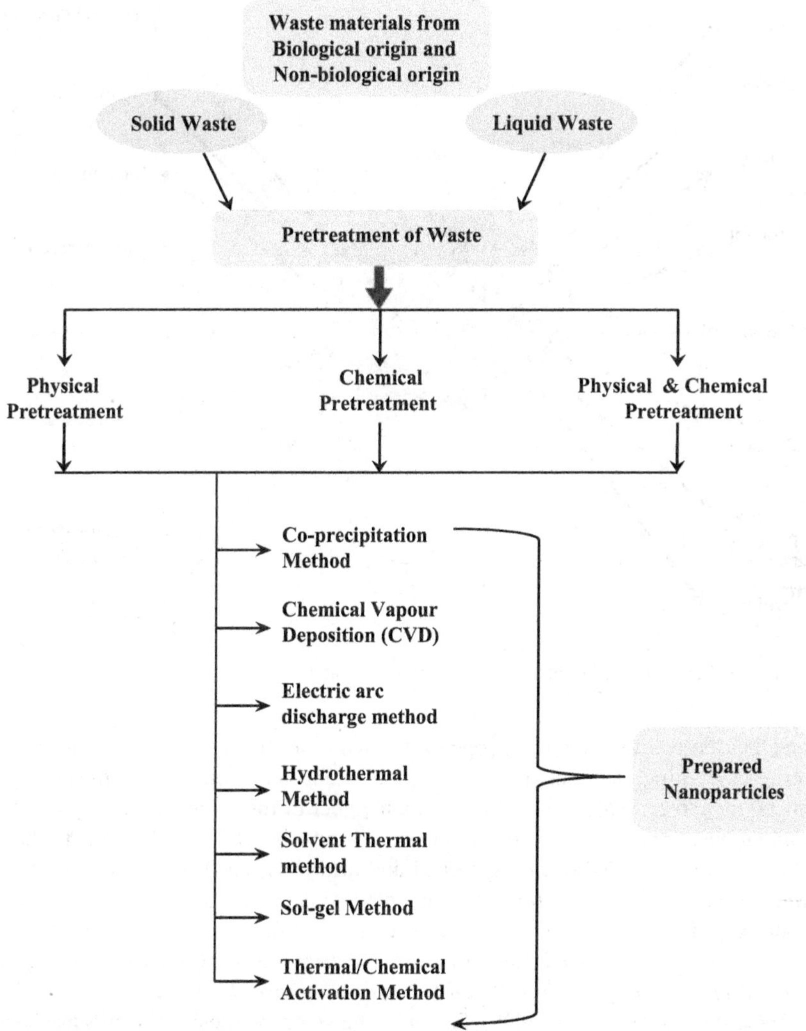

FIGURE 13.1 Schematic diagram of the synthesis of nanomaterials using wastes of biological and non-biological origin.

Quercetin is also one of the flavonoid family members that has strong chelating properties due to the presence of three sites of carbonyl and hydroxyl at C3 and C5 positions and the catechol group at C3 and C4 sites. Therefore, it is believed that several metal ions such as Al^{2+}, Co^{2+}, $Cr3^{+}$, Cu^{2+}, Fe^{2+}, Fe^{3+}, Pb^{2+} and Zn^{2+} can reduce and form chelates; consequently, the respective NPs can be formed (Makarov et al., 2014). On the other hand, the different parts of the plants, plant products or the waste generated from their residue have different kinds of biomolecules such as amino acids, enzymes, proteins, sugars and trace metals. These biomolecules are self-sufficient to reduce metal ions to stabilized NPs. It is observed by the various researchers that these metabolites can be utilized in the redox reaction in the formation of zerovalent metallic NPs (Kuppusamy et al., 2016; Maurya et al., 2016). These materials have a number of applications in different fields as shown in Figure 13.2.

13.2.1.1 Metal NPs and Non-Metal NPs Synthesis

The different kinds of wastes of biogenic origin such as fruit peels or pulp, vegetable peels or pulp and other cellulosic wastes from kitchen, restaurant and food processing industry having number of chemical

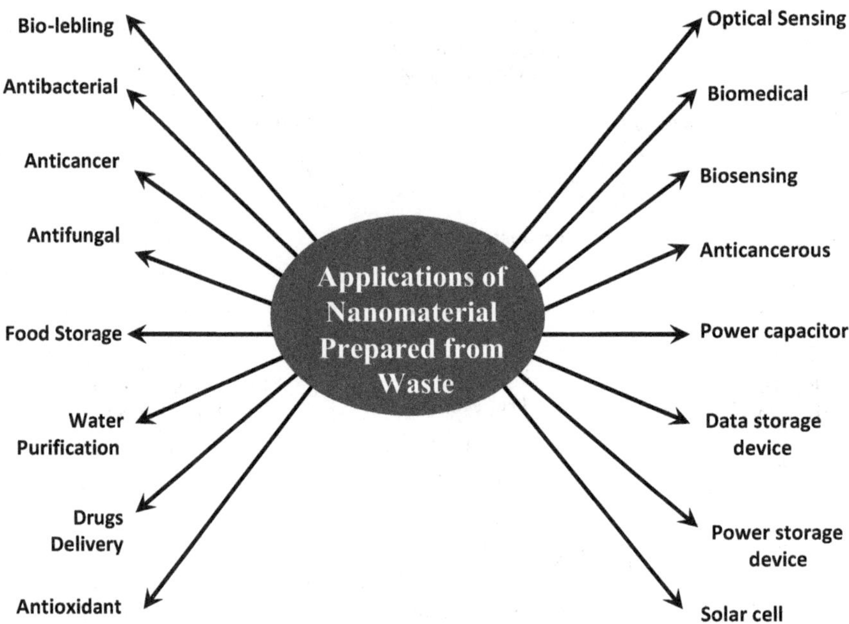

FIGURE 13.2 Applications of nanomaterials prepared from waste materials.

constituents and their derivatives that have scope for the synthesis of various kind of nanomaterials. The biogenic material containing enzymes, proteins, pigments and metabolites have intrinsic potential to reduction of metal ions to zerovalent NPs. Eugenol, a member of the terpenoid family, can be found in cinnamon (*Cinnamomum zeylanicum*) extracts for the bio-reduction of Au^{++} and Ag^{+} to the zerovalent metal NPs (Premkumar et al., 2018). The spherical and hexagonal ZnO NPs synthesized biologically using *P. granatum* showed bactericidal properties against *E. coli* and *E. faecalis*. Furthermore, it applied for human colon cancerous cell line killing and observed a effective concentration of $\geq$31.25 µg/mL (Sukri et al., 2019). AgNPs synthesized from *Tribulus terrestris* fruit were able to deactivate multi-drug-resistant bacteria *Streptococcus pyogenes, Pseudomonas aeruginosa, Escherichia coli, Bacillus subtilis* and *Staphylococcus aureus* (Gou et al., 2020). Recently, the green synthesis of FeNPs has been reported by utilizing the leaves of various plants (*Mangifera indica, Murraya koenigii, Azadirachta indica and Magnolia champaca*), wherein enzymes play an important role in synthesis. Furthermore, FeNPs were tested in domestic wastewater treatment (Devatha et al., 2016).

The production of FeNPs was reported using the extracts of green and black tea leaves, and they were utilized in the degradation of malachite green (MG) into oxidized form of iron, cleaving the bond that was connected to the benzene ring (Huang et al., 2014). Similarly, FeNPs synthesized using green tea and eucalyptus leaves were utilized in the removal of nitrate (Wang et al., 2014). Eucalyptus leaves-based magnetic NPs was prepared and attached to imprinted polymers had the ability to isolate from sample solutions when apply the external magnetic field and then template molecule selective recognize (López et al., 2020). Iron oxide NPs with different phases (Fe_3O_4, α-Fe_2O_3 and γ-Fe_2O_3) were synthesized using non-edible peel extracts of *Citrus paradisi*, and they showed antioxidant activities against 1,1-diphenyl-2-picrylhydrazyl (>15%, 100 µg) and were used in dye decolourization (methyl rose, 96.5%; methylene blue, 80.76%; methyl orange, 89.64%) (Kumar et al., 2020).

The TiO_2 NPs were prepared to reduce titanium(IV) isopropoxide using the extract of lemongrass due to the presence of reducing agents. After that, lemon peel biomass was modified with the TiO_2 nanoparticles using a polysiloxane matrix. Furthermore, it was reported that the lemon peel biomass-modified TiO_2 NPs have a maximum Ni(II) ions adsorption capacity (90%) compared to non-modified TiO_2 NPs (78%) (Herrera-Barros et al., 2020). The interesting application of waste-derived nanomaterials have found such as the nano-hydroxyapatite, prepared from Ca precursor from eggshell using reverse microemulsions at ambient temperature. The small particle size of nano-hydroxyapatite (5–20 nm) and their calcinated nano-HAP at 600°C were utilized for the removal of As(V). Finally, the experimental study revealed that the removal by non-calcinated nano-HAP was significant (up to 38.27%) compared to that by calcinated nano-HAP (0.97%) (Jahan et al., 2017).

13.2.2 NPs Prepared Using Wastes of Non-Biological Origin

The Cu and CuO NPs were successfully prepared from copper scrap using plasma arc discharge technique under the nitrogen and air atmosphere. Further, the maximum antibacterial zone of inhibition was found to be 32 and 28 mm against *S. aureus* and *K. pneumonia* (Tharchanaa et al., 2020). Rezazadeh et al. (2020) synthesized magnetite nanocomposites using raffinate waste generated from hydrometallurgical plants containing iron ions. A three-step process was developed: (i) the reduction of Fe(III) to Fe(II); (ii) The precipitation of Al^{3+} in the raffinate as Al(OH)3; and (iii) the synthesis of magnetite NPs using chemical co-precipitation technique. The prepared material was further employed as an adsorbent for the removal of Hg^{2+} ions. The adsorption capacity obtained was 73.37 mg/g, and the Hg(II) adsorption capacity of Fe_3O_4@PVA decreased from 97.2% to 88.4% after five cycles. Xiang et al. (2015) produced zinc NPs (with particle size diameter in the range of 100–300 nm) using vacuum separation and inert gas condensation method from spent zinc–manganese batteries. A zinc separation efficiency of up to 99.68% was obtained; hence, it was proved that this method is highly efficient. El-Amir et al. (2016) prepared γ- and α-aluminium nanopowders from aluminium foil wastes using co-precipitation with NH_4OH as a precipitant. Some important accounts of synthesis and applications of metal NPs, non-metal NPs and carbon nanostructure prepared from biogenic and non-biogenic waste materials are shown in Table 13.1.

13.3 Carbon Nanostructure Synthesis from Carbon Waste

The waste material as a carbon source can be of both biological origin (rice husk, coconut shell, banana peels, chicken fat, etc.) and non-biological origin (plastics, scrap rubber/tyre, etc.); further, it can also be divided into solid waste and liquid waste (leachate from municipal waste, glycerol, engine oil, heavy oil residue, etc.). These materials were utilized for the preparation of graphene by various researchers (Alfarisa et al., 2015). Carbon atoms arranged in a honeycomb manner in the form of atom-thin mesh is known as graphene, which can be consider as the parent of all graphitic materials such as carbon nanotubes (CNTs), fullerenes and graphite (Geim and Novoselov, 2010). This material is established as the world's strongest, stiffest and thinnest material, which has surprising mechanical, thermal, and electrical properties. Sun et al. (2013) prepared graphene nanosheets using biomass waste of coconut shell. The graphitic precursor ($FeCl_3$) and the activating agent ($ZnCl_2$) play a significant role in the graphitization of carbonaceous shell under heat treatment (Sun et al., 2013). Hierarchical porous nitrogen-doped carbon (HPNC) nanosheets (NS) were prepared *via* simultaneous activation and graphitization of biomass-derived natural silk. They were prepared by a one-step and facile large-scale synthesis route. The high-performance Li-ion battery and supercapacitors were developed (Hou et al., 2015).

13.3.1 Carbon Nanostructure Synthesis Using Green Waste

Carbon nanotubes (CNTs) were prepared using banana peels by the chemical activation method. Initially, the peels were carbonized at 400°C, followed by activation with phosphoric acid and then pyrolysis at 600°C to create banana peel activated carbon powder. Finally, this powder was mixed with 2% mineral oil, placed in a stainless steel tube and pyrolysed at 1,000°C–1,200°C for 1 hour to obtain CNTs. CNTs produced had outer diameters in the range of 47–10 nm and inner diameters in the range of 12–30 nm (Mopoung, 2011). Dried elm samara from plant waste has been used to produce highly porous carbon nanosheets (CNs) three-dimensional scaffold structures through carbonization and chemical activation methods. The plant waste of the elm fruit was initially treated with KOH and

TABLE 13.1

The Account of Wastes and Their Potential for the Synthesis of Various Valuable NPs

S. No.	Type of Waste	NMs	Size/Orientation	Synthesis Technique	Application & Significance	References
1.	Fruit cover, plastic waste and oil palm fibre	Graphene	Thin layered	CVD	Studied tribological potential	Tahir et al. (2020)
2.	Coconut shell	Graphene	Graphitic carbon sheet	Simultaneous activation–graphitization route	High-power supercapacitor	Sun et al. (2013)
3.	Grass waste	Silver NPs	15 nm	Solvothermal	Antibacterial activity against *P. aeruginosa, A. baumannii* and *F. solani*; anticancer activity against MCF-7	Khatami et al. (2018)
4.	Stage of a waste-based biorefinery	Iron NPs	40 nm	Biological reduction	Can be used as a wastewater treatment	Romero-Cedillo et al. (2020)
5.	Tangerine peel extract	Silver NPs	19.2±6.7 nm	Biosynthesis	Potential hybrid hydrogels, to catalyse the reduction of nitroaromatic compounds	Quadrado et al. (2019)
6.	Aluminium foil waste	Nanoscale single-crystalline γ- and α-Al powders	36–200	Co-precipitation using NH_4OH as a precipitant	Not done	El-Amir et al. (2016)
7.	Egg shell waste	Nano-hydroxyapatite	5–20 nm	Using reverse microemulsions	As(V) removal (38.27%) at different values of pH from aqueous solutions	Jahan et al. (2017)
8.	Shrimp shell waste	N-doped nano-porous carbon	>5 nm	KOH activation	Performance of lithium–sulphur battery	Qu et al. (2016)
9.	Biomass-derived natural silk	N-doped carbon nanosheets	4.6 nm	Chemical activation and graphitization	Developed supercapacitors	Hou et al. (2015)
10.	Orange peel waste	TiO_2 nanoparticles	24 nm	Solvothermal	NA	Rao et al. (2015)
11.	Tyre waste	Nano-activated carbon	Micro-porous	Chemical activation	NO reduction activity increases 75% at room temperature	Al-Rahbi and Williams (2016)
12.	Sugarcane waste (bagasse)	Nano-porous activated carbon	Nano-porous	Chemical activation technique	Used in supercapacitors	Jain and Tripathi (2015)
13.	Elm samara plant waste	Carbon nanosheets (CNs)	3D scaffold structures	Carbonization and chemical activation methods	High-performance supercapacitors	Chen et al. (2016)

(*Continued*)

TABLE 13.1 (*Continued*)
The Account of Wastes and Their Potential for the Synthesis of Various Valuable NPs

S. No.	Type of Waste	NMs	Size/Orientation	Synthesis Technique	Application & Significance	References
14.	Spent Zn–Mn from waste battery	Zink NPs	Hexagonal prisms, diameter of 100–300 nm	Vacuum separation and inert gas condensation	Zinc separation efficiency up to 99.68%	Xiang et al. (2015)
15.	Leaf extracts of *M. indica, M. Koenigii, A. indica* and *M. champaca*	Iron NPs	50–150 nm	Leaf extracts reduce to ferrous sulphate heptahydrate	Domestic wastewater treatment	Devatha et al. (2016)
16.	Ferrous sulphate waste	Nano-α-Fe_2O_3 red pigment powders	Bacteria-assisted oxidation	22–86 nm	*Acidithiobacillus ferrooxidans* bacteria-assisted synthesis	Li et al. (2016)
17.	Waste of *Citrus limetta* peel	Silver NPs	Bio-reduction	18 nm	*Found* bactericidal and antifungal *properties against M. luteus, S. mutans, S. epidermidis, S. aureus* and *E. coli*	Dutta et al. (2020)
18.	Waste of *Citrus paradisi* peel	Iron oxides Fe_3O_4, α-Fe_2O_3 and γ-Fe_2O_3	Bio-reduction	28–32 nm	Antioxidant activities found against 1,1-diphenyl-2-picrylhydrazyl (> 15%, 100 μg) and dye decolourization (m. rose, 96.5%; m. blue, 80.76; and m. orange, 89.64 %)	Kumar et al. (2020)
19.	Titanium dioxide (TiO_2) NPs	Leaf extract of lemongrass	Bio-reduction and surface modification with lemon peel extract		Lemon peel biomass-modified TiO_2 NPs have a higher adsorption capacity of Ni(II) ions (90%) than non-modified TiO_2 NPs (78%)	Herrera-Barros et al. (2020)
20.	Cu_2O	Carrot pulp waste	Bio-reduction	Hollow microspheres formed	Antibacterial activity seen	Bhardwaj et al. (2017b)
21.	Cu and CuO	Copper scrap	Plasma arc discharge	78 and 67 nm	Antibacterial activity found against *S. aureus* and *K. pneumoniae*	Tharchanaa et al. (2020)
22.	Fe_3O_4@PVA	Raffinate waste of hydrometallurgical plants	Chemical co-precipitation method	19±4	Removal of Hg^{2+} ions	Rezazadeh et al. (2020)

dried in a vacuum oven at 100°C for 10 hours. The KOH/elm samaras mixture was carbonized in a tube furnace at 700°C for 2 hours in an Ar atmosphere. The growing sample was then treated with HCl solution (10 wt%) and dried at 100°C for 12 hours to obtain the final CNs (Chen et al., 2016). The solid waste material was utilized (fruit cover, plastic waste and oil palm fibre) for the graphene synthesis. Chemical vapour deposition (CVD) technique was used for growing graphene sheet over the substrate of copper sheet. The optimal parameters were observed during the growing graphene, within 90 minutes and at 1,020°C in the presence of only argon gas using fruit cover plastic waste, whereas 30 minutes at 1,000°C temperature in the presence of argon and hydrogen gas using oil palm fibre (Tahir et al., 2020).

13.3.2 Carbon Nanostructure Using Polyethylene

To meet the global demands of plastic, the industrial scale production of plastic initiated from the last 8 decades. The world had generated nearly 300 million tons of plastic in 2016 only (Gou et al., 2020). Such a huge amount of plastics causes serious problems to the soil, various ecosystems and biodiversity mainly when disposed of in landfills. Around 8–10 million tons of plastic waste goes into the ocean directly via various point and non-point sources every year. The land-based sources contribute 80%–90%, and it is projected to double by 2030 and again double by 2050 (Wu, 2020). It can cause unexpected problems to the biodiversity of oceanic creatures; therefore, it has recently become an emerging global environmental issue drawing enormous global attention. Majority of single-use plastics are thermoset plastics that can't be reused. Scientists are constantly working on developing recycling technologies to reduce plastic waste into materials that have less environmental impact and are economically profitable. However, The carbon nanotubes were prepared by pyrolysis of granular polyethylene at 420°C–450°C in helium atmosphere using Ni plates as a catalyst (Kukovitskii et al., 1997).

13.4 Pretreatment of the Waste

The pretreatment method is the initial step in the synthesis of any nanomaterials from different kinds of wastes. These methods can vary with the type of waste and the technique opted for the synthesis. Researchers have developed different techniques of sample preparation on the basis of waste composition. These developed pretreatment techniques can be simplified into three categories: (i) physical; (ii) chemical; and (iii) combined (both physical and chemical). In the synthesis of NPs, the pretreatment of waste is normally required for getting efficiency and purity. There is number of nanomaterials generated from different waste sources along with their particle sizes and preparation techniques (Singh et al. 2021).

13.4.1 Physical Pretreatment

In this physical method, the breakdown of larger particles of any substrate into smaller to prepare extract for the reduction of metal ions here are two basic methods are given including grinding and milling technique use under this section. The mortar and pestle can be used to provide mechanical energy to the reactant, this method referred as grinding. Another important term in this section is milling in the presence of ball mills. The time and energy inputs are the most important parameters, which make the result expected more reproducible (Stolle et al., 2011).

13.4.2 Chemical Pretreatment

The chemical pretreatment removes soluble contamination or undesirable things present in the waste before NPs synthesis. The solubility can be increased with increasing temperature or treatment with suitable chemicals. Commonly, concentrated strong acids (HNO_3, HCl and H_2SO_4), alkali and other organic solvents are used to hydrolyse and treat the waste materials. However, these are vary corrosive nature acid, it always needs to neutralize and recycle acids due to minimization of toxic effect to the environment (Harmsen et al., 2010; Bhardwaj et al., 2019). For instance, aluminium NPs (γ- and α-Al nanopowders) were prepared from the waste of aluminium foils. Properly cleaned waste materials were dissolved into aqua regia initially, and then an ammonia solution was added for maintaining the solution pH in the alkaline range (El-Amir et al., 2016). The chemical and physical processes are combined together in the process of pretreatment. For example, Chen et al. (2016) performed physical and chemical pretreatment of dry waste elm samaras.

13.4.3 Combined (Both Physical and Chemical) Pretreatment

In various cases, a single pretreatment method does not successfully work out; therefore, the combined method is important. Combining chemical and physical processes one by one or simultaneously plays a significant role in the pretreatment. For instance, for the preparation of nanocomposites (NCs), first put the waste elm at 70°C in oven for drying and treat it with KOH. Then, the elm samara was dried again in a vacuum oven at 100°C for 10 hours to obtain the final sample of nanosynthesis (Chen et al., 2016).

13.5 Conclusions

This chapter is dedicated to showing the highlights of the synthesis of a number of precious metal NPs, non-metals, carbon nanostructures and other NPs using various available wastes of biological and non-biological origin, either solid waste or liquid waste. The approach, on the one hand, recovers valuable materials from waste and, on the other hand, minimizes hazardous waste. There are various kind of nanoparticles can be synthesis from the waste materials would be a big opportunity in future to utilize several kinds of waste material in useful product of nanomaterials. In this modern era, this biogenic synthesis indicates a milestone in the sustainable approach towards achieving the sustainable development goals (SDGs). Therefore, we can meet the current concern of environmentally sustainable management of waste production and at the same time can take a step towards saving our earth from the incoming 'tsunami of the waste'.

Acknowledgement

The first author (Abhishek K. Bhardwaj) is highly thankful to VBS Purvanchal University, Jaunpur, Uttar Pradesh, India, for providing Purvanchal University Postdoctoral Fellowship (PUPDF) (Grant No. 01 PUPDF/EVS).

REFERENCES

Alfarisa, S., Abu Bakar, S., Mohamed, A., Hashim, N., Kamari, A., Md Isa, I., Mamat, M.H., Rahman Mohamed, A., Rusop Mahmood, M., (2015). Carbon nanostructures production from waste materials: A review. *Advanced Materials Research* 1109, 50–54.

Al-Rahbi, A.S., Williams, P.T., (2016). Production of activated carbons from waste tyres for low temperature NOx control. *Waste Management* 49, 188–195.

Bhardwaj, A.K., Gupta, M.K., Naraian, R., (2019). Myco-nanotechnological approach for improved degradation of lignocellulosic waste: Its future aspect. *Mycodegradation of Lignocelluloses*. Springer, pp. 227–245.

Bhardwaj, A.K., Shukla, A., Maurya, S., Singh, S.C., Uttam, K.N., Sundaram, S., Singh, M.P., Gopal, R., (2018). Direct sunlight enabled photo-biochemical synthesis of silver nanoparticles and their bactericidal efficacy: Photon energy as key for size and distribution control. *Journal of Photochemistry and Photobiology B: Biology* 188, 42–49.

Bhardwaj, A.K., Shukla, A., Mishra, R.K., Singh, S., Mishra, V., Uttam, K., Singh, M.P., Sharma, S., Gopal, R., (2017a). Power and time dependent microwave assisted fabrication of silver nanoparticles decorated cotton (SNDC) fibers for bacterial decontamination. *Frontiers in Microbiology* 8, 330.

Bhardwaj, A.K., Shukla, A., Singh, S., Uttam, K.N., Nath, G., Gopal, R., (2017b). Green synthesis of Cu_2O hollow microspheres. *Advanced Materials Proceedings* 2, 132–138.

Bhardwaj, A.K., and Naraian, R., (2020). Green synthesis and characterization of silver NPs using oyster mushroom extract for antibacterial efficacy. *Journal of Chemistry, Environmental Sciences and Its Applications* 7(1), 13–18.

Bhardwaj, A.K., Sundaram, S., Yadav, K.K. and Srivastav, A.L., (2021). An overview of silver nano-particles as promising materials for water disinfection. *Environmental Technology & Innovation* 101721.

Bhardwaj, A.K. and Naraian, R., (2021). Cyanobacteria as biochemical energy source for the synthesis of inorganic nanoparticles, mechanism and potential applications: a review. *3 Biotech*, 11(10),1–16.

Chen, C., Yu, D., Zhao, G., Du, B., Tang, W., Sun, L., Sun, Y., Besenbacher, F., Yu, M., (2016). Three-dimensional scaffolding framework of porous carbon nanosheets derived from plant wastes for high-performance supercapacitors. *Nano Energy* 27, 377–389.

Devatha, C., Thalla, A.K., Katte, S.Y., (2016). Green synthesis of iron nanoparticles using different leaf extracts for treatment of domestic waste water. *Journal of Cleaner Production* 139, 1425–1435.

Dutta, T., Ghosh, N.N., Das, M., Adhikary, R., Mandal, V., Chattopadhyay, A.P., (2020). Green synthesis of antibacterial and antifungal silver nanoparticles using *Citrus limetta* peel extract: Experimental and theoretical studies. *Journal of Environmental Chemical Engineering* 104019.

El-Amir, A.A., Ewais, E.M., Abdel-Aziem, A.R., Ahmed, A., El-Anadouli, B.E., (2016). Nano-alumina powders/ceramics derived from aluminum foil waste at low temperature for various industrial applications. *Journal of Environmental Management* 183, 121–125.

Garay, A.L., Pichon, A., James, S.L., (2007). Solvent-free synthesis of metal complexes. *Chemical Society Reviews* 36, 846–855.

Geim, A.K., Novoselov, K.S., (2010). The rise of graphene. Nanoscience and technology: A collection of reviews from nature journals. *World Scientific* 11–19.

Gou, X., Zhao, D., Wu, C., (2020). Catalytic conversion of hard plastics to valuable carbon nanotubes. *Journal of Analytical and Applied Pyrolysis* 145, 104748.

Harmsen, P., Huijgen, W., Bermudez, L., Bakker, R., (2010). Literature review of physical and chemical pretreatment processes for lignocellulosic biomass. *Wageningen UR-Food & Biobased Research* 1184, 1–54.

Herrera-Barros, A., Bitar-Castro, N., Villabona-Ortíz, Á., Tejada-Tovar, C., Gonzãlez-Delgado, Á.D., (2020). Nickel adsorption from aqueous solution using lemon peel biomass chemically modified with TiO2 nanoparticles. *Sustainable Chemistry and Pharmacy* 17, 100299.

Hou, J., Cao, C., Idrees, F., Ma, X., (2015). Hierarchical porous nitrogen-doped carbon nanosheets derived from silk for ultrahigh-capacity battery anodes and supercapacitors. *ACS Nano* 9, 2556–2564.

Huang, L., Weng, X., Chen, Z., Megharaj, M., Naidu, R., (2014). Green synthesis of iron nanoparticles by various tea extracts: Comparative study of the reactivity. *Spectrochimica Acta Part A: Molecular and Biomolecular Spectroscopy* 130, 295–301.

Jahan, S.A., Mollah, M.Y.A., Ahmed, S., Susan, M.A.B.H., (2017). Nano-hydroxyapatite prepared from eggshell-derived calcium-precursor using reverse microemulsions as nanoreactor. *Materials Today: Proceedings* 4, 5497–5506.

Jain, A., Tripathi, S., (2015). Nano-porous activated carbon from sugarcane waste for supercapacitor application. *Journal of Energy Storage* 4, 121–127.

Khatami, M., Sharifi, I., Nobre, M.A., Zafarnia, N., Aflatoonian, M.R., (2018). Waste-grass-mediated green synthesis of silver nanoparticles and evaluation of their anticancer, antifungal and antibacterial activity. *Green Chemistry Letters and Reviews* 11, 125–134.

Kukovitskii, E., Chernozatonskii, L., L'vov, S., Mel'Nik, N., (1997). Carbon nanotubes of polyethylene. *Chemical Physics Letters* 266, 323–328.

Kumar, B., Smita, K., Galeas, S., Sharma, V., Guerrero, V.H., Debut, A., Cumbal, L., (2020). Characterization and application of biosynthesized iron oxide nanoparticles using *Citrus paradisi* peel: A sustainable approach. *Inorganic Chemistry Communications* 108116.

Kuppusamy, P., Yusoff, M.M., Maniam, G.P., Govindan, N., (2016). Biosynthesis of metallic nanoparticles using plant derivatives and their new avenues in pharmacological applications–An updated report. *Saudi Pharmaceutical Journal* 24, 473–484.

Li, X., Wang, C., Zeng, Y., Li, P., Xie, T., Zhang, Y., (2016). Bacteria-assisted preparation of nano α-Fe_2O_3 red pigment powders from waste ferrous sulfate. *Journal of Hazardous Materials* 317, 563–569.

López, A.S., Ramos, M.P., Herrero, R., Vilariño, J.M.L., (2020). Synthesis of magnetic green nanoparticle–Molecular imprinted polymers with emerging contaminants templates. *Journal of Environmental Chemical Engineering* 8, 103889.

Makarov, V., Love, A., Sinitsyna, O., Makarova, S., Yaminsky, I., Taliansky, M., Kalinina, N., (2014). "Green" nanotechnologies: Synthesis of metal nanoparticles using plants. *Acta Naturae* 1(20), 35–44.

Maurya, S., Bhardwaj, A., Gupta, K., Agarwal, S., Kushwaha, A., (2016). Green synthesis of silver nanoparticles using Pleurotus and its bactericidal activity. *Molecular and Cellular Biology* 62, 131.

Mopoung, S., (2011). Occurrence of carbon nanotube from banana peel activated carbon mixed with mineral oil. *International Journal of Physical Sciences* 6, 1789–1792.

Neves, A., Godina, R., Azevedo, S.G., Matias, J.C., (2020). A comprehensive review of industrial symbiosis. *Journal of Cleaner Production* 247, 119113.

Premkumar, J., Sudhakar, T., Dhakal, A., Shrestha, J.B., Krishnakumar, S., Balashanmugam, P., (2018). Synthesis of silver nanoparticles (AgNPs) from cinnamon against bacterial pathogens. *Biocatalysis and Agricultural Biotechnology* 15, 311–316.

Quadrado, R.F., Gohlke, G., Oliboni, R.S., Smaniotto, A. and Fajardo, A.R., (2019). Hybrid hydrogels containing one-step biosynthesized silver nanoparticles: Preparation, characterization and catalytic application. *Journal of Industrial and Engineering Chemistry* 79, 326–337.

Qu, J., Lv, S., Peng, X., Tian, S., Wang, J., Gao, F., (2016). Nitrogen-doped porous "green carbon" derived from shrimp shell: Combined effects of pore sizes and nitrogen doping on the performance of lithium sulfur battery. *Journal of Alloys and Compounds* 671, 17–23.

Rao, K.G., Ashok, C., Rao, K.V., Chakra, C.S., Rajendar, V., (2015). Synthesis of TiO_2 nanoparticles from orange fruit waste. *Synthesis* 1, 82–90.

Rezazadeh, L., Sharafi, S., Schaffie, M., Ranjbar, M., (2020). Application of oxidation-reduction potential (ORP) as a controlling parameter during the synthesis of Fe_3O_4@ PVA nanocomposites from industrial waste (raffinate). *Environmental Science and Pollution Research* 1–12.

Romero-Cedillo, L., Poggi-Varaldo, H.M., Santoyo-Salazar, J., Escamilla-Alvarado, C., Matsumoto-Kuwabara, Y., Ponce-Noyola, M.T., Bretón-Deval, L., García-Rocha, M., (2020). Biological synthesis of iron nanoparticles using hydrolysates from a waste-based biorefinery. *Environmental Science and Pollution Research International* 27, 28649–28669.

Shukla, A., Bhardwaj, A.K., Singh, S., Uttam, K., Gautam, N., Himanshu, A., Shah, J., Kotnala, R., Gopal, R., (2018). Microwave assisted scalable synthesis of titanium ferrite nanomaterials. *Journal of Applied Physics* 123, 161411.

Singh, M.P., Bhardwaj, A.K., Bharati, K., Singh, R.P., Chaurasia, S.K., Kumar, S., Singh, R.P., Shukla, A., Naraian, R. and Vikram, K., (2021). Biogenic and non-biogenic waste utilization in the synthesis of 2D materials (graphene, h-BN, g-C2N) and their applications. *Frontiers in Nanotechnology*, 53.

Souda, P., Sreejith, L., (2015). Magnetic hydrogel for better adsorption of heavy metals from aqueous solutions. *Journal of Environmental Chemical Engineering* 3, 1882–1891.

Stolle, A., Szuppa, T., Leonhardt, S.E., Ondruschka, B., (2011). Ball milling in organic synthesis: Solutions and challenges. *Chemical Society Reviews* 40, 2317–2329.

Sukri, S.N.A.M., Shameli, K., Wong, M.M.-T., Teow, S.-Y., Chew, J., Ismail, N.A., (2019). Cytotoxicity and antibacterial activities of plant-mediated synthesized zinc oxide (ZnO) nanoparticles using *Punica granatum* (pomegranate) fruit peels extract. *Journal of Molecular Structure* 1189, 57–65.

Sun, L., Tian, C., Li, M., Meng, X., Wang, L., Wang, R., Yin, J., Fu, H., (2013). From coconut shell to porous graphene-like nanosheets for high-power supercapacitors. *Journal of Materials Chemistry A* 1, 6462–6470.

Tahir, N.A.M., Abdollah, M.F.B., Tamaldin, N., Zin, M.R.B.M., Amiruddin, H., (2020). Optimisation of graphene grown from solid waste using CVD method. *The International Journal of Advanced Manufacturing Technology* 106, 211–218.

Tharchanaa, S., Priyanka, K., Preethi, K., Shanmugavelayutham, G., (2020). Facile synthesis of Cu and CuO nanoparticles from copper scrap using plasma arc discharge method and evaluation of antibacterial activity. *Materials Technology* 1–8.

Wang, T., Lin, J., Chen, Z., Megharaj, M., Naidu, R., (2014). Green synthesized iron nanoparticles by green tea and eucalyptus leaves extracts used for removal of nitrate in aqueous solution. *Journal of Cleaner Production* 83, 413–419.

Wu, H.-H., (2020). A study on transnational regulatory governance for marine plastic debris: Trends, challenges, and prospect. *Marine Policy* 4:103988.

Xiang, X., Xia, F., Zhan, L., Xie, B., (2015). Preparation of zinc nano structured particles from spent zinc manganese batteries by vacuum separation and inert gas condensation. *Separation and Purification Technology* 142, 227–233.

14

Use of Nanoparticles in Bioremediation of Pharmaceutical Compounds

Anushree Suresh and Jayanthi Abraham
VIT University

CONTENTS

14.1 Introduction

A wide array of pollutants, such as pharmaceutical compounds and their metabolites or residues, are continuously accumulated into the environment via hospital and municipal wastewater disposal and discharges from pharmaceutical production companies (Ebele et al. 2017). Drugs which are accumulated in the environment are mainly in the form of unchanged or partially metabolized compounds, i.e. conjugated or hydroxylated with another charged molecules such as sulphate, lysine, glutathione or glucuronic acid (Marchlewicz et al. 2015). To date, wastewater treatment plants have not been well planned for complete remediation or removal of organic micropollutants that have good water solubility and poor biodegradability, thereby prevailing in the environment for a longer period of time (El Fels et al. 2016). Furthermore, some of the conventional treatment processes, e.g. chlorination, generally releases more toxic intermediate compounds due to chemical reactions (Cao and Loh 2009). However, for some of the persistent pharmaceutical compounds, sewage treatment plants may eliminate approximately 99.9% of their initial toxic (parent) compounds by converting them to simpler non-toxic compounds (Chonova et al. 2018).

Pharmaceutical compounds have a diverse structural make-up such as saturated or unsaturated bonds, the presence of additional functional groups (e.g. halogen and sulphate), different side chain bonds, and linear or branched structure present on the parent compound; there can be no specific method followed in the wastewater sewage treatment plant for pharmaceutical compounds remediation. However, some of the advanced treatment technologies have been developed and practised, showing a high removal rate

DOI: 10.1201/9781003181224-14

of micropollutants found in wastewater to improve environmental sustainability. Presently, some of the most important remediation techniques used are adsorption using a powdered activated and granular activated carbon, biofiltration using a trickling filter, sand filtration processes and biological activated carbon, nanofiltration, reverse osmosis, membrane bioreactors and use of carbon nanocomposites having magnetic properties ((AC)/$CoFe_2O_4$) (Sun et al. 2019). The inability of the remediation process of the toxic parent pollutant compounds results in the release of other toxic compounds as active or unstable metabolites into the environment (Rozas et al. 2016). This results in the accumulation of compounds into the trophic chains having long-term adverse effects on aquatic organisms (Unger et al. 2008). Some researchers have emphasized the permissible limit of environmentally relevant concentrations (ng/L–μg/L) of residual pharmaceutical compounds released into the environment.

Though the literature data about the sensitivity of non-target organism receptors, e.g. cladocerans, cnidarians, primary producers, mussels or fish, are rare, the pharmaceutical compounds' action poses a threat to these organisms present in the aquatic system (Parolini and Binelli 2012). Oliveira et al. (2004) reported that some of various drugs have additional structural complexities that promote a negative influence on biocoenosis, e.g. the high lipophilic character of a compound, favouring the bioaccumulation and low biodegradability. Furthermore, drugs are designed to have biologically active molecules, which act on a number of different species in a slow metabolized reaction (Oliveira et al. 2004). The fate of pharmaceutical compounds in the environment is depicted in Figure 14.1.

However, Nunes et al. (2015) reported that due to the conservative nature of some of the physiological processes and similarity in target compounds in humans and many aquatic organisms, the environmental contamination of these compounds is possibly predictable and rectifiable. Parolini and Binelli (2012) also conducted another study stating that the high activity of pharmaceutical compounds can have adverse effects on the health of various biocoenosis. Additionally, Elersek et al. (2016) noticed that one compound concentration being low can also have additive or synergistic action in the presence of other pharmaceutical compounds with similar mode of action, thereby harming different aquatic life forms (Elersek et al. 2016).

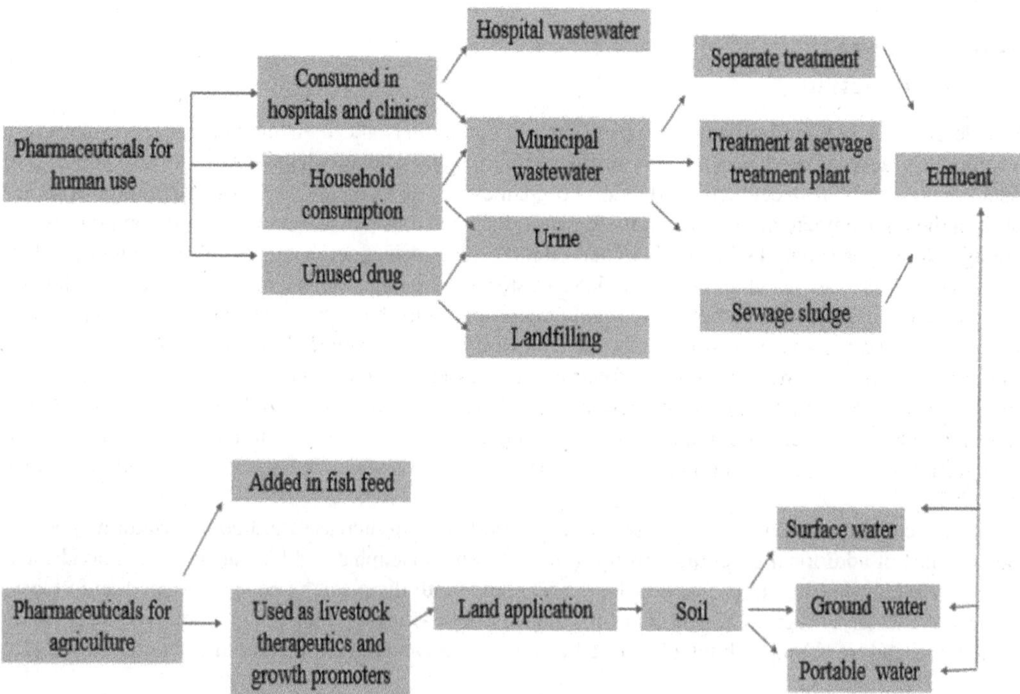

FIGURE 14.1 Fate of pharmaceutical compounds in the environment.

Biological treatment has been shown to be very effective in the removal of micropollutants (Falås et al. 2012). Even though the bioremediation processes are the most successful clean-up technology, they have some major disadvantages. Some of the disadvantages are as follows:

1. Potential interruption of bioremediation compounds by various microflora.
2. Poor susceptibility of some xenobiotic compounds to biodegradation process.
3. Inefficient removal of genetically modified microorganisms after bioremediation incubation time.
4. Formation of intermediates or metabolites that are more toxic or have longer persistence.
5. Difficulty in implementing the laboratory-scale experiments to field operations.
6. Occurrence of contaminations in all forms such as solids, gases or liquids, which often obstruct the bioremediation processes.

Microorganisms with higher degradation rate often require optimal environmental growth conditions such as temperature, pH, the presence of additional sources of carbon and appropriate levels of xenobiotics to persuade specific enzymes for metabolic processes. Additionally, some microorganisms have to be frequently acclimatized to the concentrations of pharmaceutical contaminations, which may extend the bioremediation procedure for a longer period of time (Yadav et al. 2017).

Nanotechnology has gained attention within the past decades due to its unique physical properties. Nanomaterials showed enhanced reactivity and thus better effectiveness in comparison with their bulkier counterparts because of their higher surface-to-volume ratio. The use of nanomaterials in various bioremediation processes is depicted in Figure 14.2. In addition, nanomaterials have the potential to influence unique surface chemistry as compared to traditional approaches such

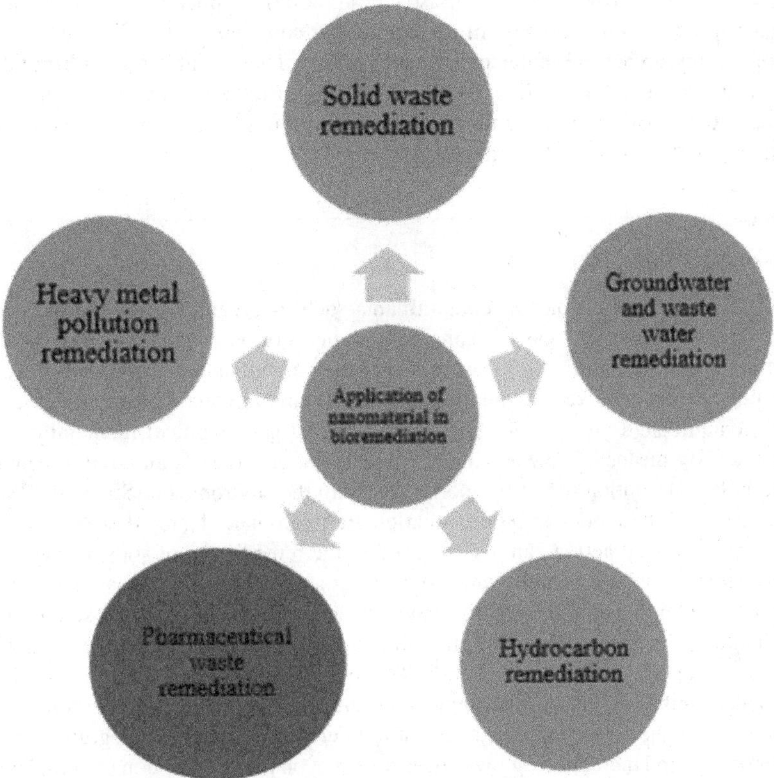

FIGURE 14.2 Use of nanomaterials in various bioremediation techniques.

that they will be functionalized or addition of functional groups which will target specific molecules of interest (pharmaceutical compounds) for efficient remediation to non-toxic and simpler metabolites.

Additional modification of physical properties of nanomaterials, such as porosity, size, structural and chemical composition, can cause supplementary and advantageous functions, which can directly affect the performance of the new material for specific remediation of contaminants. The surface modification of these nanomaterials along with the modified physical parameters can have significant advantages over conventional methods of environmental contamination remediation. Therefore, methods that are developed as a combination of several different materials such as hybrids or composites are potentially more efficient, sensitive, selective and stable than methods based upon a single nanoplatform. For instance, adhering nanoparticles to a known scaffold can be an alternative remediation to increase the stability of the material when compared to the use of nanoparticles alone. Functionalizing material with specific additional chemicals responsible for targeting contaminant molecules of interest can help increase the selectivity and efficiency of the material (Bayramoğlu et al. 2008).

The ozone and ozone or hydrogen peroxide processes are both energy-intensive and material-intensive (Pimentel et al. 2007) and are only suitable for the treatment of relatively clean surface water and groundwater with less complex structured contamination such as natural organic matter. The physical treatment processes also require the disposal of wastes such as membrane residues and spent activated carbon generated during the treatment. In addition, activated carbon adsorption has a limited ability to remove polar organic compounds due to its inefficient removal mechanism (i.e. hydrophobic interactions), especially in the presence of competitive natural organic matter (Konwarh et al. 2009), while many pharmaceutical compounds and metabolites are made up of polar compounds.

Nanotechnology uses the deliberate manipulation of matter at size less than 100 nm, which holds the promise of creating new materials and devices. These nanomaterials have high reactivity, which is caused by the large surface-to-volume ratio (Sastry et al. 2002). Application of nanomaterials for the removal of pharmaceutical contamination in wastewater has come up as an interesting area of research among most of the researchers (Schüler and Caruso 2000). They exhibit high adsorption efficiency due to their high surface area and availability of active sites for interaction with various contaminants. Furthermore, adsorbents with specific additional functional groups have been developed to improve the adsorption capacity of these materials (Mitchell et al. 2002).

14.2 Nano-Bioremediation

The process of removing environmental contaminants such as organic and inorganic pollutants and heavy metals from contaminated areas by using nanoparticles or nanomaterials synthesized by plants, bacteria and fungi using nanotechnology is called nano-bioremediation. Nano-bioremediation is an emerging technique for the removal of contaminants for the environmental clean-up process. Presently, remediation of contaminated areas is done by using physical and chemical incineration, remediation and bioremediation. By means of new advances, bioremediation suggests an environmentally friendly and economically feasible option to remove pollutants from the environment (Singh and Walker 2006). Some of the frequently used modes of bioremediation are microbial, plant and enzymatic remediation. Nanoparticles with microbial action can also be used in the remediation of soils and water containing heavy metals, organic and inorganic pollutants. Recent studies conducted by researchers have proved the removal of organic contaminants (atrazine, molinate and chlorpyrifos) using nano-sized zerovalent ions (Ghormade et al. 2011). Nanoparticles with immobilized enzyme bioremediation coupled with phytoremediation can also be an effective approach to a cleaner environment (Singh 2010). Some of the pharmaceutical compounds consist of structural complexities such as long-chain hydrocarbons and additional organochlorine and have been known to be resistant to microbial and plant biodegradation. A combination of nanotechnology and biotechnology will have an effect on the degradation of complex compounds into simpler non-toxic compounds using nano-encapsulated enzymes (Yadav et al. 2017). Some of the nanoparticles synthesized by bacteria and fungi are tabulated in Tables 14.1 and 14.2.

TABLE 14.1

Different Nanoparticles Synthesized by Bacteria Species

Nanoparticle	Bacteria Species	References
Silver nanoparticles	*Bacillus cereus*	Sunkar and Nachiyar (2012)
	Oscillatoria willei NTDMO1	Ali et al. (2011)
	Escherichia coli	Babu and Gunasekaran (2009)
	Pseudomonas stutzeri	Yadav et al. (2017)
	Bacillus subtilis	Juibari et al. (2011)
	Bacillus sp.	Mandal et al. (2006)
	Bacillus cereus	He et al. (2007)
	Bacillus thuringiensis	Kathiravan et al. (2015)
	Lactobacillus strains	Yadav et al. (2017)
	Pseudomonas stutzeri	
	Corynebacterium	
	Staphylococcus aureus	
	Ureibacillus thermosphaericus	
Magnetic nanoparticles	*Magnetospirillum magneticum*	
	Sulphate reducing bacteria	
Palladium nanoparticles	*Desulfovibrio desulfuricans* NCIMB 8307	
CdS nanoparticles	*Clostridium thermoaceticum*	
	Klebsiella aerogenes	
	Escherichia coli	
Gold nanoparticles	*Rhodopseudomonas capsulata*	
	Alkalothermophilic actinomycete	
	Thermomonospora sp.	
	Pseudomonas aeruginosa	
	Lactobacillus strain	
As-S nanotubes	*Shewanella* sp.	
ZnS nanoparticles	*Sulphate-reducing bacteria of the family* Desulfobacteraceae	

14.3 Removal of Pharmaceutical Compounds Using Different Nanomaterials

14.3.1 Carbon Nanotubes-Based Nanomaterials

Carbon-based nanomaterials are the allotropes of carbon, which include three-dimensional carbon nanotubes and two-dimensional graphene sheets. Carbon nanotubes are the rolled sheets of graphene. Carbon nanotubes are of three types depending on the number of sheets or layers folded to prepare it. Single-walled carbon nanotubes (SWCNTs) constitute only a single rolled sheet of graphene, while if two constituent sheets are present, it is called double-walled carbon nanotubes (DWCNTs), and those containing multiple sheets are called multiwalled carbon nanotubes (MWCNTs). The carbon nanotubes have a high surface area and more adsorption power in comparison with simple graphene. Hence, they have been used for the removal of various pollutants from wastewater. The types of interaction that take place between the adsorbate and adsorbent are simple hydrophobic interactions such as π–π interaction, van der Waals forces and H-bonding. These types of forces are generally used in the removal of organic non-polar contaminants from wastewater.

The carbon nanotubes need to be functionalized or modified by certain functional groups or organic moieties such as carboxylic or carbonyl groups for the removal of ionic/polar toxicants. Carbon nanotubes with their unique structural properties are suitable for the adsorptive removal of toxic pharmaceutical compounds. Zhao et al. (2013) used MWCNTs for the removal of tetracycline from aqueous solutions. Tetracycline is an antibiotic used to kill or inhibit the disease-producing bacteria. The maximum

TABLE 14.2

Different Nanoparticles Synthesized by Yeasts and Fungi

Nanoparticles	Yeast and Fungus	References
PbS nanoparticles	*Torulopsis species*	Seshadri et al. (2011)
	Rhodosporidium dibovatum	Sanghi and Verma (2009)
CdS quantum dots	*Candida glabrata*	Yadav et al. (2017)
	Schizosaccharomyces pombe	Yadav et al. (2017)
Ag nanoparticles	*Silver-tolerant yeast strains* MKY3	Mourato et al. (2011)
	Cladosporium cladosporioides	Yadav et al. (2017)
	Coriolus versicolor	Popescu et al. (2010)
	Fusarium semitectum	
	Fusarium oxysporum	
	Phanerochaete chrysosporium	
	Aspergillus flavus	
	Extremophilic yeast	
	Aspergillus niger	
	Aspergillus oryzae	
	Fusarium solani	
	Pleurotus sajor-caju	
	Trichoderma viride	
Stable silver nanoparticles	*Aspergillus flavus*	
	A. fumigatus	
	A. terreus	
	A. nidulans	
Gold nanoparticles	*Verticillium* sp.	
	Fusarium oxysporum	
Bioactive nanoparticles	*Lichen fungi (Usnea longissima)*	

adsorption capacity obtained in 80 minutes at pH 5 and 25°C was 269.54 mg/g for tetracycline (Zhao et al. 2013). In another study done by Yu et al. (2014) on the adsorption of drug ciprofloxacin on MWCNTs having different oxygen contents, the maximum adsorption occurred at pH 4 and contact time 240 minutes. Although oxygen content has less effect on adsorption capacity, the main role was played by π–π electron donor–acceptor interaction (Yu et al. 2014b). Different carbon nanotubes used for the remediation of various antibiotics are tabulated in Table 14.3.

14.3.2 Silica-Based Nanomaterials for the Removal of Pharmaceutical Compounds

Due to their inert nature, silica nanoparticles are used in drug delivery applications and can also be used in drug removal. Nassar et al. (2019) synthesized pure silica nanoparticles from rice husk and used them as an adsorbent for the adsorptive removal of ciprofloxacin from polluted water. The concentration of the drug was determined by UV–visible spectrophotometer (Nassar et al. 2019). The adsorption of ciprofloxacin onto the surface of silica nanoparticles occurred at pH 7, in 24 hours. The silica nanoparticles showed an adsorption capacity of 24.1 mg/g (Nassar et al. 2019).

Silica nanoparticles being cheap, non-toxic and biocompatible are a good candidate to be used as an adsorbent for the removal of toxic substances from water. Silica nanoparticles have many advantages such as high surface area, transparent appearance, mechanical resistance and, most importantly, high porosity, which not only provides enough space for any type of contaminants either large or small, but also helps in their modification with a suitable functional group. The modification enhances its selectivity, adsorption capacity and reusability. Although silica nanoparticles have many advantages, they also have disadvantages and certain limitations; for example, their surface modification is a complex and tedious process (Saxena et al. 2020).

TABLE 14.3

Different Carbon Nanotubes used for Remediation of Various Antibiotics

S. No.	Carbon Nanotubes	Pharmaceutical Pollutants	pH	Contact Time (min)	Adsorption Capacity/mg/g	References
1.	SWCNTs	Oxytetracycline	7	30	554.0	Yu et al. (2016)
2.	DWCNTs	Oxytetracycline			507.0	Zhou et al. (2019)
3.	MWCNTs	Ciprofloxacin			724.0	Shan et al. (2016)
4.	MWCNTs	Tetracycline	5	80	260.2	Yang et al. (2015)
		Ciprofloxacin	4	240	20.0	
		Sulphamethoxazole and sulphapyridine	5.6		67.9	
5.	MWCNTs/Al_2O_3	Carbamazepine	6	72	157.4	
		Diclofenac sodium (DS)	-	102	106.5	
6.	O-MWCNTs	Pefloxacin	6	240	61.1	
7.	MWCNTs/$CoFe_2O_4$	Sulphamethoxazole	5.5	120	6.9	
8.	CNTs–CoFe–chitosan composite	Tetracycline	6	110	104.0	
9.	Granular MWCNTs	Carbamazepine	6	72	369.5	
		Tetracycline		85	284.2	
		Diclofenac sodium		120	203.1	
10.	Pristine MWCNTs	Sulphamethazine	5	24	24.7	
11.	Hydroxylated MWCNTs				13.3	
12.	MWNTs and MWNTs-COOH	2-Nitrophenol	4	25	310	
13.	AC, CNT and carbon xerogel	Cefazolin	5	25	100	
14.	SWCNTs, MWCNTs and powdered AC	Iopromide Lincomycin Sulphamethoxazole	5	25	12000	
15.	Hydroxylated, carboxylate, graphitized MWCNTs and SWCNTs	Cefazolin	7	80	35	
16.	MWCNTs	Ofloxacin	7	25	700	
		Norfloxacin			60	
17.	Graphitized MWCNTs, carboxylate MWCNTs, hydroxylated MWCNTs and AC	Norfloxacin	7.2	120	40	
18.	Hydroxylated, carboxylate and graphitized MWCNTs	Sulphamethoxazole	2-12	25	50	

14.3.3 Metal Oxide-Based Nanomaterials

Metal oxides in nano-size are potential adsorbents for the removal of various pollutants present in the wastewater. The nano-sized metal oxide includes nano-ferric oxide, nano-titanium oxide, nano-zinc oxides and nano-aluminium oxides. These nanoparticles have a large surface area with excellent adsorption capacity due to the presence of charges on their surface. The metal oxide nanoparticles have a high tendency to attract the toxicants such as ionic organic pollutants with great selectivity. The metal oxide nanoparticles are easy to synthesize and are of low cost. However, these particles have some limitations: when the size of the metal oxide nanoparticles decreases to nanometre scale, the increased surface energy leads to poor stability. As a result, the nano-metal oxide tends to agglomerate due to van der Waals forces or other types of interaction.

The consequences of this agglomeration are the decrease in or even loss of adsorption capability and selectivity. Metal oxide nanoparticles are also unstable in stationary beds or any flow through the system due to excessive pressure drops and their low mechanical strength. So, to improve their capability, they are modified with activated carbon or polymeric support, etc. Recently, the production of magnetic nanoparticles has attracted the interest of many researchers as they have many advantages over the conventional metal oxide nanoparticles. The magnetic nanoparticles can be easily separated and collected with the help of external magnetic fields, and to further enhance their adsorption capability and selectivity, they can be easily modified with a suitable moiety (Saxena et al. 2020).

Pharmaceutical drugs are designed to be selective and target–specific, which creates a more dangerous effect on living organisms and the environment. They have the capability to enter the cell membrane and thus harm the vital tissues if they enter the body. The disposal of the pharmaceutical drugs in the water bodies leads to various kinds of diseases in different organisms, and their removal is vital. Although not much work has been done on magnetic nanoparticles for the removal of pharmaceuticals, some research work has been carried out by Soares et al. (2016). They prepared a novel biosorbent of magnetite nanoparticles functionalized by k-carrageenan hybrid siliceous shells and used it for the removal of metoprolol tartrate from aqueous solutions. The structure of modified nanoparticles was spheroidal, and the specific surface area was 38.4 m^2/g. The high adsorption capacity of 447 mg/g was obtained at pH 7 and in 15 minutes (Soares et al. 2016). The FTIR results revealed that the adsorption of metoprolol on the surface of nano-adsorbent occurred via electrostatic interaction between the sulphonate group of nano-adsorbent and the protonated amine group of metoprolol (Soares et al. 2016).

Similarly, Fakhri and his co-workers reported MgO nanoparticles for the removal of cephalosporins (antibiotic) from aqueous solutions. The adsorption capacity obtained was 500 mg/g in just 10 minutes at pH 9 (Fakhri and Adami 2014). Different metal oxides used for the remediation of pharmaceutical compounds are tabulated in Table 14.4.

TABLE 14.4
Different Metal Oxides Used for Remediation of Pharmaceutical Compounds

S. No.	Metal Nanoparticles	Pharmaceutical Pollutants	pH	Contact Time (minutes)	Adsorption Capacity / mg/g	References
1.	TiO_2/UV (photocatalytic oxidation)	Chloramphenicol	6.5	48	20	Jodat and Jodat (2014)
		Tartrazine	6.5	32	20	
2.	ZnS	β-Lactam antibiotics	4.5	60	500	
3.	$Sn/Zn/TiO_2$	Amoxicillin	5.6	28	20	Mohammadi et al. (2015)
4.	TiO_2/UV, TiO_2/H_2O_2/UV and Fe^{2+}/H_2O_2/UV	Trimethoprim	6	48	5	Dias et al. (2014)
		Sulphamethoxazole	2.8	27	5	
5.	$TiO_2/SiO_2/Fe_3O_4$	Diclofenac	10	48	150	Hu et al. (2011)
6.	S-nZVI	Florfenicol	7	28	200	

14.3.4 Chitosan-Based Nanomaterials

Chitosan is a biodegradable polysaccharide that is prepared by deacetylation of chitin. Its basic structure has β-1,4-linked 2-amino-2-deoxy-D-glucose. It has three different functional groups: an amine group and primary and secondary hydroxyl groups. The functional groups improve the adsorption of ionic or polar contaminants on the surface of chitosan. Chitosan also has special properties such as viscosity, solubility in various solvents, polyelectrolytic behaviour, polyoxin salt formation, mucoadhesive nature, and ability to form metal chelates and to form films, which make it a good candidate to be used in various fields including food, agriculture, pharmacy and cosmetics. Advantages as it is biodegradable, biocompatible, cheap (as manufactured from naturally abundant animal shells) and is environmentally friendly which makes it useful in biomedical as well as to work as environmental remediation agent (Kaur and Dhillon 2014).

Chitosan-based nanoparticles showed good results in the removal of pharmaceuticals from wastewater. Raeiatbin et al. (2017) prepared magnetic chitosan nanoparticles and used them for the removal of tetracycline (a frequently used antibiotic in hospitals) from medical wastewater. The magnetic chitosan nanoparticles showed a maximum adsorption capacity of 78.11 mg/g at pH 5 and temperature 25°C. The advantage of using magnetic material is the ease in separation of adsorbent (Raeiatbin and Açıkel 2017). Similarly, Liang (2019) and his co-workers modified the magnetic chitosan nanoparticles with amine functionality and used them for the adsorptive removal of diclofenac sodium. The modified magnetic chitosan nanoparticles showed a much better adsorption capacity (469.48 mg/g) obtained from Langmuir model at pH 4.5 and 30°C, which showed that amine modification improved the removal tendency of pharmaceuticals/drugs. Different chitosan nanomaterials for pharmaceutical compounds remediation are tabulated in Table 14.5.

14.3.5 Nanofilters

Nanofiltration is a recently developed technique involving pressure-driven membrane separation process. It has extensively been used in aqueous system treatment plants to determine the concentration of antibiotics in aqueous solutions (Wang et al. 2010a). Removal of hardness and dissolved organics in water, antibiotic removal from drinking water (Vrijenhoek and Waypa 2000), heavy metal ions recovery from electroplated wastewater (Hafiane et al. 2000) and separation of pharmaceutical compounds from fermentation broths (Christy and Vermant 2002) are some of the examples of nanofiltration used in industries. Nanofiltration shares some functions with ultrafiltration and reverse osmosis. Generally, reverse osmosis membranes cannot process both salts and organic matter (Salahi et al. 2010), whereas ultrafiltration membranes freely pass most of the salts and most organic matter.

TABLE 14.5
Different Chitosan Nanomaterials for Pharmaceutical Compounds Remediation

S. No.	Chitosan Nanomaterial	Pharmaceutical Pollutants	pH	Contact Time (min)	Adsorption Capacity /mg/g	References
1.	Magnetic chitosan nanoparticles	Tetracycline	5	-	78.1	Raeiatbin and Açıkel (2017)
2.	Magnetic amine functionalization chitosan nanoparticles	Diclofenac	4.5	60	469.5	Soares et al. (2019)
3.	Magnetic quaternary chitosan hybrid nanoparticles	Diclofenac	6	5	240.0	
4.	Glutaraldehyde cross-linked magnetic chitosan nanocomposites	Sulphamethazine	5	70	112.4	

Nanofiltration membrane allows the passage of some low molecular weight micropollutants unlike ultrafiltration and reverse osmosis membranes. Furthermore, nanofiltration membranes are non-permeable to some of the bivalent ions, but are comparatively permeable for monovalent ions present in pharmaceuticals (Zhu et al. 2003). Most of the nanofiltration membranes have their function similar to reverse osmosis membranes, and they are also called loose reverse osmosis membranes or tight ultrafiltration membrane in accordance with their permeability, flux and separation performance rate (Rama et al. 1996). Hence, nanofilters have served the purpose as an attractive economic alternative to the existing reverse osmosis membranes due to their lower operating pressure application and higher permeability of water.

There are many studies reported on the application of nanofiltration for the efficient removal of pharmaceutical compounds from wastewater. In all the studies, the removal percentage obtained for nanofilters was higher than 90% for all the pharmaceutical compounds studied (Košutić et al. 2007; Radjenović et al. 2008). Some results obtained showed the lowest values for the removal of tetracycline (50%–80%) and sulphonamides (11%–20%) (Koyuncu et al. 2008). Naproxen is a non-steroidal anti-inflammatory drug frequently used for treating fever, inflammation and different health problems and has recently been detected in sewage effluents, groundwater and occasionally in drinking water in high concentration. A study of a wastewater treatment plant using ultrafiltration, activated carbon filters and reverse osmosis after preliminary biological treatment by Qurie et al. (2014) showed that both nano-ultrafiltration and micro-ultrafiltration were not efficient enough for removing high concentration of naproxen to a permissible level; however, reverse osmosis membrane was successful in the removal process. Naproxen was successfully degraded using biological treatment within a period of three days in activated sludge releasing the metabolite O-desmethyl-naproxen (Qurie et al. 2014).

Jurecska et al. (2015) showed a new pathway for the removal of persistent inorganic or organic pollutants in wastewater. In their study, the removal of antibiotic–cyclodextrin with nanofilters of varying thickness (1.5–3.5 mm) and chemical composition was investigated. The adsorption capacity of cyclodextrin was determined by applying an ibuprofen antibiotic containing a model solution with total organic carbon analyser. The regeneration of nanofilters with the application of ethanol in conjugation with some of the inorganic additives such as NaCl, $NaHCO_3$ and NH_4HCO_3 have tremendously increased the nanofilters adsorption capacity used in the process of remediation. The best results were achieved with a chemical composition containing 30 m/m% β-cyclodextrin polymer beads along with 70 m/m% ultra-high molecular weight polyethylene submerged in 12 mmol ammonium hydrogen carbonate or nanofilter (Jurecska et al. 2015). Some of the different nanofilters used for pharmaceuticals remediation are tabulated in Table 14.6.

14.4 Role of Single-Enzyme Nanoparticles in Bioremediation

Enzymes are biological molecules (typically proteins), which are specific, sensitive and effective and function as catalysts in the bioremediation process. However, the instability and relatively short catalytic lifetime of most of the enzymes in the metabolic pathway limit their utility as cost-effective alternatives to existing synthetic catalysts. An effective alternative is to maximize the steadiness and reusability of these enzymes by connecting them to magnetic nanoparticles made of iron. Enzymes with magnetic iron nanoparticles can be used for the separation of enzymes from reactant compounds or end products by applying appropriate amounts of magnetic flux. With this purpose, two different catabolic enzymes, trypsin and peroxides, were used to build uniform core–shell magnetic nanoparticles. A study conducted showed that the enzyme activity and lifetime increase intensely from a couple of hours to days, thereby making magnetic nanoparticle enzyme conjugates more stable, efficient and economical (Qiang et al. 2007). A list of immobilized enzymes with nanoparticles for pharmaceuticals remediation is given in Table 14.7.

TABLE 14.6
Different Nanofilters for Pharmaceutical Compounds Remediation

S. No.	Nanofilters	Pharmaceutical Pollutants	Operating Conditions	Adsorption Capacity/ mg/g	References
1.	Nanofiltration and reverse osmosis	Clarithromycin Ciprofloxacin Diclofenac Ibuprofen Metronidazole Moxifloxacin Telmisartan Tramadol Bezafibrate Bisoprolol	Pump pressure at 7 bar for NF and 14 bar for RO	20–250	Beier et al. (2010)
2.	Nanofiltration membrane	Amoxicillin	Pressure at 3–15 bar	24225	
3.	Combined nanofiltration and mild solar photo-Fenton	Carbamazepine Flumequine Ibuprofen Ofloxacin Sulphamethoxazole	Fenton operated at low iron and hydrogen peroxide	15–150	Miralles-Cuevas et al. (2014)
4.	Nanofiltration combined with advanced tertiary treatments (solar photo-Fenton, photo-Fenton-like Fe(III)–EDDS complex and ozonation)	Carbamazepine Flumequine Ibuprofen Ofloxacin Sulphamethoxazole	pH around 8–8.5	300	
5.	Nanofiltration combined with ozone-based advanced oxidation processes	Norfloxacin Ofloxacin Roxithromycin Azithromycin	pH 7.9	600	Liu et al. (2014)

14.5 Future Prospects

More work is required in neutralizing the toxic effects of organic compounds in pharmaceutical compounds accumulated in wastewater. The researchers have to develop an effective sensing mechanism or tools that can selectively and accurately detect pollutants. The sensor should be portable and less expensive. The approach is to make a portable, low-cost and effective sensor of chitosan nanoparticles so that it can be used in the field for rapid, sensitive, selective and reliable determination of organic pollutants in water and soil environment.

A slight variation in the surface of various nanomaterials such as TiO_2 and nZVI by the addition of second catalytic metal results in enhanced wastewater quality, thereby increasing the reactivity and selectivity of the contaminant. Surface modification has led to the enhanced photocatalytic activity of the contaminant due to its short lifetime of reactive oxygen radicals and increase in the affinity for modified nanomaterials. Bimetallic nanoparticles have extensively been used for the remediation of various water contaminants. Nevertheless, further studies are yet to be conducted for understanding the mode of action of bimetallic nanoparticles degradation of various pharmaceuticals in wastewater. For current applications, an improved understanding of the mode of action of these nanocomposites is very important for successful wastewater treatment.

TABLE 14.7

List of Immobilized Enzymes with Nanoparticles for Pharmaceuticals Remediation (Qiang et al. 2007)

S. No.	Enzyme	Nanoparticle	Application
1.	β-Glucosidase (BGL) from *Aspergillus niger*	Iron oxide	Amoxicillin breakdown
2.	Superoxide dismutase (SOD)	Nano-Fe_3O_4	Functional group breakdown in β-lactam antibiotic series
3.	α-Amylase	Silica nanoparticles	Penicillin
4.	Trypsin	Nano-diamond prepared by detonation (dND)	Diclofenac sodium
5.	Lysozyme	Chitosan nanofibres	Breaking of chained hydrogen bonds in most antibiotics
6.	*Mucor javanicus* lipase	Nano-sized magnetite	Carbamazepine
7.	Lipases from *C. rugosa* and *Pseudomonas cepacia*	Zirconia nanoparticles	Naproxen
8.	Horseradish peroxidase (HRP)	Magnetite silica nanoparticles	Ibuprofen
9.	Alcohol dehydrogenase from *T. brockii* (TbADH)	Gold and silver	Trimethoprim
10.	Cholesterol oxidase	Fe_3O_4 nanoparticles	Fluoroquinolone compounds
11.	Haloalkane dehalogenase	Silica-coated iron oxide nanoparticles	Sulphide antibiotics
12.	Laccase	Chitosan magnetic nanoparticles	Ciprofloxacin
13.	Keratinase	Fe_3O_4 nanoparticles	Most of the hormones, especially oestradiol and ethinyloestradiol compounds
14.	α-Amylase	Cellulose-coated magnetite nanoparticles	Metoprolol
15.	β-Galactosidase	Con A layered ZnO nanoparticles	Diclofenac sodium
16.	α-Chymotrypsin	Polystyrene nanoparticles	Ofloxacin and Norfloxacin
17.	Diastase	Silica-coated nickel nanoparticles	Tetracycline
18.	Bitter gourd peroxidase (BGP)	TiO_2 nanoparticles	Iopromide
19.	Cellulase		Sulphadiazine
20.	Alpha amylase		Diclofenac sodium
21.	Trypsin		Sulphapyridine
22.	Alpha amylase	Silver nanoparticles	Nalidixic acid

14.6 Conclusions

It is generally recognized that nanotechnology and its application may play an important role in resolving issues relating to water quality. The great potential of nanomaterials is obviously due to the large surface area, small dimension, great adsorption capacity, catalytic activity and sometimes microbial effectiveness. But a lot is still unknown about the behaviour of these materials in the environment. There has been research on the development and potential benefits of nanomaterials in water treatment processes and very little insight into the human and environmental toxicity of these nanomaterials. The shape, the instability and the possible transformation after aggregation and utilization of nanomaterials still raise controversy on their suitability in water remediation. To this effect, improvements in the field of nanomaterials characterization are also necessary. The equivalent materials involved at the nanoscale are still under investigation. It would be beneficial to practice the development of new strategies for the use of these nanoparticles in the remediation of pharmaceutical compounds accumulated in the environment.

REFERENCES

Ali DM, Sasikala M, Gunasekaran M, et al (2011) Biosynthesis and characterization of silver nanoparticles using marine cyanobacterium, *Oscillatoria willei* NTDM01. *Dig J Nanomater Biostruct* 6:385–390.

Babu MMG, Gunasekaran P (2009) Production and structural characterization of crystalline silver nanoparticles from *Bacillus cereus* isolate. *Colloids Surfaces Biointerfac* 74:191–195.

Bayramoğlu G, Yılmaz M, Şenel AÜ, et al (2008) Preparation of nanofibrous polymer grafted magnetic poly(GMA-MMA)-g-MAA beads for immobilization of trypsin via adsorption. *Biochem Eng J* 40:262–274. doi: 10.1016/j.bej.2007.12.013.

Beier S, Köster S, Veltmann K, et al (2010) Treatment of hospital wastewater effluent by nanofiltration and reverse osmosis. *Water Sci Technol* 61:1691–1698.

Cao B, Loh K-C (2009) Physiological comparison of Pseudomonas putida between two growth phases during cometabolism of 4-chlorophenol in presence of phenol and glutamate: a proteomics approach. *J Chem Technol Biotechnol* 84:1178–1185. doi: 10.1002/jctb.2155.

Chonova T, Labanowski J, Cournoyer B, et al (2018) River biofilm community changes related to pharmaceutical loads emitted by a wastewater treatment plant. *Environ Sci Pollut Res* 25:9254–9264. doi: 10.1007/s11356-017-0024-0.

Christy C, Vermant S (2002) The state-of-the-art of filtration in recovery processes for biopharmaceutical production. *Desalination* 147:1–4.

Dias IN, Souza BS, Pereira JHOS, et al (2014) Enhancement of the photo-Fenton reaction at near neutral pH through the use of ferrioxalate complexes: a case study on trimethoprim and sulfamethoxazole antibiotics removal from aqueous solutions. *Chem Eng J* 247:302–313.

Ebele AJ, Abou-Elwafa Abdallah M, Harrad S (2017) Pharmaceuticals and personal care products (PPCPs) in the freshwater aquatic environment. *Emerg Contam* 3:1–16. doi: 10.1016/j.emcon.2016.12.004.

El Fels L, Lemee L, Ambles A, et al (2016) Identification and biotransformation of aliphatic hydrocarbons during co-composting of sewage sludge-date palm waste using pyrolysis-GC/MS technique. *Environ Sci Pollut Res* 23:16857–16864. doi: 10.1007/s11356-016-6670-9.

Elersek T, Milavec S, Korošec M, et al (2016) Toxicity of the mixture of selected antineoplastic drugs against aquatic primary producers. *Environ Sci Pollut Res* 23:14780–14790. doi: 10.1007/s11356-015-6005-2.

Fakhri A, Adami S (2014) Adsorption and thermodynamic study of Cephalosporins antibiotics from aqueous solution onto MgO nanoparticles. *J Taiwan Inst Chem Eng* 45:1001–1006.

Falås P, Baillon-Dhumez A, Andersen HR, et al (2012) Suspended biofilm carrier and activated sludge removal of acidic pharmaceuticals. *Water Res* 46:1167–1175. doi: 10.1016/j.watres.2011.12.003.

Ghormade V, Deshpande MV, Paknikar KM (2011) Perspectives for nano-biotechnology enabled protection and nutrition of plants. *Biotechnol Adv* 29:792–803.

Hafiane A, Lemordant D, Dhahbi M (2000) Removal of hexavalent chromium by nanofiltration. *Desalination* 130:305–312. doi: 10.1016/S0011-9164(00)00094-1.

He S, Guo Z, Zhang Y, et al (2007) Biosynthesis of gold nanoparticles using the bacteria *Rhodopseudomonas capsulata*. *Mater Lett* 61:3984–3987. doi: 10.1016/j.matlet.2007.01.018.

Hu X, Yang J, Zhang J (2011) Magnetic loading of TiO2/SiO2/Fe3O4 nanoparticles on electrode surface for photoelectrocatalytic degradation of diclofenac. *J Hazard Mater* 196:220–227.

Jodat A, Jodat A (2014) Photocatalytic degradation of chloramphenicol and tartrazine using Ag/TiO2 nanoparticles. *Desalin Water Treat* 52:2668–2677.

Juibari MM, Abbasalizadeh S, Jouzani GS, et al (2011) Intensified biosynthesis of silver nanoparticles using a native extremophilic *Ureibacillus thermosphaericus* strain. *Mater Lett* 65:1014–1017. doi: 10.1016/j.matlet.2010.12.056.

Jurecska L, Dobosy P, Barkács K, et al (2015) Reprint of "characterization of cyclodextrin containing nanofilters for removal of pharmaceutical residues." *J Pharm Biomed Anal* 106:124–128.

Kathiravan V, Ravi S, Ashokkumar S, et al (2015) Green synthesis of silver nanoparticles using Croton sparsiflorus morong leaf extract and their antibacterial and antifungal activities. *Spectrochim Acta Part A Mol Biomol Spectrosc* 139:200–205. doi: 10.1016/j.saa.2014.12.022.

Kaur S, Dhillon GS (2014) The versatile biopolymer chitosan: potential sources, evaluation of extraction methods and applications. *Crit Rev Microbiol* 40:155–175.

Klaus-Joerger T, Joerger R, Olsson E, et al (2001) Bacteria as workers in the living factory: metal-accumulating bacteria and their potential for materials science. *Trends Biotechnol* 19:15–20.

Konwarh R, Karak N, Rai SK, et al (2009) Polymer-assisted iron oxide magnetic nanoparticle immobilized keratinase. *Nanotechnology* 20:225107. doi: 10.1088/0957-4484/20/22/225107.

Košutić K, Dolar D, Ašperger D, et al (2007) Removal of antibiotics from a model wastewater by RO/NF membranes. *Sep Purif Technol* 53:244–249.

Koyuncu I, Arikan OA, Wiesner MR, et al (2008) Removal of hormones and antibiotics by nanofiltration membranes. *J Memb Sci* 309:94–101.

Liu P, Zhang H, Feng Y, et al (2014) Removal of trace antibiotics from wastewater: a systematic study of nanofiltration combined with ozone-based advanced oxidation processes. *Chem Eng J* 240:211–220.

Mandal D, Bolander ME, Mukhopadhyay D, et al (2006) The use of microorganisms for the formation of metal nanoparticles and their application. *Appl Microbiol Biotechnol* 69:485–492. doi: 10.1007/s00253-005-0179-3.

Marchlewicz A, Guzik U, Wojcieszyńska D (2015) Over-the-counter monocyclic non-steroidal anti-inflammatory drugs in environment-sources, risks, biodegradation. *Water, Air, Soil Pollut* 226:355. doi: 10.1007/s11270-015-2622-0.

Miralles-Cuevas S, Oller I, Aguirre AR, et al (2014) Removal of pharmaceuticals at microg L– 1 by combined nanofiltration and mild solar photo-fenton. *Chem Eng J* 239:68–74.

Mitchell DT, Lee SB, Trofin L, et al (2002) Smart nanotubes for bioseparations and biocatalysis. *J Am Chem Soc* 124:11864–11865. doi: 10.1021/ja027247b.

Mohammadi R, Massoumi B, Eskandarloo H (2015) Preparation and characterization of Sn/Zn/TiO2 photocatalyst for enhanced amoxicillin trihydrate degradation. *Desalin Water Treat* 53:1995–2004.

Mourato A, Gadanho M, Lino AR, et al (2011) Biosynthesis of crystalline silver and gold nanoparticles by extremophilic yeasts. *Bioinorg Chem Appl* 2011:546074. doi: 10.1155/2011/546074.

Nassar MY, Ahmed IS, Raya MA (2019) A facile and tunable approach for synthesis of pure silica nanostructures from rice husk for the removal of ciprofloxacin drug from polluted aqueous solutions. *J Mol Liq* 282:251–263.

Nunes B, Verde MF, Soares AMVM (2015) Biochemical effects of the pharmaceutical drug paracetamol on *Anguilla anguilla*. *Environ Sci Pollut Res* 22:11574–11584. doi: 10.1007/s11356-015-4329-6.

Oliveira LCA, Petkowicz DI, Smaniotto A, et al (2004) Magnetic zeolites: a new adsorbent for removal of metallic contaminants from water. *Water Res* 38:3699–3704. doi: 10.1016/j.watres.2004.06.008.

Parolini M, Binelli A (2012) Sub-lethal effects induced by a mixture of three non-steroidal anti-inflammatory drugs (NSAIDs) on the freshwater bivalve *Dreissena polymorpha*. *Ecotoxicology* 21:379–392. doi: 10.1007/s10646-011-0799-6.

Pimentel MCB, Leao ABF, Melo EHM, et al (2007) Immobilization of *Candida rugosa* lipase on magnetized dacron: kinetic study. *Artif Cells Blood Subst Biotechnol* 35:221–235. doi: 10.1080/10731190601188380.

Popescu M, Velea A, Lorinczi A (2010) Biogenic production of nanoparticles. *Dig J Nanomater Biostructures* 5:1035–1040.

Qiang Y, Sharma A, Paszczynski A, et al (2007) Conjugates of magnetic nanoparticle-enzyme for bioremediation. In: Proceedings of the 2007 NSTI Nanotechnology Conference and Trade Show. pp 656–659

Qurie M, Khamis M, Malek F, et al (2014) Stability and removal of naproxen and its metabolite by advanced membrane wastewater treatment plant and Micelle–C lay complex. *Clean Soil Air Water* 42:594–600.

Radjenović J, Petrović M, Ventura F, et al (2008) Rejection of pharmaceuticals in nanofiltration and reverse osmosis membrane drinking water treatment. *Water Res* 42:3601–3610.

Raeiatbin P, Açıkel YS (2017) Removal of tetracycline by magnetic chitosan nanoparticles from medical wastewaters. *Desalin Water Treat* 73:380–388. doi: 10.5004/dwt.2017.20421.

Rama LP, Cheryan M, Rajagopalan N (1996) Solvent recovery and partial deacidification of vegetable oils by membrane technology. *Lipid/fett* 98:10–14.

Rozas O, Vidal C, Baeza C, et al (2016) Organic micropollutants (OMPs) in natural waters: oxidation by UV/H2O2 treatment and toxicity assessment. *Water Res* 98:109–118. doi: 10.1016/j.watres.2016.03.069.

Salahi A, Mohammadi T, Rekabdar F, et al (2010) Reverse osmosis of refinery oily wastewater effluents. *Iran J Environ Health Sci Eng* 7(5):413–422.

Sanghi R, Verma P (2009) Biomimetic synthesis and characterisation of protein capped silver nanoparticles. *Bioresour Technol* 100:501–504. doi: 10.1016/j.biortech.2008.05.048.

Sastry M, Rao M, Ganesh KN (2002) Electrostatic assembly of nanoparticles and biomacromolecules. *Acc Chem Res* 35:847–855. doi: 10.1021/ar010094x.

Saxena R, Saxena M, Lochab A (2020) Recent progress in nanomaterials for adsorptive removal of organic contaminants from wastewater. *ChemistrySelect* 5:335–353. doi: 10.1002/slct.201903542.

Schüler C, Caruso F (2000) Preparation of enzyme multilayers on colloids for biocatalysis. *Macromol Rapid Commun* 21:750–753. doi: 10.1002/1521-3927(20000701)21:11<750::AID-MARC750>3.0.CO;2-3.

Seshadri S, Saranya K, Kowshik M (2011) Green synthesis of lead sulfide nanoparticles by the lead resistant marine yeast, *Rhodosporidium diobovatum*. *Biotechnol Prog* 27:1464–1469 doi: 10.1002/btpr.651.

Shan D, Deng S, Zhao T, et al (2016) Preparation of regenerable granular carbon nanotubes by a simple heating-filtration method for efficient removal of typical pharmaceuticals. *Chem Eng J* 294:353–361.

Singh BK (2010) Exploring microbial diversity for biotechnology: the way forward. *Trends Biotechnol* 28:111–116.

Singh BK, Walker A (2006) Microbial degradation of organophosphorus compounds. *FEMS Microbiol Rev* 30:428–471.

Soares SF, Fernandes T, Sacramento M, et al (2019) Magnetic quaternary chitosan hybrid nanoparticles for the efficient uptake of diclofenac from water. *Carbohydr Polym* 203:35–44.

Soares SF, Simões TR, António M, et al (2016) Hybrid nanoadsorbents for the magnetically assisted removal of metoprolol from water. *Chem Eng J* 302:560–569.

Sun Y, Liu Z, Fei Z, et al (2019) Synergistic effect and degradation mechanism on Fe-Ni/CNTs for removal of 2,4-dichlorophenol in aqueous solution. *Environ Sci Pollut Res* 26:8768–8778. doi: 10.1007/s11356-019-04394-w.

Sunkar S, Nachiyar CV (2012) Microbial synthesis and characterization of silver nanoparticles using the endophytic bacterium *Bacillus cereus*: a novel source in the benign synthesis. *Glob J Med Res* 12:43–50.

Unger KK, Skudas R, Schulte MM (2008) Particle packed columns and monolithic columns in high-performance liquid chromatography-comparison and critical appraisal. *J Chromatogr A* 1184:393–415. doi: 10.1016/j.chroma.2007.11.118.

Vrijenhoek EM, Waypa JJ (2000) Arsenic removal from drinking water by a "loose" nanofiltration membrane. *Desalination* 130:265–277. doi: 10.1016/S0011-9164(00)00091-6.

Wang Y, Shu L, Jegatheesan V, et al (2010a) Removal and adsorption of diuron through nanofiltration membrane: the effects of ionic environment and operating pressures. *Sep Purif Technol* 74:236–241. doi: 10.1016/j.seppur.2010.06.011.

Yadav KK, Singh JK, Gupta N, et al (2017) A review of nano-bioremediation technologies for environmental cleanup: a novel biological approach. *J Mater Environ Sci* 8:740–757.

Yang Q, Chen G, Zhang J, et al (2015) Adsorption of sulfamethazine by multi-walled carbon nanotubes: effects of aqueous solution chemistry. *RSC Adv* 5:25541–25549.

Yu F, Sun S, Han S, et al (2016) Adsorption removal of ciprofloxacin by multi-walled carbon nanotubes with different oxygen contents from aqueous solutions. *Chem Eng J* 285:588–595.

Yu J-G, Zhao X-H, Yang H, et al (2014b) Aqueous adsorption and removal of organic contaminants by carbon nanotubes. *Sci Total Environ* 482:241–251.

Zhao D, Zhang W, Chen C, et al (2013) Adsorption of methyl orange dye onto multiwalled carbon nanotubes. *Procedia Environ Sci* 18:890–895.

Zhou Y, He Y, Xiang Y, et al (2019) Single and simultaneous adsorption of pefloxacin and Cu (II) ions from aqueous solutions by oxidized multiwalled carbon nanotube. *Sci Total Environ* 646:29–36.

Zhu A, Zhu W, Wu Z, et al (2003) Recovery of clindamycin from fermentation wastewater with nanofiltration membranes. *Water Res* 37:3718–3732.

15

Heavy Metal Remediation through Nanoparticles

Ankita Chatterjee and Jayanthi Abraham
VIT University

CONTENTS

15.1 Introduction

A pollutant is any substance which causes hazardous or undesirable effects in the environment, thereby causing an imbalance in the natural ecosystem. As described by Hill in 2020, a pollutant can be defined as 'a chemical out of place' (Hill 2020). The primary causes of environmental pollution are large-scale eradication of nature replaced by industrialization (Singh and Prasad 2015). Pollution caused by heavy metals has slowly increased with time and is considered as one of the chief environmental concerns across the globe. Contrary to several organic pollutants, heavy metals cannot be decomposed or degraded by natural processes. It has been reported that 99% of the heavy metals that are released into the aquatic environment reach the rivers and are trapped in the sediments. Apart from residing in sediments, heavy metals also exist in different forms in nature, such as colloids, water-soluble forms and suspended forms. The toxicity of heavy metals that enter the food chain affects human beings and other life forms. Common toxic heavy metals found in the environment are mercury, chromium, cadmium, lead, arsenic and certain radioactive metals. The heavy metals might mimic an important element in the human body and thus disturb the metabolism of the human body (Baby et al. 2019).

15.2 Sources of Heavy Metals

Heavy metals are introduced into the environment via natural and anthropogenic activities. Anthropogenic activities that cause heavy metal pollution include agricultural processes, combustion of fossil fuel, mining and smelting, disposal of wastes and corrosion. The anthropogenic sources and the permissible limit of heavy metals in the environment are tabulated in Table 15.1. A recent study suggests that street dust has been a major carrier of heavy metals in the environment, thus increasing the heavy metal pollution.

DOI: 10.1201/9781003181224-15

TABLE 15.1

Anthropogenic Sources and Permissible Limits of Pollution-Causing Heavy Metals

S. No.	Heavy Metal	Anthropogenic Sources	Permissible Limit in Groundwater as Suggested by WHO	References
1.	Arsenic	Effluents of pesticide and fertilizer industries, mining, smelting, processing of metal ores	10 μg/L	Ali et al. (2019)
2.	Cadmium	Battery industries, electroplating industries, animal breeding, aquaculture, wastewater treatment, crop farming	30 μg/L	Yuan et al. (2019)
3.	Chromium	Industrial process (dust from cement industries), metal processing factories, combustion of coal, organic composts, phosphate fertilizers	50 μg/L	Poznanović Spahić et al. (2019)
4.	Cobalt	Tannery, leather factory, metal processing factories, combustion of coal	500 μg/L	Noorka et al. (2019)
5.	Copper	Electroplating industries, fungicides, storage of ash, mining, smelting	2000 μg/L	Łanocha-Arendarczyk and Kosik-Bogacka (2019)
6.	Lead	Discharge from wastewater treatment plants, mining and smelting, combustion of fossil fuel, urban run-off	100 μg/L	Hernández et al. (2020)
7.	Mercury	Battery industries	10 μg/L	Nasir et al. (2019)
8.	Nickel	Dust released from cement industries, tannery, leather factory, metal processing factories, combustion of coal, phosphate fertilizers, organic composts	200 μg/L	Poznanović Spahić et al. (2019)
9.	Selenium	Combustion of biomass, coal, oil and wood; manufacturing and implementation of agricultural products; smelting of non-ferrous metals	400 μg/L	Pilarczyk et al. (2019)

Street dusts consist of a mixture of airborne particles, substances released from construction sites, fumes and soot released from vehicles along with soil. The heavy metals which are present in the street dusts are transported in the atmosphere, contributing to its prolonged existence in the environment.

15.3 Remediation Strategies of Heavy Metals Pollution

The common methods of remediation of heavy metals include electrochemical separation, chemical precipitation, flocculation, coagulation, membrane separation and ion exchange. However, these methods are not very suitable for industrial purposes as they have drawbacks. However, electrochemical separation is an efficient way of remediation of heavy metals, but the requirements of high energy and operational costs during the process add to the disadvantages, making the process unfit for application. In case of chemical precipitation, production of excess amounts of wastes as end products makes it difficult to be used on a large scale. Production of high amounts of sludge restrains the application of coagulation and flocculation methods. Membrane separation technique is not suitable to be used in large scale due to their high cost of generation and difficulties in the removal of the residues produced during the process. The inadequacy of regeneration capacity or recyclability in case of ion exchange technique adds to the drawback. Adsorption is considered to be a better alternative in the remediation of heavy metals compared to the above-mentioned techniques. Adsorbents can be prepared at low costs and show a much higher efficiency. Removal of heavy metals even at very low concentrations can be achieved using adsorption

TABLE 15.2

Comparison between the Conventional Techniques of Heavy Metal Remediation

S. No.	Conventional Technique	Heavy Metals Treated	Advantages	Disadvantages	References
1.	Chemical precipitation	Copper, chromium, nickel, zinc, lead, iron	Easy procedure, inexpensive	Production of toxic sludge	Mirbagheri and Hosseini (2005)
2.	Floatation	Copper, cadmium, lead, zinc, nickel	Efficient separation process, recovery of selective metals, production of comparatively less sludge	Interference of intensity of ionic strength of the floatation agent solution, requirement of high amount of energy and electricity, production of bigger fragments	Salmani et al. (2013)
3.	Ion exchange	Arsenic, copper, cadmium, lead, nickel, mercury, zinc	Production of limited quantity of sludge, selective recovery of heavy metals, regeneration of the resins	Efficacy of the process reduced due to existence of acids, sensitive towards pH, high cost of the process	Budak (2013)
4.	Solvent extraction	Chromium, copper, nickel, zinc	Rapid procedure, easy separation technique	Economically infeasible, production of toxic compounds	Lewinsky (2007)
5.	Membrane filtration	Copper, nickel, zinc, chromium, lead	Efficient in the removal of heavy metals treatment, requirement of less energy, nature-friendly technique, requirement of compressed space for activity	Fouling of membrane resulting in a decrease in flux and deterioration of membrane quality, increased cost of operation	Huang et al. (2015)
6.	Adsorption	Chromium, cadmium, copper, lead, zinc, nickel	Less operational expenditure, can occur in a broad range of pH, easy application and strong affinity for metals	Problematic for application in large scale, unable to accomplish selective separation, waste products are formed by the end of the process	Bobade and Eshtiagi (2015)

methods (Chatterjee et al. 2020). The advantages and disadvantages of the conventional techniques of heavy metals remediation are listed in Table 15.2.

15.4 Importance of Nanoparticles in Remediation of Heavy Metals

Nanotechnology and its application in various fields, such as medicine, plant genomics and animal science, have been proven to be of much importance in recent research. Encouraging developments have been noticed by applying nanotechnology in the remediation of environmental contaminants and are considered to be a feasible approach to the remediation of heavy metals from the environment. Nanoparticles can act as catalysts or can be incorporated in sensors in order to improve the air and soil quality. Treatment of water contaminated with heavy metals using nanoparticles can be achieved by approaches such as filtration and bioremediation. Filtration process involves straining techniques to eliminate heavy metal pollutants. Nanoparticles, when used in filtration, attract the heavy metals towards them and thus enhance the quality of filtration. Hybrid membranes for filtration have been accepted among researchers due to the efficient sorption ability of the membranes, which in turn increases the quality of filtration. The hybrid membranes are prepared by impregnating various types of inorganic nanoparticles in certain polymeric matrix. Studies revealed that the poor filtration capacity of polysulphone matrix can be enhanced by incorporation of inorganic nanoparticles in them (Yurekli et al. 2017).

Magnetic iron oxide nanoparticles are reported to remove heavy metals from wastewater by following filtration methods of treatment. Nanoparticles are used in bioremediation due to various reasons; for example, the exposure of increased surface area per unit mass helps in the availability of a larger portion of the particle to react with the pollutants. Application of nanoparticles in bioremediation is effective due to their immense diffusion or penetration capacity and high reactivity to pollutants. In

TABLE 15.3

List of Few Sources of Biosynthesis of Nanoparticles Used in Heavy Metal Remediation

S. No.	Nanoparticles	Source of Synthesis	References
1.	Silver	Bacteria: *E. coli*, *Pseudomonas* sp. Fungi: *Penicillium fellutanum*, *Thraustochytrium* sp. *Aspergillus niger*, *Fusarium oxysporum* Algae: *Sargassum wightii*, *Ulva fasciata* Plant and plant products: *Capsicum annuum*, *Azadirachta indica*, *Ocimum sanctum*	Asmathunisha and Kathiresan (2013)
2.	Iron	Bacteria: *Bacillus subtilis*, cytoplasmic extract of *Lactobacillus fermentum*, *Bacillus megaterium* Fungi: *Pleurotus* sp., *Alternaria alternata* Algae: *Sargassum muticum* Plant and plant products: sorghum bran extract, leaf extract of *Sageretia thea*	
3.	Titanium dioxide	Bacteria: *Bacillus subtilis*, *Bacillus amyloliquefaciens*, *Aeromonas hydrophila* Fungi: *Trichoderma viride* Algae: *Chlamydomonas reinhardtii* Plant and plant products: seed extracts of *Vigna radiata*, seed extracts of *Vigna unguiculata*, leaf of *Sesbania grandiflora*, extract of *Trigonella foenum-graecum*	
4.	Manganese oxide	Bacteria: *Bacillus* sp., *Acinetobacter* sp., *Shewanella loihica* Fungi: *Fusarium oxysporum* Plant and plant products: aqueous leaf extract of *Cucurbita pepo*, *Kalopanax pictus* extract	
5.	Magnesium oxide	Fungi: melanin obtained from *Penicillium chrysogenum* Plant and plant products: *Citrus limon* leaf extract, leaf extract of *Bauhinia purpurea*, peels of *Nephelium lappaceum*, *Trigonella foenum-graecum* leaf extract, aqueous extract of *Swertia chirayita*	El-Sayyad et al. (2018)
6.	Zinc oxide	Bacteria: *Lactobacillus plantarum*, *Lactobacillus sporogenes* Fungi: *Aspergillus fumigatus*, culture filtrates of *Aspergillus niger*, *Pichia kudriavzevii*, *Alternaria alternate* Algae: *Ulva lactuca* Plant and plant products: leaf extract of *Hibiscus sabdariffa*, extract of the flower obtained from *Nyctanthes* ***arbor-tristis***, leaves of *Camellia sinensis*	Yusof et al. (2019)

certain cases, nanoparticles are biosynthesized from plants or microbes and are used in the remediation of heavy metals (termed as bioadsorption). Table 15.3 shows the different sources of biosynthesis of commonly used nanoparticles. Bioadsorption has been proven to be an effective approach to removing heavy metals maintaining low cost of the process. The efficiency of a nanomaterial in adsorption technique is also detected by its desorption and regeneration ability. Desorption is the phenomenon of extracting the heavy metals from the surface of the nanomaterials after the completion of adsorption. Desorption of heavy metals from nanomaterial adsorbents helps in determining the reusability of the nanomaterials and their loss of effectiveness. An increased regeneration ability helps in reducing the cost of the adsorption process by reusing an adsorbent for several cycles of adsorption–desorption (Chatterjee and Abraham 2019).

15.5 Conventional Techniques of Synthesis of Nanoparticles

The mechanism of synthesis of nanoparticles can be broadly classified into two types: top-down method and bottom-up method. The top-down approach to nanoparticles synthesis involves the process of reducing the size of bulk material by splitting them into fine nanoparticles (physical techniques). Examples of top-down approaches include ball milling, sputtering, evaporation–condensation, nanolithography, pulsed laser ablation, pulsed wire discharge and thermal decomposition. However, in the case of bottom-up mechanisms, biological and chemical processes of synthesis are involved. In these types, atoms assemble to form clusters of nanoparticles. Examples of bottom-up methods of nanoparticles synthesis are sol–gel technique, spinning, pyrolysis, chemical vapour deposition and biosynthesis. Figure 15.1 shows the pictorial representation of the basic concept of top-down and bottom-up approaches to the synthesis of nanoparticles.

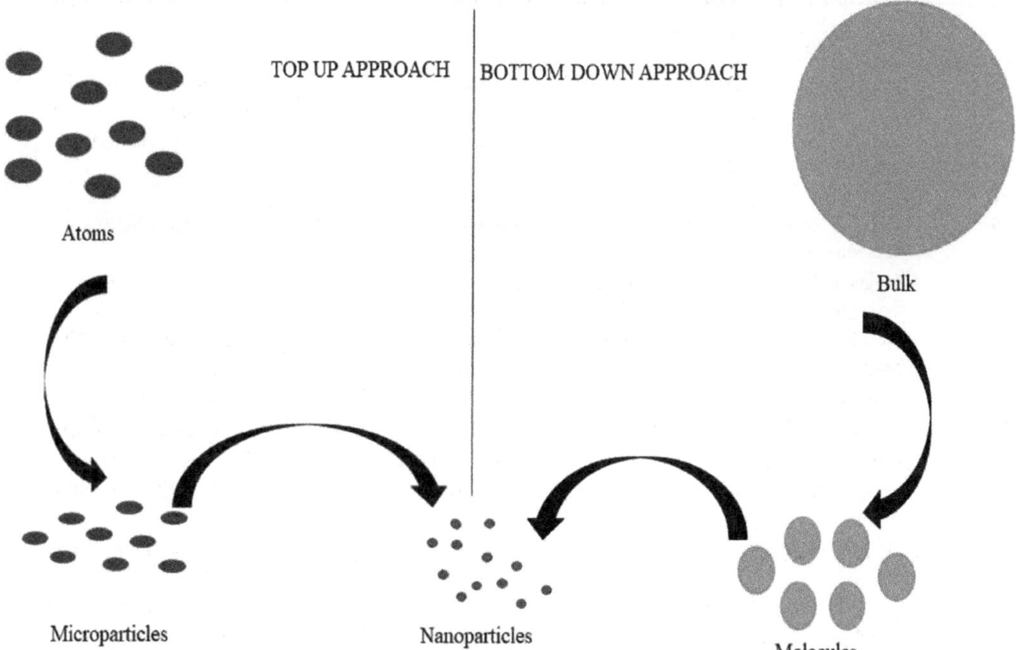

FIGURE 15.1 Pictorial representation of top-down and bottom-up approaches.

15.6 Types of Nanoparticles and Nanomaterials Used in Heavy Metal Remediation

Numerous nanoparticles have been experimented in the remediation of heavy metals from contaminated soil or water, some of which are briefly explained below.

15.6.1 Carbon-Based Nanoparticles

Carbon-based nanoparticles are widely accepted in various industrial fields. The exceptional electrical and thermal characteristics of carbon-based nanoparticles result in their usage in electronics industries. However, the capability of carbon-based nanoparticles in the remediation of both organic and inorganic contaminants along with their feasible physical or chemical modifications are the primary reasons why carbon-based nanoparticles are chosen in the remediation of heavy metals. Carbon nanotubes are used by several researchers in the removal of heavy metals due to their higher surface area, rapid adsorption kinetics and higher adsorption ability (Gupta et al. 2016). The first report regarding carbon nanotubes was revealed in 1991 (Iijima 1991). Carbon nanotubes are found to be of two types, which are single-walled carbon nanotubes and multiwalled carbon nanotubes. Carbon nanotubes have been found to be effective in the remediation of heavy metals causing severe pollution, such as arsenic, lead, mercury, manganese, copper and chromium. Table 15.4 shows the various types of carbon nanotubes and the heavy metals they are capable of adsorbing. The heavy metals are adsorbed on the outer surface, external groove sites, internal sites and interstitial channels of a carbon nanotube. Modifications in the

TABLE 15.4

Application of Carbon-Based Nanoparticles and Nanocomposites in Heavy Metal Remediation

S. No.	Carbon-Based Nanoparticles	Heavy Metal Remediated	References
1.	Carbon nanotubes	Arsenic, lead	Naghizadeh et al. (2012)
2.	Iron-coated ethylenediamine-functionalized multiwalled carbon nanotubes (e-MWCNTs)	Arsenate	Veličković et al. (2012)
3.	Iron oxide coated with MWCNTs	Arsenic	Addo Ntim and Mitra (2011)
4.	Nanocomposite of poly(amidoamine) dendrimer (PANAM) and carbon nanotubes	Zinc, cobalt, arsenic	Hayati et al. (2018)
5.	Goethite-incorporated graphene oxide–carbon nanotubes aerogel	Arsenic	Fu et al. (2017)
6.	Magnetic MWCNTs and iron oxide composite	Arsenic	Chen et al. (2014)
7.	Nanohybrid of MWCNTs and zirconia	Arsenic	Ntim and Mitra (2012)
8.	MWCNTs	Cadmium, lead, copper, zinc	Oloumi et al. (2018)
9.	Lignin-grafted CNTs	Lead	Li et al. (2017)
10.	Magnetic CNTs	Lead	Alijani et al. (2014)
11.	Cup-stacked CNTs	Lead, cadmium, chromium	Gong et al. (2014)
12.	Oxidized MWCNTs	Mercury, cadmium, lead, copper, nickel, zinc	Mubarak et al. (2013)
13.	Graphene oxide	Copper, cadmium, Cobalt, zinc, lead	Lee and Yang (2012)
14.	EDTA-based graphene oxide	Lead	Madadrang et al. (2012)
15.	Graphene oxide nanocomposite modified with magnetic water-soluble hyperbranched polyol	Lead	Hu et al. (2016)
16.	Magnetic graphene nanocomposites	Chromium	Zhu et al. (2013)
17.	Carbon shell magnetic adsorbent	Chromium	Zhu et al. (2013)

conventional carbon nanotubes can be conducted to increase the efficiency of the remediation process by introduction of functional groups.

According to previous studies, HNO_3, $KMNO_4$, NaOCl and H_2SO_4 aid in increasing the adsorption efficiency of the carbon nanotubes (El-Sheikh et al. 2011). Carbon nano-onions (commonly known as multilayered fullerenes) have been studied to be able to remove lead, copper, cadmium, nickel and zinc. The sorption kinetics analysis revealed that the nano-onions possess a high adsorption capacity and can be reused after several cycles of adsorption–desorption process (Seymour et al. 2012). However, the increased cost of carbon nanotubes and the difficulties faced while separating the carbon nanotubes from the wastewater after the remediation process hinder the wide utilization of the carbon nanotubes in industries.

Graphene nanoparticles are also well-known carbon-based nanomaterials that are used in the removal of heavy metals from wastewater and soil. The elasticity, stiffness, strong mechanical properties and effective electric and thermal characteristics make them suitable in various applications. The common graphene nanoparticles that are used in heavy metals remediation are graphene oxide and reduced graphene oxide nanoparticles. Graphene oxide nanoparticles have stable functional groups on their surface, which help in the adsorption of metal ion complexes via both coordinate and electrostatic mechanisms (Wang et al. 2013). Heavy metals adsorbed by graphene oxide nanoparticles are listed in Table 15.4. In certain cases, modifications of graphene oxide nanoparticles were conducted to obtain better adsorption of heavy metals. Graphene oxide nanoparticles modified with thiol groups are reported to adsorb mercury at much higher rates (around six times) when compared to only graphene oxide or activated carbon. Reports suggest that impregnation of chitosan with graphene oxide nanoparticles increased the sorption percentage when experimented with the adsorption of lead (He et al. 2011). Apart from graphene oxide, reduced graphene oxide is also used in heavy metals remediation.

Recent studies suggest that reduced graphene oxide successfully treats mercury-contaminated water by adsorption of mercury (Nhlane et al. 2020). Immobilization of zinc oxide nanorods and reduced graphene oxide showed removal of copper and cobalt with a higher efficiency than other adsorbents such as titanium dioxide nanoparticles, kaolin, magnetite–reduced graphene oxide and amine-functionalized magnetic nanoadsorbents. Reduced graphene oxide modified with 4-sulphophenylazo groups showed efficient adsorption of copper, chromium, cadmium, lead and nickel (Zhang et al. 2018).

15.6.2 Zerovalent Metal Nanoparticles

Application of metal nanoparticles in heavy metal remediation has gained the attention of researchers due to the transformation of contaminants in laboratory scale at a faster rate. The toxicity of the heavy metals in the environment depends on the surrounding redox reactions, adsorption and desorption conditions and precipitation. The oxidation states of the metal ions firmly regulate the toxicity of the heavy metals. Iron, silver, gold and palladium are the mostly used in their nanoforms. However, zerovalent iron nanoparticles are widely accepted for remediation purposes. Zerovalent iron nanoparticles remediate heavy metals depending on the standard redox potential of the heavy metal pollutant. The heavy metals having standard redox potential more negative as compared to standard redox potential of iron are remediated using adsorption technique. However, in case of metal ions that have standard redox potential more positive than that of iron, they are removed via both adsorption and reduction (O'Carroll et al. 2013). The particle size of zerovalent iron nanoparticle can be controlled, and the presence of numerous reactive sites on their surface along with their extreme reducing capacity enhances the chances of adsorption. Figure 15.2 shows the different types of remediation technique followed by zerovalent iron for specific metal ions. Iron nanoparticles supported with reduced graphene oxide are reported to remove uranium from contaminated groundwater.

Few studies have reported about the application of silver and gold nanoparticles in the remediation of heavy metals. Silver and gold nanoparticles are capable of removing mercury (Ojea-Jiménez et al. 2012). Green synthesis of silver and gold nanoparticles has been studied thoroughly. Green synthesis reduces the cost of the production of nanoparticles and thus adds value to application of these nanoparticles. Nanocomposites prepared using silver nanoparticles along with chitosan, carbon nanotubes and copper nanoparticles have been found to show promising adsorption capacity against copper, lead and cadmium (Alsabagh et al. 2015).

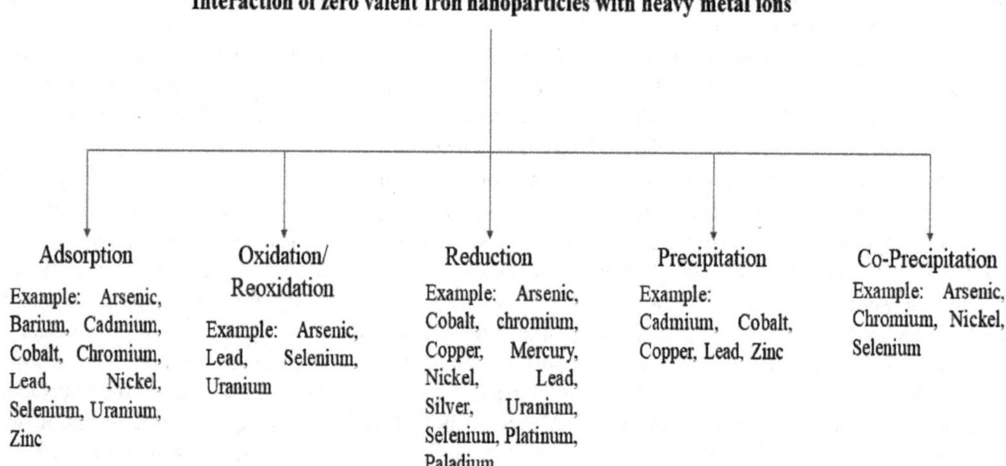

FIGURE 15.2 Heavy metals remediation techniques followed by zerovalent iron nanoparticles.

15.6.3 Metal Oxide Nanoparticles

Metal oxide nanoparticles are considered as promising adsorbents for heavy metals due to their enhanced remediation capability and selectivity towards a certain heavy metal (Yang et al. 2019). Some of the well-known metal oxides that are used in the remediation of heavy metals are manganese oxides, iron oxides, zinc oxides, aluminium oxides, titanium oxides and magnesium oxides. Hydrous manganese oxide nanoparticles was reported to remove mercury from water suspension in 1973. Hydroxyl group present on the surface of hydrous manganese oxides interacts with the metal ions via adsorption followed by intraparticle diffusion. Manganese oxide nanoparticles have been used in modifying the aggregates of graphene oxide, which showed effective sorption of cadmium and copper. The higher dispersion capacity of manganese nanoparticles helps in the increased sorption activity of the metal ions (Wan et al. 2018). Lead and chromium have successfully been removed from wastewater using manganese oxide nanoparticles (Dargahi et al. 2016).

Nanocomposites prepared using manganese oxide nanoparticles biochar derived from peanut shells showed effective results in the removal of cadmium and lead. The adsorbent was found to have a high regeneration ability (Wan et al. 2018). Combinations of manganese oxide nanoparticles with cellulose were experimented in the removal of lead, which showed efficient outcome in lead remediation. These studies indicated the firm selectivity of manganese oxide nanoparticles towards lead. Application of iron oxide nanoparticles in the remediation of arsenic in the form of both arsenite and arsenate has been confirmed by several studies. Along with arsenic, haematite and magnetite nanoparticles are also able to remove chromium, cobalt and molybdenum. Magnetite and haematite are forms of iron oxide nanoparticles that are commonly used in remediation approaches. The unique properties of magnetite and haematite include magnetic, biochemical and catalytic characteristics. The superparamagnetism of these nanoparticles enhances their sorption capability. Lead, copper, chromium, nickel and mercury can be removed from contaminated wastewater or soil using these iron oxide nanoparticles as adsorbents (Bhateria and Singh 2019).

In certain cases, organic or inorganic substances are incorporated with iron oxide nanoparticles in order to increase the stability of the material, thereby helping in metal ion uptake process. Modification of iron oxide nanoparticles with synthetic polymers has been found to be effective in the removal of chromium and mercury. Sorption of heavy metals using iron oxide nanoparticles fulfils the important aspects to be considered for large-scale remediation treatment, such as easy treatment, efficient outcome, regeneration ability of the adsorbent, harmlessness to the environment and low cost of the process. The

TABLE 15.5
Metal Oxide Nanoparticles and Composites in Remediation of Heavy Metals

S. No.	Metal Oxide Nanoparticles	Heavy Metal Remediated	References
1.	Manganese dioxide nanoparticles immobilized with activated carbon	Cadmium, lead	Lee and Tiwari (2013)
2.	Manganese oxide nanoparticles immobilized on graphene oxide	Cadmium, copper	Wan et al. (2018)
3.	Hybrid of ZnO nanorods with reduced graphene oxide	Cobalt, copper	
4.	Magnesium oxide	Copper, nickel, lead, cadmium	Prasad et al. (2007)
5.	Iron oxide nanoparticles with synthetic resins support	Copper	Min and Hering (1998)
6.	Iron oxide nanoparticles immobilized with alginate	Arsenic, copper	Lim et al. (2008)
7.	Zirconium oxide nanoparticles impregnated with quaternary ammonium resin	Copper, arsenic	Mohan and Pittman (2007)
8.	Nanocomposite of iron oxide and gum arabic	Copper	Banerjee and Chen (2007)
9.	Cobalt ferrite nanoparticles, titanium nanotubes, nanocomposite of cobalt ferrite, cobalt ferrite nanoparticles supported by alginate, nanocomposite of titanium nanotubes supported by alginate	Copper, iron, arsenic	Esmat et al. (2017)

cost of preparation of iron oxide nanoparticles can also be reduced using the green synthesis method. El-Kassas et al. (2016) revealed that iron oxide nanoparticles synthesized using seaweeds are effective in the remediation of lead.

Zinc oxide and tin oxide nanoparticles are reported to remove chromium from aqueous solution following intraparticle diffusion mechanism. A study revealed that photo-remediation of chromium can be conducted using zinc oxide nanoparticles (Banerjee et al. 2012). Copper has been remediated from wastewater released from electroplating industries using zinc oxide and magnesium oxide nanoparticles (Rafiq et al. 2014). Titanium oxide, magnesium oxide, aluminium oxide and zirconium oxide nanoparticles are also found to be effective in heavy metals remediation. Table 15.5 shows the applications of various metal oxide nanoparticles in the remediation of heavy metals.

15.7 Conclusions

The application of different types of nanoparticles in the remediation of heavy metals has been discussed in this chapter. Extensive exploitation of nanoparticles for remediation purposes is due to their unique properties. The synthesis of nanomaterials is carried out using top-down and bottom-up approaches. Nanoparticles as adsorbents can be used to replace various conventionally used adsorbents. Nanoparticles, in different forms, are applied in the remediation of heavy metals, which has been explained in this chapter. Nanocomposites are considered to be better adsorbents than nanoparticles because nanocomposites can be easily separated from the aqueous solution. However, as the application of nanomaterials in wastewater management increases, the toxic effects of the materials on the environment should be checked regularly to avoid any hazardous conditions. Considering the ecological condition and human health, implementation of nanoparticles in wastewater treatment is having a great impact. The factors that have been taken into consideration while using nanoparticles as adsorbents in large scale are their efficiency, regeneration capacity, cost of synthesis and cost of the treatment process.

Acknowledgement

The authors would like to express their gratitude to the management of VIT, Vellore.

REFERENCES

Addo Ntim S, Mitra S (2011) Removal of trace arsenic to meet drinking water standards using iron oxide coated multiwall carbon nanotubes. *J Chem Eng Data* 56(5):2077–83. https://doi.org/10.1021/je1010664.

Ali W, Rasool A, Junaid M, Zhang H (2019) A comprehensive review on current status, mechanism, and possible sources of arsenic contamination in groundwater: a global perspective with prominence of Pakistan scenario. *Environ Geochem Health* 41(2):737–60. https://doi.org/10.1007/s10653-018-0169-x.

Alijani H, Beyki MH, Shariatinia Z, Bayat M, Shemirani F (2014) A new approach for one step synthesis of magnetic carbon nanotubes/diatomite earth composite by chemical vapor deposition method: application for removal of lead ions. *Chem Eng* 253:456–63. https://doi.org/10.1016/j.cej.2014.05.021.

Alsabagh AM, Fathy M, Morsi RE (2015) Preparation and characterization of chitosan/silver nanoparticle/copper nanoparticle/carbon nanotube multifunctional nano-composite for water treatment: heavy metals removal; kinetics, isotherms and competitive studies. *RSC Adv* 5(69):55774–83. https://doi.org/10.1039/C5RA07477K.

Asmathunisha N, Kathiresan K (2013) A review on biosynthesis of nanoparticles by marine organisms. *Colloids Surf B Biointerfaces* 103:283–7. https://doi.org/10.1016/j.colsurfb.2012.10.030.

Baby R, Saifullah B, Hussein MZ (2019) Carbon nanomaterials for the treatment of heavy metal-contaminated water and environmental remediation. *Nanoscale Res Lett* 14(1):341. https://doi.org/10.1186/s11671-019-3167-8.

Banerjee P, Chakrabarti S, Maitra S, Dutta BK (2012) Zinc oxide nano-particles–sonochemical synthesis, characterization and application for photo-remediation of heavy metal. *Ultrason Sonochem* 19(1):85–93. https://doi.org/10.1016/j.ultsonch.2011.05.007.

Banerjee SS, Chen DH (2007) Fast removal of copper ions by gum Arabic modified magnetic nano-adsorbent. *J Hazard Mater* 147(3):792–9. https://doi.org/10.1016/j.jhazmat.2007.01.079.

Bhateria R, Singh R (2019) A review on nanotechnological application of magnetic iron oxides for heavy metal removal. *J Water Process Eng* 31:100845. https://doi.org/10.1016/j.jwpe.2019.100845.

Bobade V, Eshtiagi N (2015) Heavy metals removal from wastewater by adsorption process: A review. In *Asia Pacific Confederation of Chemical Engineering Congress 2015: APCChE 2015, incorporating CHEMECA 2015* (p. 312). Engineers Australia.

Budak TB (2013) Removal of heavy metals from wastewater using synthetic ion exchange resin. *Asian J Chem* 25(8):4207–10.

Chatterjee A, Abraham J (2019) Desorption of heavy metals from metal loaded sorbents and e-wastes: a review. *Biotechnol Lett* 41(3):319–33. https://doi.org/10.1007/s10529-019-02650-0.

Chatterjee A, Das R, Abraham J (2020) Bioleaching of heavy metals from spent batteries using *Aspergillus nomius* JAMK1. *Int J Environ Sci Technol* 17(1):49–66. https://doi.org/10.1007/s13762-019-02255-0.

Chen B, Zhu Z, Ma J, Yang M, Hong J, Hu X, Qiu Y, Chen J (2014) One-pot, solid-phase synthesis of magnetic multiwalled carbon nanotube/iron oxide composites and their application in arsenic removal. *J Colloid Interface Sci* 434:9–17. https://doi.org/10.1016/j.jcis.2014.07.046.

Dargahi A, Golestanifar H, Darvishi P, Karam A (2016) An investigation and comparison of removing heavy metals (lead and chromium) from aqueous solutions using magnesium oxide nanoparticles. *Pol J Environ Stud* 25:557–67. https://doi.org/10.15244/pjoes/60281.

El-Kassas HY, Aly-Eldeen MA, Gharib SM (2016) Green synthesis of iron oxide (Fe_3O_4) nanoparticles using two selected brown seaweeds: characterization and application for lead bioremediation. *Acta Oceanol Sin* 35(8):89–98. https://doi.org/10.1007/s13131-016-0880-3.

El-Sayyad GS, Mosallam FM, El-Batal AI (2018) One-pot green synthesis of magnesium oxide nanoparticles using *Penicillium chrysogenum* melanin pigment and gamma rays with antimicrobial activity against multidrug-resistant microbes. *Adv Powder Technol* 29(11):2616–25. https://doi.org/10.1016/j.apt.2018.07.009.

El-Sheikh AH, Al-Degs YS, Al-As'ad RM, Sweileh JA (2011) Effect of oxidation and geometrical dimensions of carbon nanotubes on Hg (II) sorption and preconcentration from real waters. *Desalination* 270(1–3):214–20. https://doi.org/10.1016/j.desal.2010.11.048.

Esmat M, Farghali AA, Khedr MH, El-Sherbiny IM (2017) Alginate-based nanocomposites for efficient removal of heavy metal ions. *Int J Biol Macromol* 102:272–83.

Fu D, He Z, Su S, Xu B, Liu Y, Zhao Y (2017) Fabrication of α-FeOOH decorated graphene oxide-carbon nanotubes aerogel and its application in adsorption of arsenic species. *J Colloid Interface Sci* 505: 105–14.

Gong J, Feng J, Liu J, Jiang Z, Chen X, Mijowska E, Wen X, Tang T (2014) Catalytic carbonization of polypropylene into cup-stacked carbon nanotubes with high performances in adsorption of heavy metallic ions and organic dyes. *Chem Eng J* 248:27–40. https://doi.org/10.1016/j.cej.2014.01.107.

Gupta VK, Moradi O, Tyagi I, Agarwal S, Sadegh H, Shahryari-Ghoshekandi R, Makhlouf AS, Goodarzi M, Garshasbi A (2016) Study on the removal of heavy metal ions from industry waste by carbon nanotubes: effect of the surface modification: a review. *Crit Rev Env Sci Tech* 46(2):93–118. https://doi.org/10.1080/10643389.2015.1061874.

Hayati B, Maleki A, Najafi F, Gharibi F, McKay G, Gupta VK, Puttaiah SH, Marzban N (2018) Heavy metal adsorption using PAMAM/CNT nanocomposite from aqueous solution in batch and continuous fixed bed systems. *Chem Eng J* 346:258–70. https://doi.org/10.1016/j.cej.2018.03.172.

He YQ, Zhang NN, Wang XD (2011) Adsorption of graphene oxide/chitosan porous materials for metal ions. *Chin Chem Lett* 22(7):859–62. https://doi.org/10.1016/j.cclet.2010.12.049.

Hernández E, Obrist-Farner J, Brenner M, Kenney WF, Curtis JH, Duarte E (2020) Natural and anthropogenic sources of lead, zinc, and nickel in sediments of Lake Izabal, Guatemala. *J Environ Sci* 96:117–26. https://doi.org/10.1016/j.jes.2020.04.020.

Hill MK (2020) *Understanding Environmental Pollution* 327–352. Cambridge University Press, United Kingdom.

Hu L, Li Y, Zhang X, Wang Y, Cui L, Wei Q, Ma H, Yan L, Du B (2016) Fabrication of magnetic water-soluble hyperbranched polyol functionalized graphene oxide for high-efficiency water remediation. *Sci Rep* 6:28924. https://doi.org/10.1038/srep28924.

Huang Y, Du JR, Zhang Y, Lawless D and Feng X (2015) Removal of mercury (II) from wastewater by polyvinylamine-enhanced ultrafiltration. *Sep Purif Technol* 154:1–10. https://doi.org/10.1016/j.seppur.2015.09.003.

Iijima S (1991) Helical microtubules of graphitic carbon. *Nature* 354(6348):56-8. https://doi.org/10.1038/354056a0

Łanocha-Arendarczyk N, Kosik-Bogacka DI (2019) Copper, Cu. In *Mammals and Birds as Bioindicators of Trace Element Contaminations in Terrestrial Environments* (pp. 125–161). Springer, Cham.

Lee SM, Tiwari D (2013) Manganese oxide immobilized activated carbons in the remediation of aqueous wastes contaminated with copper (II) and lead (II). *Chem Eng* 225:128–37. https://doi.org/10.1016/j.cej.2013.03.083.

Lee YC, Yang JW (2012) Self-assembled flower-like TiO2 on exfoliated graphite oxide for heavy metal removal. *J Ind Eng Chem* 18(3):1178–85. https://doi.org/10.1016/j.jiec.2012.01.005.

Lewinsky AA (2007) *Hazardous Materials and Wastewater: Treatment, Removal and Analysis.* Nova Publishers, 1–44.

Li Z, Chen J, Ge Y (2017) Removal of lead ion and oil droplet from aqueous solution by lignin-grafted carbon nanotubes. *Chem Eng* 308:809–17. https://doi.org/10.1016/j.cej.2016.09.126.

Lim SF, Zheng YM, Zou SW, Chen JP (2008) Characterization of copper adsorption onto an alginate encapsulated magnetic sorbent by a combined FT-IR, XPS, and mathematical modeling study. *Environ Sci Technol* 42(7):2551–6. https://doi.org/10.1021/es7021889.

Madadrang CJ, Kim HY, Gao G, Wang N, Zhu J, Feng H, Gorring M, Kasner ML, Hou S (2012) Adsorption behavior of EDTA-graphene oxide for Pb (II) removal. *ACS Appl Mater Interfaces* 4(3):1186–93. https://doi.org/10.1021/am201645g.

Min JH, Hering JG (1998) Arsenate sorption by Fe (III)-doped alginate gels. *Water Res* 32(5):1544–52. https://doi.org/10.1016/S0043-1354(97)00349-7.

Mirbagheri SA, Hosseini SN (2005) Pilot plant investigation on petrochemical wastewater treatment for the removal of copper and chromium with the objective of reuse. *Desalination* 171(1):85–93. https://doi.org/10.1016/j.desal.2004.03.022.

Mohan D, Pittman Jr CU (2007) Arsenic removal from water/wastewater using adsorbents—a critical review. *J Hazard Mater* 142(1–2):1–53. https://doi.org/10.1016/j.jhazmat.2007.01.006.

Mubarak NM, Alicia RF, Abdullah EC, Sahu JN, Haslija AA, Tan J (2013) Statistical optimization and kinetic studies on removal of Zn2+ using functionalized carbon nanotubes and magnetic biochar. *J Environ Chem Eng* 11(3):486–95. https://doi.org/10.1016/j.jece.2013.06.011.

Naghizadeh A, Yari AR, Tashauoei HR, Mahdavi M, Derakhshani E, Rahimi R, Bahmani P, Daraei H, Ghahremani E (2012) Carbon nanotubes technology for removal of arsenic from water. *Arch Hyg Sci* 1(1):6–11.

Nasir AM, Goh PS, Abdullah MS, Ng BC, Ismail AF (2019) Adsorptive nanocomposite membranes for heavy metal remediation: recent progresses and challenges. *Chemosphere* 232:96–112. https://doi.org/10.1016/j.chemosphere.2019.05.174.

Nhlane D, Richards H, Etale A (2020) Facile and green synthesis of reduced graphene oxide for remediation of Hg (II)-contaminated water. In *Materials Today: Proceedings*, 38:737–42. https://doi.org/10.1016/j.matpr.2020.04.163.

Noorka IR, Ahmad K, Nadeem M, Ugulu I (2019) 15. Biotransfer of cobalt along a soil-plant-chicken food chain: implication for public health. *Pure Appl Biol* 8(3):2015–27.

Ntim SA, Mitra S (2012) Adsorption of arsenic on multiwall carbon nanotube–zirconia nanohybrid for potential drinking water purification. *J Colloid Interface Sci* 375(1):154–9. https://doi.org/10.1016/j.jcis.2012.01.063.

O'Carroll D, Sleep B, Krol M, Boparai H, Kocur C (2013) Nanoscale zero valent iron and bimetallic particles for contaminated site remediation. *Adv Water Res* 51:104–22. https://doi.org/10.1016/j.advwatres.2012.02.005.

Ojea-Jiménez I, López X, Arbiol J, Puntes V (2012) Citrate-coated gold nanoparticles as smart scavengers for mercury (II) removal from polluted waters. *ACS Nano* 6(3):2253–60. https://doi.org/10.1021/nn204313a.

Oloumi H, Mousavi EA, Nejad RM (2018) Multi-wall carbon nanotubes effects on plant seedlings growth and cadmium/lead uptake in vitro. *Russ J Plant Physl* 65(2):260–8. https://doi.org/10.1134/S102144371802019X.

Pilarczyk B, Tomza-Marciniak A, Pilarczyk R, Marciniak A, Bąkowska M, Nowakowska E. (2019). Selenium, Se. In *Mammals and Birds as Bioindicators of Trace Element Contaminations in Terrestrial Environments* (pp. 301–362). Springer, Cham.

Poznanović Spahić MM, Sakan SM, Glavaš-Trbić BM, Tančić PI, Škrivanj SB, Kovačević JR, Manojlović DD (2019) Natural and anthropogenic sources of chromium, nickel and cobalt in soils impacted by agricultural and industrial activity (Vojvodina, Serbia). *J Environ Sci Health A* 54(3):219–30. https://doi.org/10.1080/10934529.2018.1544802.

Prasad PV, Pisipati S, Nagisetti G, Kirkham MB, Reddi L (2007) Application of metal oxide nanoparticles for phytostabilization of heavy metals in soil. In *Abstracts. Annual Meeting of American Society of Agronomy* (pp. 4–8), New Orleans, Louisiana, USA

Rafiq Z, Nazir R, Shah MR, Ali S (2014) Utilization of magnesium and zinc oxide nano-adsorbents as potential materials for treatment of copper electroplating industry wastewater. *J Environ Chem Eng* 2(1):642–51. https://doi.org/10.1016/j.jece.2013.11.004.

Salmani MH, Davoodi M, Ehrampoush MH, Ghaneian MT, Fallahzadah MH (2013) Removal of cadmium (II) from simulated wastewater by ion flotation technique. *Iran J Environ Health Sci Eng* 10(1):16. https://doi.org/10.1186/1735-2746-10-16.

Seymour MB, Su C, Gao Y, Lu Y, Li Y (2012) Characterization of carbon nano-onions for heavy metal ion remediation. *J Nanopart Res* 14(9):1087. https://doi.org/10.1007/s11051-012-1087-y.

Singh A, Prasad SM (2015) Remediation of heavy metal contaminated ecosystem: an overview on technology advancement. *Int J Environ Sci Technol* 12(1):353–66. https://doi.org/10.1007/s13762-014-0542-y.

Veličković Z, Vuković GD, Marinković AD, Moldovan MS, Perić-Grujić AA, Uskoković PS, Ristić MĐ (2012) Adsorption of arsenate on iron (III) oxide coated ethylenediamine functionalized multiwall carbon nanotubes. *Chem Eng J* 181:174–81. https://doi.org/10.1016/j.cej.2011.11.052.

Wan S, Wu J, Zhou S, Wang R, Gao B, He F (2018) Enhanced lead and cadmium removal using biochar-supported hydrated manganese oxide (HMO) nanoparticles: behavior and mechanism. *Sci Total Environ* 616:1298–306. https://doi.org/10.1016/j.scitotenv.2017.10.188.

Wang S, Sun H, Ang HM, Tadé MO (2013) Adsorptive remediation of environmental pollutants using novel graphene-based nanomaterials. *Chem Eng J* 226:336–47. https://doi.org/10.1016/j.cej.2013.04.070.

Yang J, Hou B, Wang J, Tian B, Bi J, Wang N, Li X, Huang X (2019) Nanomaterials for the removal of heavy metals from wastewater. *Nanomaterials* 9(3):424. https://doi.org/10.3390/nano9030424.

Yuan Z, Luo T, Liu X, Hua H, Zhuang Y, Zhang X, Zhang L, Zhang Y, Xu W, Ren J (2019) Tracing anthropogenic cadmium emissions: from sources to pollution. *Sci Total Environ* 676:87–96. https://doi.org/10.1016/j.scitotenv.2019.04.250.

Yurekli Y, Yildirim M, Aydin L, Savran M (2017) Filtration and removal performances of membrane adsorbers. *J Hazard Mater* 332:33–41. https://doi.org/10.1016/j.jhazmat.2017.02.061.

Yusof HM, Mohamad R, Zaidan UH (2019) Microbial synthesis of zinc oxide nanoparticles and their potential application as an antimicrobial agent and a feed supplement in animal industry: a review. *J Animal Sci Biotechnol* 10(1):57. https://doi.org/10.1186/s40104-019-0368-z.

Zhang CZ, Chen B, Bai Y, Xie J (2018) A new functionalized reduced graphene oxide adsorbent for removing heavy metal ions in water via coordination and ion exchange. *Sep Sci Technol* 53(18):2896–905. https://doi.org/10.1080/01496395.2018.1497655.

Zhu J, Wei S, Chen M, Gu H, Rapole SB, Pallavkar S, Ho TC, Hopper J, Guo Z (2013) Magnetic nanocomposites for environmental remediation. *Adv Powder Technol* 24(2):459–67.

16

Applicability of Plants in Detoxification of Dyes

Pankaj Kumar Chaurasia
B.R.A. Bihar University

Shashi Lata Bharati
North Eastern Regional Institute of Science and Technology, Nirjuli (Itanagar)

CONTENTS

16.1 Introduction

Plants on the earth are actually boon for humans, and moreover, humans can exist on the earth only in the fairly green environment of the plants. In spite of several incredible significances, various types of plants are known for their great values in the fields of food and nutrition, medicines, industries, nanotechnology, scientific uses, research purposes and so on (Chaurasia and Bharati, 2022). In spite of several great values of plants in the life of humans, which are not going to be described here, plants are also greatly valuable in protecting the environment from various types of pollution by stabilizing them, converting them into non-pollutant form, destroying them or deactivating them.

Plants can be significantly used either individually or in combination with microbes and other biological sources for the treatment of different types of pollutants such as heavy metals (Adiloğlu, 2017), petroleum wastes (Ite and Ibok, 2019), water pollutants (Luqman et al., 2013) and dyes (Tahir et al., 2016; Dogra et al., 2018). This chapter is based on the use of plant sources for the bioremediation of poisonous dyes.

As we know, dyes are being used extensively in the modern time for colouring various materials of use, such as textiles, plastics, fibres, woods and food materials. Actually, dye may be synthetic or natural according to their origin. If dyes are synthesized purely in laboratory using chemical substances/reagents, then they are called synthetic dyes (which means they are man-made), and if they are isolated or extracted from natural sources such as different parts of the plants, fungi and lichens (Burgess, 2017), they are called natural dyes. But in the modern time of extensive industrialization, most of the dyes being used for various purposes are synthetic dyes, which means man-made dyes using petrochemicals.

DOI: 10.1201/9781003181224-16

Besides their use in the field of colouring; they are also used for various other purposes such as organic dye lasers (Silfvast, 2008), optical media and camera sensors (colour filter array).

Dyes may be classified into various categories according to their chemical properties (Booth, 2000): acid dyes (water-soluble anionic dyes which are used for colouring the fibres of nylon, silk, wool and modified acrylic fibres by the reaction between the cationic group of dyes and anionic group of the fibres), basic dyes (water-soluble cationic dyes which are mainly used for dyeing the acrylic fibres), vat dyes (water-insoluble dyes which cannot be used to dye the fibres directly; colouring of fibres by this dye is based on the chemical reaction of reduction–oxidation), direct dyes (such dyes are used in neutral or slightly alkaline medium for dyeing cotton, silk, leather, wool, etc. Such dyes also work as indicator and biological stains), reactive dyes (chromophores attached to substituent are able to directly react with fibres), mordant dyes (such dyes use mordant for the fastness against water and light), disperse dye (these are water-insoluble and mainly developed for colouring the cellulose acetate), azo dyes (formation of such insoluble dyes occurs on/within fibres directly) and sulphur dyes (these inexpensive dyes are used to give dark colour to fibres).

There are another important class of dyes called food dyes, which are known as food additives and may be direct, mordant and vat dyes. Since food dyes are related to health, their production takes place with high standards and rules.

16.2 Pollution and Harmful Effects of Dyes

Dyes are highly useful in the present scenario of the world, and it is not possible to ban the use and synthesis of dyes from the environment safety point of view. Synthetic dyes which are being continuously released into the soil as well as into the aquatic environment are extremely harmful, toxic and carcinogenic for human beings as well as for aquatic lives. Textile industries using dyes abundantly for colouring the textiles release huge amounts of by-products and waste dyes into the aquatic environment and sewages. Adverse and poisonous effects of textile dyes on human health as well as aquatic organisms are the major challenges facing the scientific community. Several researches are continuously running all around the world in order to demonstrate the influences of industrial dyes on human health and to find out solutions to these serious challenges (Lellis et al., 2019; Gita et al., 2017; Kant, 2012). Different types of dyes have different natures of pollution. For example, in case of disperse dyes, the nature of pollution is carriers, reducing agents and organic acids; in case of acid dyes, the nature of pollution is unfixed dyes and organic dyes; in case of direct dyes, the nature of pollution is salts, unfixed dyes, copper salts and cationic fixing agents; in case of vat dyes, the nature of pollution is alkali, oxidizing agents and reducing agents; in case of sulphur dyes, the nature of pollution is alkali, oxidizing agents, reducing agents and unfixed dyes; and in case of reactive dyes, the nature of pollution is salts, unfixed dyes, alkali and so on (Ramachandran and Gnanadoss, 2013).

In the present time of speedy developments of the world, everyone searches for cheap products in the modern fashionable world under suitable range of their pocket. Industries are also working in this direction of fashionable world to meet the latest trends in the field of fashion world by rapid production of cheap clothing. The fast-growing fashion is majorly responsible for the production of chemical wastes in the aqueous medium. In this way, ultimately due to the unfettered rise in the growth of production and demands, there is also a huge growth in the production of pollution and chemical wastes. An important article titled 'How polluting is the fashion industry?' (Boggon, 2019) How polluting is the fashion industry?, https://www.ekoenergy.org/how-polluting-is-the-fashion-industry/) shows that how, in the modern time, rapidly growing fashions are responsible for the growth of such pollution and how we can tackle these problems using renewable energies.

Dyes released from textile industries in the form of wastewater are responsible for the major environmental pollution. The wastewater contains various chemical constituents extremely toxic and carcinogenic to lives. These may be in the form of whole unused dye or its by-products. The characteristics of wastewaters released from the textile industries and the different constituents of synthetic effluents have nicely been reviewed by Yaseen and Scholz (2019).

16.3 Bioremediation

Processes that involve the green techniques using microbes, enzymes, plants and other natural resources for the detoxification and degradation of toxic, harmful or unsafe pollutants present in polluted site, water system, soil and the environment in the form of chemicals, metals, polymers, effluents or others are called bioremediation. In other words, bioremediation is a process used to treat contaminated media, including water, soil and subsurface materials, by altering the environmental conditions to stimulate growth of microorganisms and degrade the target pollutants. In comparison with other remediation techniques, bioremediation processes are safer, green and less expensive in many cases (Green Remediation Best Management Practices: Sites with Leaking Underground Storage Tank Systems. EPA 542-F-11-008. EPA. June 2011).

Oxidation–reduction reactions are involved in most of the bioremediation processes. Reduced pollutants such as hydrocarbons undergo oxidation with the addition of an electron acceptor such as oxygen, or oxidized pollutants (such as nitrate, chlorinated solvents and explosives) undergo reduction with the addition of an electron donor such as an organic substrate (US Environmental Protection Agency, 2013, p. 30, EPA 542-R-13-018). In order to optimize the conditions for microorganisms, different additional nutrients and others may be added and, for enhancing the process of biodegradation further, specialized microbial cultures also may be added (bioaugmentation) in some cases. Phytoremediation, mycoremediation, rhizofiltration, bioaugmentation, composting, landfarming, bioleaching, etc., are a few examples of technologies related to bioremediation.

16.4 Biological Remediation and Its Resources

For the purpose of bio-treatment of environmental pollutants, different microbes and microbial enzymes as well as plants in different ways have been utilized by various researchers. Several fungi, bacteria, enzymes, archaea (less studied) (Bioremediation, https://microbewiki.kenyon.edu/index.php/Bioremediation) and plants are well known for their biodegradation ability. There are many publications that provide much information on remediation (Pal et al., 2020; Chaurasia and Bharati, 2017, 2019, 2020; Chaurasia et al., 2015, 2018, 2019, 2020a, 2020b; Bharati and Chaurasia, 2018; Agarwal et al., 2018).

Fungi are one of the most valuable sources that are used in the bio-treatment of various environmental pollutants. Fungi and fungal-based enzymes play significant roles in these processes. Biodegradation of harmful pollutants with the help of fungi are known as mycoremediation, a type of bioremediation. In other words, the removal, destruction or suppression of active contaminants causing adverse or any types of harmful effects to the environment by the use of fungi or fungal technology is known as mycoremediation (myco = fungus; remediation = to restore/cure/treat) (Chaurasia et al., 2019). Simply, it can also be defined as decontamination of the environment by the use of fungal-based technology (Sujata and Bharagava 2016; Kumari et al., 2016; Goutam et al., 2018). Fungi have been found to have a very valuable role in detoxifying the environment or wastewater contaminated with toxins including heavy metals, persistent organic pollutants, textile dyes, leather tanning industries, petroleum fuels, polycyclic aromatic hydrocarbons, pharmaceuticals and personal care products, pesticides and herbicides (Deshmukh et al., 2016). They are very cheap, effective and environmentally friendly techniques. The enzymes secreted by fungi, such as laccase, may also be highly valuable in the bioremediation process (Chaurasia et al., 2019).

Seeing the potency of fungi in soil bioremediation, it can easily be understood that they play a great role in soil bioremediation, and white rot fungi are now a very hot topic in this area (Gadd, 2001; Harms et al., 2011) (Bioremediation, https://microbewiki.kenyon.edu/index.php/Bioremediation). Interestingly, Chaurasia and Bharati (2021) described the role of fungi in agriculture and the sustainability of environment. The use of *Phanerochaete chrysosporium,* a white rot fungus, was reported in 1985 for metabolizing multiple key environmental pollutants, and the potency of fungi in treating pollutants was shown (Bioremediation, https://microbewiki.kenyon.edu/index.php/Bioremediation; Fragoeiro, 2005).

Enzymatic functionalities of these fungi are special characteristics of them, due to which they have the strong ability to metabolize the complex forms of chemical compounds. White rot fungi play a significant role in the degradation of complex compounds such as lignin *via* hyphal growth and extension, extracellularly. Extracellular degradation of complex pollutants and chemicals is an advantage of white rot fungi. Hyphal extension makes these fungi more useful by maximizing the area of surface for interaction by enzymes of these fungi. One another major advantage of the use of such types of fungi is that they can be directly used for bioremediation purposes, while pre-conditioning is the major requirement for the use of other types of microbes in biodegradation processes because they cannot survive in unconditioned environment (Bioremediation, https://microbewiki.kenyon.edu/index.php/Bioremediation; Fragoeiro, 2005). For the degradation of organic pollutants, *P. chrysosporium* was the first used fungi, which shows the potential for biodegradation of pollutants such as polycyclic aromatic hydrocarbons, carbon tetrachloride and pesticides. In this way, this fungus is a model for bioremediation processes among fungal systems. *Pleurotus ostreatus* and *Trametes versicolor* are some other species under white rot fungi, which also show their potential for bioremediation process (Bioremediation, https://microbewiki.kenyon.edu/index.php/Bioremediation; Fragoeiro, 2005).

In the present time of the rapidly growing demands of food and nutrition for heavily growing populations, extensive uses of pesticides in the field of agriculture and other places for pest control are of serious concern for the world because they are directly associated with public health in the form of food products. Their pollution is not only up to the food product; rather, they pollute and harm the environment very badly by entering the water system from the places of use. Thus, their use should be under special rules and regulations with proper knowledge of the techniques and methods that can minimize their harmful effects. There are several technologies for cleaning these pesticides, but they are very expensive and inefficient for their general use. In this direction, white rot fungi may be a good option for the bioremediation of pesticides-polluted soils because they have shown their potency in studies. They have great ability to degrade specific pesticides in soil (Bioremediation, https://microbewiki.kenyon.edu/index.php/Bioremediation; Fragoeiro, 2005). Table 16.1 shows some latest works on bioremediation of contaminated soil (Chaurasia et al., 2019).

Similarly, bacteria also play an important role in the remediation of wastes and pollutants (Bilal and Iqbal, 2020; d'Errico et al., 2020; Brim et al., 2000; Jessica et al., 1999). Several enzymes are known to have significant applications in the field of bioremediation. Laccases, peroxidases and other enzymes are significantly used for treating several pollutants (Osuoha and Nwaichi, 2020; Thakur et al., 2019; Bharati and Chaurasia, 2018; Agarwal et al., 2018; Chaurasia et al., 2015, 2018; Chaurasia and Bharati, 2017, Kushwaha et al., 2017).

16.5 Microbial Dye Detoxification/Decolourization

Decolourization done by fungi is known as mycodyedecolourization. Fungal-based decolourization may be performed in two ways. In the first way, a fungus or a group of fungi may be applied directly for the decolourization of dyes, and in the second way, their enzymatic liquid culture medium may also be applied to dyes. There are significant numbers of researches being performed per year all over the world. Some latest works on mycodyedecolourization are given in Table 16.2 (Chaurasia and Bharati, 2019).

16.6 Significances of Plants in Bioremediation

The use of plants for the remediation of environmental pollutants is known as phytoremediation. In other words, it may be defined as follows: phytoremediation deals with the clean-up of organic pollutants and heavy metal contaminants using plants and rhizospheric microorganisms (Ojuederie and Babalola, 2017; Dixit et al., 2015; Ali et al., 2013; Jan and Parray, 2016). The main advantage of plant-based remediation is that it is completely inexpensive and eco-friendly for the purpose of restoration of heavy

TABLE 16.1

Some Selected Fungi or Association for Bioremediation Processes

S. No.	Fungi/Fungal Association	Functions	References
1.	*Phlebia* species	Aerobic degradation of lindane	Xiao and Kondo (2020)
2.	*Aspergillus niger*-MK452260.1 (F1), *A. fumigatus*-KU321562.1 (F2), *A. flavus*-MH270609.1 (F4) and *Penicillium chrysogenum*-MK696383.1 (F3).	Bioremediation process	Ahmad and Ganjo (2020)
3.	*Autochthonous fungi*	Bioremediation of different types of leachate	Zegzouti et al. (2020)
4.	Two selected saprotrophic fungal strains	Bioremediation of DDT (dichlorodiphenyltrichloroethane)	Russo et al. (2019)
5.	*Fomitopsis meliae*, *Trichoderma ghanense* and *Rhizopus microsporus*	Metal tolerance	Oladipo et al. (2018)
6.	*Pleurotus pulmonarius*	Bioremediation of crude oil-contaminated soil	Stanley et al. (2017)
7.	Combined bacteria–white rot fungi	Bioremediation of petroleum-contaminated soil	Liu et al. (2017)
8.	*Oenothera picensis*	Bioremediation of copper-contaminated soil	Cornejo et al. (2017)
9.	*Aspergillus niger* and *Lichtheimia ramosa*	Reduction of oil spill	Moustafa (2016)
10.	*Acremonium* sp. P0997 and *Bacillus subtilis*	Bioremediation of fluoranthene-contaminated sand	Ma et al. (2016)
11.	*Aspergillus niger, Candida glabrata, Candida krusei* and *Saccharomyces cerevisiae*	Crude oil remediation	Burghal et al. (2016)
12.	*Phanerochaete velutina*	Bioremediation of PAH-contaminated soil	Winquist et al. (2014)
13.	*Penicillium sp.*	Biodegradation of petrol	Vanishree et al. (2014)
14.	*Bionectria ochroleuca*	Degradation of more than 70% of C12–C28 hydrocarbons, and 100% degradation of C12 and C28 hydrocarbons	Kota et al. (2014)

metal-contaminated environments. The level of efficiency of phytoremediation depends mainly on the level of contamination of polluted sites, amount of metal-based soil contamination and ability of plant in remediating it (Tak et al., 2013). Hyperaccumulators and non-hyperaccumulators are the types of plants used in phytoremediation. Hyperaccumulators have a very high potential for heavy metal accumulation and little biomass yield efficiency, while non-hyperaccumulators possess lesser extraction capacity than hyperaccumulators, but the total biomass yield is substantially higher (Ojuederie and Babalola, 2017; Abbaszadeh-Dahaji et al., 2016; Choudhary et al., 2017). Figure 16.1 shows the role of plants in bioremediation (Ojuederie and Babalola, 2017).

Plants are also effective against several organic contaminants and, thus, can be used to remediate organic pollutants (Rada et al., 2019; Donati et al., 2019; Ahmad et al., 2017; Cole, 1998). The use of plants in bioremediation processes has several benefits such as lower cost, possible recovery and reuse of metals, maintains the fertility of soil (Ali et al., 2013), helps to increase the health of soil, yield, plant phytochemical (Othman and Leskovar, 2018), reduces leaching of metals and erosion of soils (Ali et al., 2013), but the processes also have some limitations. There are various processes mediated by plants in the treatment of environmental issues, such as phytoextraction or phytoaccumulation or phytosequestration (phytoextraction exploits the capability of plants or algae to eliminate pollutants from soil or water by converting them into harvestable plant biomass. Substances are taken up by roots from either soil or water, and then, they concentrate above ground in the form of plant biomass) (Ali et al., 2013), phytostabilization (in this process, the mobility of substances is decreased in the environment, for example, by limiting the leaching of substances from the soil) (Lone et al., 2008),

TABLE 16.2
Dye Decolourization by Fungi

S. No.	Fungi/Fungal Enzymes	Name of Dyes Decolourized	Percentage Decolourization	References
1.	*Trametes versicolour* and *Phanerochaete chrysosporium* co-cultures	Azo dye decolourization	87.6%	Singh et al. (2020)
2.	Yeast strain *Sterigmatomyces halophilus* SSA-1575	Azo dye decolourization	Excellent (up to 100%)	Al-Tohamy et al. (2020)
3.	White rot fungi – *Stereum ostrea* and *Phanerochaete chrysosporium*	Triphenylmethane dye (crystal violet)	90% and 67%	Usha et al. (2019)
4.	*Aspergillus sp.* XJ-2 (with microalgae *Chlorella sorokiniana* XJK)	Disperse Red 3B	98.09%	Tang et al. (2019)
5.	Secretome of *Myrothecium roridum* IM6482	Acid Blue 113, Acid Red 27, Direct Blue 14 and Acid Orange 7	66%, 91%, 79% and 80%	Jasińska et al. (2019)
6.	Extracts of *C. versicolor* and *P. ostreatus*	Remazol Brilliant Blue Royal	80.42% and 70.42%	Afiya et al. (2019)
7.	*Aspergillus niger* D2-1	Reactive yellow and reactive red (4BL) dyes (azo dyes)	98.62% and 92.42%	Salem et al. (2019)
8.	*Aspergillus niger* LAG	Thiazole Yellow G	98%	Bankole et al. (2019)
9.	Mutant PIE_5 (laccase from *Coprinopsis cinerea)*	Indigo dye	87.1%–90.9%	Yin et al. (2019)
10.	Crude laccase from *Ganoderma lucidum*	Remazol Brilliant Blue R dye	50.3%	Qin et al. (2019)
11.	Laccase from *Trametes hirsuta* EDN084	Remazol Brilliant Blue R, Reactive Blue 4, Acid Blue 129 and Direct Blue 71	50%, 47%, 51% and 85%	Yanto et al. (2019)
12.	Isolated *Pleurotus ostreatus*	Congo red and brilliant green	98.82% and 95.74%	Vantamuri et al. (2019)
13.	Recombinant TaDyP (from *Pleurotus ostreatus)*	Acid Blue 129 and Acetyl Yellow G	77% and 34%	Cuamatzi-Flores et al. (2019)
14.	*Aspergillus niger*	Textile effluents containing direct red dye	Above 97%	Verma et al. (2019)
15.	Selected fungal isolates	Bromophenol blue, bromothymol blue and Remazol Brilliant Blue R	100%	Rao et al. (2019)

phytodegradation or phytotransformation (in which the plant degrades the organic pollutants in soil or plant's body), phytostimulation or rhizodegradation (in this process soil, microbial activity enhancement takes place for organic pollutant's degradation) (Phytoremediation Processes. www.unep.or.jp.), phytovolatilization (this is the process in which toxic/harmful substances are removed from soil or water and released into air, sometimes as a result of phytotransformation into less harmful substances and/or more volatile form. Plants take up the contaminants and evaporate them into air by transpiration) (Phytoremediation Processes. www.unep.or.jp.) and rhizofiltration (in this process, plants filter water for the removal of pollutants or excess nutrients using their mass of roots. The pollutants remain absorbed in or adsorbed to the roots (Phytoremediation Processes. www.unep.or.jp.)

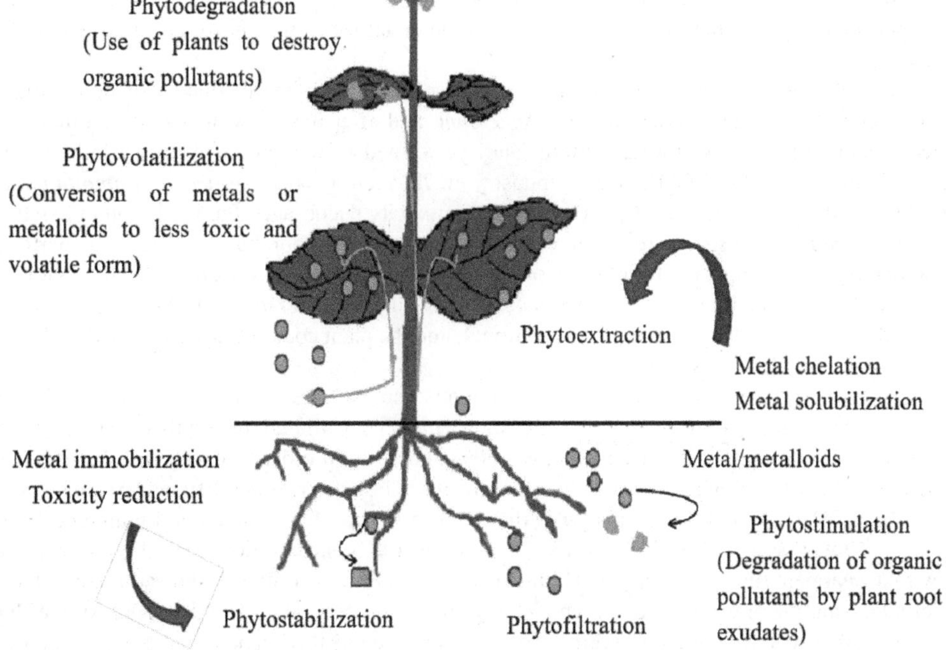

FIGURE 16.1 Plants in phytoremediation of metals. (From Ojuederie, O.B., Babalola, O.O., Int. J. Environ. Res. Public Health, 14, 1504, 2017.)

16.7 Significances of Plants in Dye Degradation

Plants either individually or in proper combination with microbes can be effectively utilized for the treatment of various dyes, which are continuously being discharged from various dye industries as well as industries using dyes directly or indirectly. Many research works on the phytoremediation of dyes are going on worldwide, and many works have already been performed by researchers.

Bacopa monnieri (L.) Pennell is a potential plant that has been used actively in the treatment of textile azo dyes. An important research performed by Shanmugam et al. (2020) showed the potential of this plant species for the treatment of 14 different azo dyes. They obtained promising results, and the decolourization varies between 90% and 100% after incubation of 2 weeks in in vitro hydroponic cultures. In the presence of dyes R. Magenta MB, R. Navy Blue M2R, Dt. Orange RS, Dt. T Blue GLL and Dt. Sky Blue FF, alone and together in the medium, there were no significant alteration in the growth of *B. monnieri*. On the other hand, the authors also observed toxic effects on the growth of plant after using higher concentrations of dyes (60–100 mg/L) along with decline in the rate of decolourization. Their experiments and result suggest that L-proline, N-valeryldecyl ester, 3,5-di-tert-butyl-4-trimethylsilyloxytoluene, 1,2-benzenedicarboxylic acid and diisooctyl ester were degraded intermediates of Dt. Blue GLL. According to them, due to the presence of oxidative and antioxidative enzymes in leaves and roots and their higher enzymatic activities in the presence of dye over control, dyes degradation occurred, which means these enzymes were involved in the decolourization of dyes. Their results suggest that this plant species may be a potential and practical tool for the phytoremediation of such textile dyes (Shanmugam et al., 2020).

Shah et al. (2019) performed a study on the role and ability of plant *Tinospora cordifolia* for the degradation of malachite green. They performed different experiments using different concentrations of this dye and used UV–visible spectroscopy for the study of degradation of malachite green. During experiments of before and after the treatment with *Tinospora cordifolia*, they also studied the parameters such as pH and conductivity. They found a decrease in these values by *Tinospora cordifolia* within 72 hours. Their experiment suggests that this species may have potential for its use in dye decolourization (Shah et al., 2019).

Priyanka and Krishnaswamy (2019) performed a study on the phytoremediation of mixed reactive azo dyes using plant species *Ceratophyllum demersum*, and after this, they also analysed their toxicity effects on other life forms in water system. They performed a study on the toxicity effects of mixed azo dyes obtained from local fabric dyeing industry on *Trigonella foenum-graecum, Lampito mauritii* and bacterial pathogens. Their results showed that the toxicity encouraged the inhibition of *Trigonella foenum-graecum* root and shoot growth at a concentration of 1,000 ppm, but when this was performed on remediated soil with higher concentration of the effluents, the effect was decreased. They studied the phytoremediation of mixed azo dyes-contaminated water using *C. demersum* at different concentrations. They found that dye removal percentage was minimal, and the plant couldn't sustain more than 500 ppm of dye (Priyanka and Krishnaswamy, 2019).

For treating the textile effluent based toxic pollutants using *Chara vulgaris*, Mahajan et al. (2019) assessed the phytoremediation efficacy of this plant species. They diluted the highly concentrated toxic effluents to different concentrations: 10%, 25%, 50% and 75%, in order to check the accessibility of macroalgae to bear the pollutant load of textile effluents. They determined different parameters such as pH, TDS, BOD, COD and EC for analysing the textile effluents. They found maximum reductions in TDS (68%), COD (78%), BOD (82%) and EC (86%) when 10% concentrations of textile effluents were used with a treatment time of 120 hours. High concentration of toxic effluents showed toxic influences on microalgae and there were no remarkable change in parameters on 75% and 100% textile effluent. Their UV–visible spectroscopic observation confirmed the removal of toxic organic pollutants by the use of *C. vulgaris.* They also found important information from this experiment that the typical X-ray spectra recorded using EDXRF technique showed Cd, a heavy metal, in dried samples of microalgae after treatment, which indicates its efficiency in the treatment of heavy metals also (Mahajan et al., 2019).

Imron et al. (2019) performed a study on the phytoremediation of methylene blue by using *Lemna minor.* They used 2 g of *L. minor* in 50 mg/L of methylene blue dyes for 24 hours. They measured the absorbance values at 0, 0.5, 1, 2, 3, 4, 5, 6 and 24 hours with a maximum wavelength of 665 nm and observed the percentage of dye removal and relative growth rate of *L. minor* during exposure to methylene blue. They found a removal percentage 80.56% ± 0.44% at 24 hours with a relative growth rate of 0.006/hour. Their experiment showed the potency of *L. minor* in the phytoremediation of dyes from wastewater, and thus, this can be used as a phytoremediation agent (Imron et al., 2019).

Tan et al. (2016) performed a study in order to check the potential of water hyacinth (*Eichhornia crassipes*) in decolourization of dyes. It was used for the treatment of methylene blue (50 mg/L) and methyl orange (50 mg/L) for 20 days. For both types of dyes, they found interesting result; however, dye decolourization was higher for methylene blue (98.42%) compared to methyl orange (66.80%) (Tan et al., 2016).

Nisha and Emilia (2016) performed studies on potential tissue culture approach for the remediation of dyes using callus cultures of the plant, *Tecoma stans.* In this study, they focused on effective degradation of brilliant green using this plant. Callus culture of this plant showed good peroxidase activity. By using batch culture, they initially optimized the dye degradation parameters. Their result shows the effective role of peroxidase of *Tecoma* in the degradation of brilliant green, and it was found that further use of immobilized callus for seven cycles of dye degradation in packed bed reactor was also possible (Nisha and Emilia, 2016).

A study was done on aquatic fern (*Salvinia molesta*) for observing its potential in the degradation of an azo dye (Rubine GFL). Chandanshive et al. (2016) showed the potential role of root biomass (60 ± 2) in the degradation of this dye to be up to 97% in 72 hours. At the time of decolourization of Rubine GFL, induction of different types of enzymatic activities (lignin peroxidase, veratryl alcohol oxidase, laccase, tyrosinase, catalase, DCIP reductase and superoxide dismutase) was observed in root and stem tissues. They analysed the metabolites and found that lignin peroxidase and laccase degraded Rubine GFL and formed 2-methyl-4-nitroaniline and N-methylbenzene-1,4-diamine by attacking at its azo bond. Their toxicity work on seeds of *Triticum aestivum* and *Phaseolus mungo* also showed decreased toxicity of metabolites (Chandanshive et al., 2016).

The plant–microbe interaction also plays a significant role in the remediation of dyes and heavy metals. Such a valuable interaction may have several applications in phytoremediation technology. Rhizorestitution is a specific type of phytoremediation technique that may be helpful for the remediation of sites polluted with dyes and heavy metals. Further, plant's system in connection with rhizospheric and endophytic microbiome has nice ability to degrade organic molecules in places polluted by organic matter. The potential of plants in combination with microbes for the purpose of dye remediation (phytoremediation) becomes enhanced. This type of interaction alters the dyes/organic pollutants and heavy metals accumulation in plant tissues and thus enhances the process of phytoremediation (Dogra et al., 2018). Dogra et al. (2018) very nicely discussed the different types of plant–microbe interactions and their efficiency in the phytoremediation of dyes and heavy metals. Readers can also visit some other reviews and research works on phytoremediation (Bharathiraja et al., 2018; Kimmatkar et al., 2017; Tahir et al., 2016; Khandare and Govindwar, 2015; Vafaei et al., 2013; Kabra et al., 2011).

16.8 Future Perspectives

Since the plants are omnipresent and easily available to everyone, future researches in the field of bioremediation using plants or using combination of plants with other sources may be very fruitful from the point of cost-effectiveness. Further, easily available plants may be highly valuable for the researches on the removal of pollution caused by specific types of toxic and carcinogenic organic dyes. Proper attention of the scientific community to work on plant-based detoxification and decolourization of dyes and the exploitation of the capability of various plants for such purposes is much needed.

16.9 Conclusions

Thus, plants can play a significant role in removing the toxic and adverse effects of different types of dyes being released into water systems from textile and other types of industries. Regarding other literature studies, many more works have been performed by various researchers working worldwide. In this chapter, only a few recent works have been shown.

Acknowledgements

Dr P.K. Chaurasia is thankful to the P.G. Department of the Chemistry, L.S. College, Muzaffarpur, and Dr S.L. Bharati is thankful to the Department of Chemistry, NERIST, Nirjuli, Arunachal Pradesh.

REFERENCES

Abbaszadeh-Dahaji P, Omidvari M, Ghorbanpour M (2016) Increasing phytoremediation efficiency of heavy metal-contaminated soil using PGPR for sustainable agriculture. In: Choudhary DK, Varma A, Tuteja N (eds) *Plant-Microbe Interaction: An Approach to Sustainable Agriculture*, Springer: New Delhi, India, pp. 187–204. doi: 10.1007/978-981-10-2854-0_9.

Adiloğlu S (2017) *Heavy Metal Removal with Phytoremediation*. doi: 10.5772/intechopen.70330.

Afiya H, Ahmet EE, Shah MM (2019) Enzymatic decolorization of remazol brilliant blue royal (RB 19) textile dye by white rot fungi. *J Appl Adv Res* 4(1):11–15. doi: 10.21839/jaar.2019.v4i1.260.

Agarwal S, Gupta KK, Chaturvedi VK, et al (2018) The potential application of peroxidase enzyme for the treatment of industry wastes. In: Bharati SL and Chaurasia PK (eds) *Research Advancements in Pharmaceutical, Nutritional and Industrial Enzymology*, IGI Global: USA. doi: 10.4018/978-1-5225-5237-6.ch012, ISBN 9781522552376.

Ahmad I, Imran M, Hussain MB, et al (2017) Remediation of organic and inorganic pollutants from soil: The role of plant bacteria partnership. In: Naser A. Anjum (ed) *Chemical Pollution Control with Microorganisms*, Nova Sci. Publisher, pp. 197–243.

Ahmad TA, Ganjo DGA (2020) Fungal isolates and their bioremediation for pH, chloride, tph and some toxic heavy metals. *Eurasia J Biosci* 14:149–160.

Ali H, Khan E, Sajad MA (2013) Phytoremediation of heavy metals - Concepts and applications. *Chemosphere* 91(7):869–881. doi: 10.1016/j.chemosphere.2013.01.075.

Al-Tohamy R, Kenawy E-R, Sun J, et al (2020) Performance of a newly isolated salt-tolerant yeast strain *Sterigmatomyces halophilus* SSA-1575 for azo dye decolorization and detoxification. *Front Microbiol* 11:1163. doi: 10.3389/fmicb.2020.01163.

Bankole PO, Adekunle AA, Govindwar SP (2019) Demethylation and desulfonation of textile industry dye, Thiazole Yellow G by *Aspergillus niger* LAG. *Biotechnology Reports* 23:e00327. doi: 10.1016/j.btre.2019.e00327.

Bharathiraja B, Jayamuthunagai J, Praveenkumarand R, et al (2018) Phytoremediation techniques for the removal of dye in wastewater. In: Varjani S, Agarwal A, Gnansounou E, Gurunathan B (eds) *Bioremediation: Applications for Environmental Protection and Management. Energy, Environment, and Sustainability*, Springer: Singapore. doi: 10.1007/978-981-10-7485-1_12.

Bharati SL, Chaurasia PK (eds) (2018) *Research Advancements in Pharmaceutical, Nutritional and Industrial Enzymology*, IGI Global: USA. 1–549. doi: 10.4018/978-1-5225-5237-6. ISBN: 978-1522552376 (hardcover), ISBN: 9781522552383 (ebook).

Bilal M, Iqbal HMN (2020) Microbial bioremediation as a robust process to mitigate pollutants of environmental concern. *Case Stud Chem Environ Eng*. doi: 10.1016/j.cscee.2020.100011.

Bioremediation, https://microbewiki.kenyon.edu/index.php/Bioremediation.

Boggon C (2019) How Polluting Is the Fashion Industry. https://www.ekoenergy.org/how-polluting-is-the-fashion-industry/.

Booth G (2000) Dyes, general survey. *Ullmann's Encyclopedia of Industrial Chemistry*, Wiley-VCH. doi: 10.1002/14356007.a09_073. ISBN 3527306730.

Brim H, McFarlan SC, Fredrickson JK, et al (2000) Engineering *Deinococcus radiodurans* for metal remediation in radioactive mixed waste environments. *Nat Biotechnol* 18: 85–90.

Burgess R (2017) *Harvesting Color: How to Find Plants and Make Natural Dyes*, Artisan Books. ISBN 9781579654252.

Burghal AA, Abu-Mejdad NMJA, Al-Tamimi WH (2016) Mycodegradation of crude oil by fungal species isolated from petroleum contaminated soil. *Int J Innov Res Sci Eng Technol* 5(2):1517–1524.

Chandanshivea VV, Rane NR, Gholave AR, et al (2016) Efficient decolorization and detoxification of textile industry effluent by *Salvinia molesta* in lagoon treatment. *Environ Res* 150:88–96. doi: 10.1016/j.envres.2016.05.047.

Chaurasia PK, Bharati SL (2017) Significance of laccases in food chemistry and related bioremediation. In: Grumezescu and Holban (eds) *Multi Volume Handbook of Food Bioengineering*, Vol. 3, *Soft Chemistry and Food Fermentation*, Elsevier, 299–335. doi: 10.1016/B978-0-12-811412-4.00011-4. ISBN: 978-0-12-811412-4.

Chaurasia PK, Bharati SL (2019) Recent myco-dye decolorization studies (mini-review). *J Biotechnol Bioeng* 3(4):27–31.

Chaurasia PK, Bharati SL (eds) (2020) *Research Advances in the Fungal World: Culture, Isolation, Identification, Classification, Characterization, Properties and Kinetics*, Nova Science Publisher, Inc: USA. ISBN: 978-1-53617-197-6.

Chaurasia PK, Bharati SL (eds) (2022) The Chemistry inside Spices and Herbs: Research and Development (Part I & Part II), Bentham Science Publishers (In Press). ISBN: 978-981-5039-57-3 (ISBN of Part I).

Chaurasia PK, Bharati SL, Kumar S (2020a) Recent Studies on Biotechnological Roles of *Pleurotus* spp. J Biotechnol Bioprocess 1(3). doi:10.31579/2766-2314/018.

Chaurasia PK, Bharati SL, Kumar S (2020b) Microbes as Remediating Agent against Toxicity of Dyes. J Appl Biotechnol Bioeng 7(6), 242-244. doi: 10.15406/jabb.2020.07.00240.

Chaurasia PK, Bharati SL, Mani A (2018) Enzymatic treatment of petroleum based hydrocarbons. In: Pathak and Navneet (eds) *Microbial Tools and Techniques for Environmental Waste Management*, IGI Global Publisher: USA, 396–408. doi: 10.4018/978-1-5225-3540-9.ch019.

Chaurasia PK, Bharati SL, Mani A (2019) Significances of fungi in bioremediation of contaminated soil. In: Singh JS and Singh DP (eds) *New and Future Developments in Microbial Biotechnology and Bioengineering*, Elsevier, 281–294. doi: 10.1016/B978-0-444-64191-5.00020-1.

Chaurasia PK, Bharati SL, Sharma M, et al (2015) Fungal laccases and their biotechnological significances in the current perspective: A review. *Curr Org Chem* 19(19): 1916–1934. doi: 10.2174/1385272819666150629175237.

Chaurasia, PK, Bharati SL (2021) Applicability of fungi in agriculture and environmental sustainability. In: Singh JS, Tiwari S, Singh C and Singh AK (eds) *Microbes in Land Use Change Management*,Elsevier, 155–172. doi: 10.1016/B978-0-12-824448-7.00010-3.

Choudhary DK, Varma A, Tuteja N (2017) *Plant-Microbe Interaction: An Approach to Sustainable Agriculture*, Springer: New Delhi, India.

Cole MA (1998) Remediation of soils contaminated with toxic organic compounds. In: Brown S, Angle JS, Jacobs L (eds) *Beneficial Co-Utilization of Agricultural, Municipal and Industrial by-Products*, Springer: Dordrecht. doi: 10.1007/978-94-011-5068-2_15.

Cornejo P, Meier S, García S, et al (2017) Contribution of inoculation with arbuscular mycorrhizal fungi to the bioremediation of a copper contaminated soil using *Oenothera picensis*. *J Soil Sci Plant Nutr* 17(1):14–21.

Cuamatzi-Flores J, Esquivel-Naranjo E, Nava-Galicia S, et al (2019) Differential regulation of *Pleurotus ostreatus* dye peroxidases gene expression in response to dyes and potential application of recombinant Pleos-DyP1 in decolorization. *PLoS One* 14(1):e0209711. doi: 10.1371/journal.pone.0209711.

d'Errico G, Aloj V, Ventorino V et al (2020) Methyl t-butyl ether-degrading bacteria for bioremediation and biocontrol purposes. *PLoS One* 15(2):e0228936. doi: 10.1371/journal.pone.0228936.

Deshmukh R, Khardenavis AA, Purohit HJ (2016) Diverse metabolic capacities of Fungi for bioremediation. *Indian J Microbiol* 56(3):247–264.

Dixit R, Malaviya D, Pandiyan K, et al (2015) Bioremediation of heavy metals from soil and aquatic environment: An overview of principles and criteria of fundamental processes. *Sustainability* 7:2189–2212.

Dogra V, Kaur G, Kumar R, Prakash C (2018) The importance of plant-microbe interaction for the bioremediation of dyes and heavy metals. In: Kumar V, Kumar M, Prasad R (eds) *Phytobiont and Ecosystem Restitution*, Springer: Singapore. doi: 10.1007/978-981-13-1187-1_22.

Donati ER, Sani RK, Goh KM, et al (2019) Editorial: Recent advances in bioremediation/biodegradation by extreme microorganisms. *Front Microbiol* 10:1851. doi: 10.3389/fmicb.2019.01851.

Fragoeiro SIDS (2005) *Use of fungi in bioremediation of pesticides. Applied Mycology Group Institute of Bioscience and Technology*, Cranfield University.

Gadd GM (ed) (2001) *Fungi in Bioremediation* (No. 23), Cambridge University Press.

Gita S, Hussan A, Choudhury TG (2017) Impact of textile dyes waste on aquatic environments and its treatment. *Environ Ecol* 35 (3C): 2349–2353.

Goutam SP, Saxena G, Singh V, et al (2018) Green synthesis of TiO2 nanoparticles using leaf extract of Jatropha curcas L. for photocatalytic degradation of tannery wastewater. *Chem Engin J* 336: 386–396.

Green Remediation Best Management Practices: Sites with Leaking Underground Storage Tank Systems. EPA 542-F-11-008. EPA. June 2011.

Harms H, Schlosser D, Wick LY (2011) Untapped potential: Exploiting fungi in bioremediation of hazardous chemicals. *Nat Rev Microbiol* 9(3):177–192.

Imron MF, Kurniawan SB, Soegianto A, et al (2019) Phytoremediation of methylene blue using duckweed (*Lemna minor*). *Heliyon* 5:e02206. doi: 10.1016/j.heliyon.2019.e02206.

Ite AE and Ibok UJ (2019) Role of plants and microbes in bioremediation of petroleum hydrocarbons contaminated soils. *Int J Environ Bioremediat Biodegrad* 7(1): 1–19. doi: 10.12691/ijebb-7-1-1.

Jan S, Parray JA (2016) *Approaches to Heavy Metal Tolerance in Plants*, Springer: New Delhi, India.

Jasińska A, Soboń A, Góralczyk-Bińkowska A et al (2019) Analysis of decolorization potential of *Myrothecium roridum* in the light of its secretome and toxicological studies. *Environ Sci Pollut Res* 26:26313–26323. doi: 10.1007/s11356-019-05324-6.

Jessica R, Ackerman CE, Scow KM (1999) Biodegradation of methyl tert-butyl ether by a bacterial pure culture. *Appl Environ Microbiol* 11: 4788–4792.

Kabra AN, Khandare RV, Kurade MB et al (2011) Phytoremediation of a sulphonated azo dye Green HE4B by *Glandularia pulchella* (Sweet) Tronc. (Moss Verbena). *Environ Sci Pollut Res* 18:1360–1373. doi: 10.1007/s11356-011-0491-7.

Kant R (2012) Textile dyeing industry an environmental hazard. *Nat Sci* 4:22–26. doi: 10.4236/ns.2012.41004.

Khandare RV, Govindwar SP (2015) Phytoremediation of textile dyes and effluents: Current scenario and future prospects. *Biotechnol Adv* 33(8):1697–1714. doi: 10.1016/j.biotechadv.2015.09.003.

Kimmatkar KS, Purohit AV, Sanyal AJ (2017) Phytoremediation techniques and species for combating contaminants of textile effluents – An overview. *Int J Sci Res* 6(3):ART20171923.

Kota MF, Hussaini AASA, Zulkharnain A, et al (2014) Bioremediation of crude oil by different fungal genera. *Asian J Plant Biol* 2(1):11–18.

Kumari V, Yadav A, Haq I, et al (2016) Genotoxicity evaluation of tannery effluent treated with newly isolated hexavalent chromium reducing *Bacillus cereus*. *J Environ Manag* 183:204–211.

Kushwaha A, Agarwal S, Gupta KK, Maurya S, Chaurasia PK *et al.* (2017) Laccase enzyme from white rot fungi: An overview and its' application. In: Singh MP, Verma V, Singh AK (eds), Incredible World of Biotechnology, Nova Science Publishers, Inc., N.Y., USA, 25-41.

Lellis B, ZaniFávaro-Polonio C, Pamphile JA, et al (2019) Effects of textile dyes on health and the environment and bioremediation potential of living organisms. *Biotechnol Res Innov* 3(2):275–290. doi: 10.1016/j.biori.2019.09.001.

Liu B, Liu J, Ju M, et al (2017) Bacteria-white-rot fungi joint remediation of petroleum-contaminated soil based on sustained release of laccase. *RSC Adv* 7:39075.

Lone MI, He, Z-l, Stoffella PJ, et al (2008) Phytoremediation of heavy metal polluted soils and water: Progresses and perspectives. *J Zhejiang Univ Sci B* 9(3):210–220. doi:10.1631/jzus.B0710633.

Luqman M, Butt TM, Tanvir A, et al (2013) *Afr J Agric Res* 8(17):1591–1595. doi: 10.5897/AJAR11.1111.

Ma X-K, Ding N, Peterson EC, et al (2016) Heavy metals species affect fungal-bacterial synergism during the bioremediation of fluoranthene. *Appl Microbiol Biotechnol* 100:7741–7750.

Mahajan P, Kaushal J, Upmanyu A, et al (2019) Assessment of phytoremediation potential of *Chara vulgaris* to treat toxic pollutants of textile effluent. *J Toxicol* Article ID 8351272, 11 p. doi: 10.1155/2019/8351272.

Moustafa AM (2016) Bioremediation of oil spill in Kingdom of Saudi Arabia by using fungi isolated from polluted soils. *Int J Curr Microbiol App Sci* 5(5):680–691.

Nisha RD, Emilia AT (2016) A potential tissue culture approach for the phytoremediation of dyes in aquaculture industry. *Biochem Eng J* 115:23–29. doi: 10.1016/j.bej.2016.08.001.

Ojuederie OB, Babalola OO (2017) Microbial and plant-assisted bioremediation of heavy metal polluted environments: A review. *Int J Environ Res Public Health* 14:1504. doi: 10.3390/ijerph14121504.

Oladipo OG, Awotoye OO, Olayinkac A, et al (2018). Heavy metal tolerance traits of filamentous fungi isolated from gold and gemstone mining sites. *Braz J Microbiol* 49:29–37.

Osuoha JO, Nwaichi EO (2020) Enzymatic technologies as green and sustainable techniques for remediation of oil-contaminated environment: State of the art. *Int J Environ Sci Technol* doi: 10.1007/s13762-020-02876-w.

Othman YA, Leskovar D (2018) Organic soil amendments influence soil health, yield, and phytochemicals of globe artichoke heads. *Biol Agric Hortic* 34(4):1–10. doi: 10.1080/01448765.2018.1463292.

Pal AK, Singh J, Soni R, et al (2020) The role of microorganism in bioremediation for sustainable environment management. In: Pandey and Singh (eds) *Bioremediation of Pollutants: From Genetic Engineering to Genome Engineering*, 227–249. doi: 10.1016/B978-0-12-819025-8.00010-7.

Phytoremediation processes, www.unep.or.jp.

Priyanka JV, Krishnaswamy VG (2019) Phytoremediation of mixed reactive azo dyes in contaminated water by *Ceratophylum demersum* and its toxicity analysis on other life forms. *SPC J Environ Sci* 1(2):16–25.

Qin, P., Wu, Y., Adil, B., et al (2019) Optimization of laccase from *Ganoderma lucidum* decolorizing remazol brilliant blue R and GLAC1 as main laccase-contributing gene. *Molecules* 24:3914. doi: 10.3390/molecules24213914.

Rada EC, Andreottola G, Istrate IA, et al (2019) Remediation of soil polluted by organic compounds through chemical oxidation and phytoremediation combined with DCT. *Int J Environ Res Public Health* 16:3179. doi: 10..3390/ijerph16173179.

Ramachandran R, Gnanadoss JJ (2013) Mycoremediation for the treatment of dye containing effluents. *Int J Comput Algor* 02(02):101–105.

Rao RG, Ravichandran A, Kandalam G, et al (2019) Screening of wild basidiomycetes and evaluation of the biodegradation potential of dyes and lignin by manganese peroxidases. *BioRes* 14(3): 6558–6576.

Russo F, Ceci A, Pinzari F, et al (2019) Bioremediation of dichlorodiphenyltrichloroethane (DDT)-contaminated agricultural soils: Potential of two autochthonous saprotrophic fungal strains. *Appl Environ Microbiol* 85:e01720-19. doi: 10.1128/AEM.01720-19.

Salem SS, Mohamed A, El-Gamal M, et al (2019) Biological decolorization and degradation of azo dyes from textile wastewater effluent by *Aspergillus niger. Egypt J Chem* 62(10):1799–1813. doi: 10.21608/ejchem.2019.11720.1747.

Shah CN, George L-B, Upadhyaya AM (2019) Phytoremediation potential of an azo dye-malachite green. *Int J Pharm Biol Sci* 9(1): 792–796. Doi: 10.21276/ijpbs.2019.9.1.100.

Shanmugam L, Ahire M, Nikam T (2020) *Bacopa monnieri* (L.) Pennell, a potential plant species for degradation of textile azo dyes. *Environ Sci Pollut Res* 27:9349–9363. doi: 10.1007/s11356-019-07430-x.

Silfvast WT (2008) *Laser Fundamentals*, Cambridge University Press. ISBN 9781139855570.

Singh J, Das A, Yogalakshmi KN (2020) Enhanced laccase expression and azo dye decolourization during co-interaction of *Trametes versicolor* and *Phanerochaete chrysosporium. SN Appl Sci* 2:1095. doi: 10.1007/s42452-020-2832-y.

Stanley HO, Offorbuike OM, Stanley CN (2017) Bioremediation of crude oil contaminated soil using *Pleurotus pulmonarius*, a white-rot fungus. *IOSR J Environ Sci Tox Food Technol* 11(4):122–128. Ver. II.

Strong PJ, Burgess JE (2007) Bioremediation of a wine distillery wastewater using white rot fungi and the subsequent production of laccase. *Water Sci Technol* 56(2):179–186.

Sujata M, Bharagava RN (2016) Exposure to crystal violet, its toxic, genotoxic and carcinogenic effects on environment and its degradation and detoxification for environmental safety. *Rev Environ Contaminat Toxicol* 237:71–104.

Tahir U, Yasmin A, Khan UH (2016) Phytoremediation: Potential flora for synthetic dyestuff metabolism. *J King Saud Univ Sci* 28:119–130. doi: 10.1016/j.jksus.2015.05.009.

Tak HI, Ahmad F, Babalola OO (2013) Advances in the application of plant growth-promoting rhizobacteria in phytoremediation of heavy metals. In: Whitacre, DM (ed) *Reviews of Environmental Contamination and Toxicology*, Springer: New York, NY, USA, Vol. 223, pp. 33–52.

Tan KA, Morad N, Ooi JQ (2016) Phytoremediation of methylene blue and methyl orange using *Eichhornia crassipes. Int J Environ Sci Dev* 7(10):724–728. doi: 10.18178/ijesd.2016.7.10.869.

Tang W, Xu X, Ye BC, et al (2019) Decolorization and degradation analysis of disperse red 3B by a consortium of the fungus *Aspergillus* sp. XJ-2 and the microalgae *Chlorella sorokiniana* XJK. *RSC Adv* 9:14558. doi: 10.1039/C9RA01169B.

Thakur M, Medintz IL, Walper SA (2019) Enzymatic bioremediation of organophosphate compounds—progress and remaining challenges. *Front Bioeng Biotechnol* 7:289. doi: 10.3389/fbioe.2019.00289.

US Environmental Protection Agency. (2013) *Introduction to In Situ Bioremediation of Groundwater.* 30, EPA 542-R-13-018.

Usha KY, Dhar A, Ridham, Patil SJ, Kumar KP (2019) Decolorization of triphenylmethane dye by white-rot fungi. *Indian J Adv Chem Sci* 7(3): 65–69.

Vafaei F, Movafeghi A, Khataee AR, Zarei M, Lisar SSY (2013) Potential of *Hydrocotyle vulgaris* for phytoremediation of a textile dye: Inducing antioxidant response in roots and leaves. *Ecotoxicol Environ Saf* 93:128–134. doi: 10.1016/j.ecoenv.2013.03.035.

Vanishree M, Thatheyus AJ, Ramya D (2014) Biodegradation of petrol using the fungus *Penicillium* sp. *Sci Int* 2:26–31. doi: 10.17311/sciintl.2014.26.31.

Vantamuri AB, Kotresh O, Kannur LS, Bhat PV (2019) Decolorization of different dyes by using white rot fungus. *World J Pharmaceut Life Sci* 5(2):174–176.

Verma A, Agarwal M, Bhati D, Garg H (2019) Investigation on the removal of direct red dye using *Aspergillus niger. J Mater Sci Surface Eng* 6(6):881–883.

Winquist E, Björklöf K, Schultz E, Räsänen M, Salonen K, Anasonye F, Cajtham T, Steffen KT, Jørgensen KS, Tuomela M (2014) Bioremediation of PAH-contaminated soil with fungi- from laboratory to field scale. *Int Biodeter Biodegrad* 86:238–247.

Xiao P, Kondo R (2020) Potency of *Phlebia* species of white rot fungi for the aerobic degradation, transformation and mineralization of lindane. *J Microbiol* 58:395–404. doi: 10.1007/s12275-020-9492-x.

Yanto DHY, Auliana N, Anita SH, Watanabe T (2019) Decolorization of synthetic textile dyes by laccase from newly isolated *Trametes hirsuta* EDN084 mediated by violuric acid. The 8th International Symposium for Sustainable Humanosphere. IOP Conf. Series: Earth and Environmental Science 374:012005. doi: 10.1088/1755-1315/374/1/012005.

Yaseen DA, Scholz M (2019) Textile dye wastewater characteristics and constituents of synthetic effluents: A critical review. *Int J Environ Sci Technol* 16:1193–1226. doi: 10.1007/s13762-018-2130-z.

Yin Q, Zhou G, Peng C, Zhang Y, Kües U, Liu J, Xiao Y, Fang Z (2019) The first fungal laccase with an alkaline pH optimum obtained by directed evolution and its application in indigo dye decolorization. *AMB Exp* 9:151. doi: 10.1186/s13568-019-0878-2.

Zegzouti Y, Boutafda A, El Fels L, El Hadek M, Ndoye F, Mbaye N, Kouisni L, Hafidi M (2020) Screening and selection of autochthonous fungi from leachate contaminated-soil for bioremediation of different types of leachate. *Environ Eng Res* 25(5): 722–734. doi: 10.4491/eer.2019.317.

Index

Note: **Bold** page numbers refer to tables and *italic* page numbers refer to figures.